Current Progress in Tectonics

Current Progress in Tectonics

Edited by **Agnes Nolan**

R Callisto
Reference

New York

Published by Callisto Reference,
106 Park Avenue, Suite 200,
New York, NY 10016, USA
www.callistoreference.com

Current Progress in Tectonics
Edited by Agnes Nolan

International Standard Book Number: 978-1-63239-141-4 (Hardback)

Printed in the United States of America.

Contents

Preface

In my initial years as a student, I used to run to the library at every possible instance to grab a book and learn something new. Books were my primary source of knowledge and I would not have come such a long way without all that I learnt from them. Thus, when I was approached to edit this book; I became understandably nostalgic. It was an absolute honor to be considered worthy of guiding the current generation as well as those to come. I put all my knowledge and hard work into making this book most beneficial for its readers.

Tectonics is concerned with the large-scale processes which affect the structure of the earth's crust. This book focuses on distinct characteristics of tectonic researches, primarily on modern geodynamic procedures. A combination of existing and new works, along with novel outcomes and interpretations, has been elucidated in this book for diverse tectonic settings. The book provides updated content on descriptive geological-geophysical investigations, which can help readers to comprehend the essence of mechanisms of distinct tectonic processes more vividly. Descriptive procedures in axes of slow-spreading mid-ocean ridges with an example of central part of Mid-Atlantic Ridge and in continental collision zones are some of the general problems of tectonics that have been presented in this book. Composition of sedimentary basins has been analyzed with examples of Niger Delta, Mesozoic and Cenozoic basins of the Alpine margin (Tunisia), and Triassic Cuyana Basin (Argentina). Neotectonic procedures in Morocco and Turkey have been evaluated; tectonic evolution of the southern margin of Laurasia in the Paleozoic has been discussed about; and interconnection of western Troms-Lofoten and the Lewisian complexes in the Middle Paleoproterozoic has also been elucidated.

I wish to thank my publisher for supporting me at every step. I would also like to thank all the authors who have contributed their researches in this book. I hope this book will be a valuable contribution to the progress of the field.

Editor

General Problems of Tectonics

Features of Caucasian Segment of the Alpine-Himalayan Convergence Zone: Geological, Volcanological, Neotectonical, and Geophysical Data

E.V. Sharkov, V.A. Lebedev, A.G. Rodnikov,
A.V. Chugaev, N.A. Sergeeva and L.P. Zabarinskaya

Additional information is available at the end of the chapter

1. Introduction

The Caucasus Mountain System (Fig. 1) is the part of the largest Alpine-Himalayan collision zone, stretching out for 16,000 kilometers across Eurasia from the Western Mediterranean to the Western Pacific. In geological terms, it is represented by huge area of the Trans-Eurasian Belt (TEB) of Late Cenozoic activation, which was formed after the closure of the Mesozoic-Early Cenozoic ocean Neotetis (Sharkov, 2011). TEB is characterized by powerful modern processes of mountain building, appearance of rift structures, numerous intraplate basaltic lava plateaus and chains of intracontinental andesite-latite volcanic arcs that trace suture zone of continental plates collision. In that both types of magmatism developed almost simultaneously with each other and with the tectonic processes of the entire length of the belt. Large amagmatic blocks (North-Eurasian and Indian) are located on its both sides.

Despite of long-standing geological and geophysical investigations, the Caucasus is still insufficiently studied region, especially in terms of interrelation of its deep-seated structure with neotectonics and manifestations of late Cenozoiv volcanism. However, numerous diverse data obtained during the recent years, allow us to consider this region as a testing place for a comprehensive study of modern geodynamic and petrologic processes in zone of continental plates collision.

The aim of this chapter is synthesis of geological, petrological and geophysical information about the deep-seated processes in the Caucasian parts of the collision zone and their expression in the geological processes in the crust.

Figure 1. Location of the Caucasus Mountain System in World map of volcanoes, earthquakes and plate tectonics (World map of volcanoes, earthquakes and plate tectonics;. Compiled by T. Simkin, R.L. Tilling, J.N. Taggart, W.J. Jones and H. Spall. Smithsonian Institution and US Geological Survey, 1989).

2. Alpine-Caucasian part of the convergence zone

The most complicated structure within of the Trans-Eurasian Belt of the late Cenozoic activization has Alpine-Caucasian segment, where systems of andesite-latite volcanic arcs and back-arc basins (Alboran, Tyrrhenian, and Aegean seas as well as Pannonian depression) occur (Fig. 2). Despite the differences in the morphology of these structures, they have several common features. Along their periphery developed fold-thrust zone, a kind of "accretionary prism" that form the ridges of the Alps, Carpathians, Apennines, Gibraltar arc, etc. Among tectonic slices are often observed deep-water sedimentary rocks of the Tethys, ophiolite complexes, and sometimes blocks of the lower crust and upper mantle rocks like Ivrea-Verbano, Ronda, Beni Bousera and others (Magmatic ..., 1988).

In the rear of these structures, repeating their configuration, andesite-latite volcanic arc of three types occur: (i) island (Aegean), (ii) "semicontinental", located partly on the continent and in part - directly adjacent to it (Alboranian, South-Italian, etc.), and (iii) "intracontinental" (Carpathian, Anatolian-Caucasian-Elburssian, etc.). Behind these arcs newly-formed depressions occur with thinned at the expense of lower high-velocity layers (Gize, Pavlenkova, 1988) crust of intermediate or even oceanic type, within which basaltic volcanism develops. Oceanic crust in the Western Mediterranean has evolved in the Late Cenozoic due

Features of Caucasian Segment of the Alpine-Himalayan Convergence Zone: Geological, Volcanological, Neotectonical, and Geophysical Data

5

to continental crust the African plate (Ricou et al., 1986; Ziegler et al., 2006) as a result of back-arc spreading (Sharkov, Svalova, 2011).

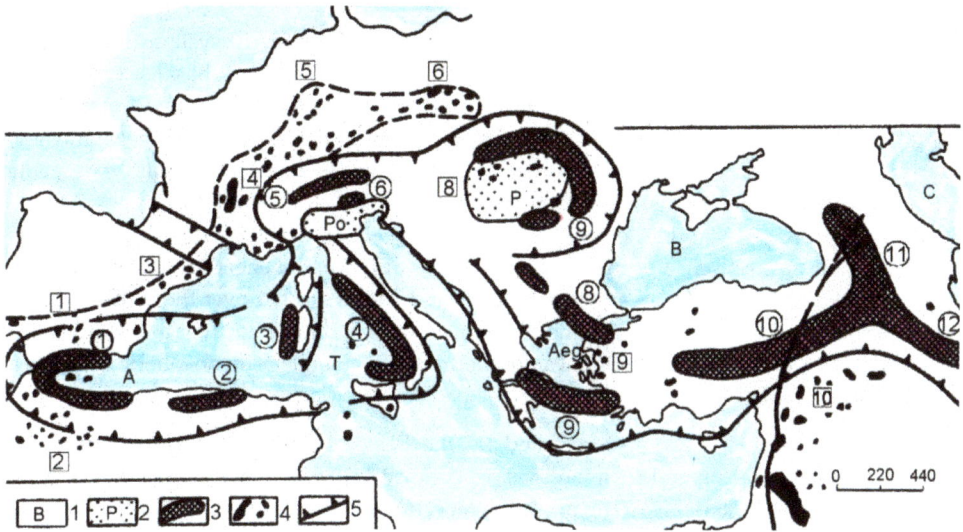

Figure 2. Distribution of the late Cenozoic igneous rocks within the Alpine Belt
1 – back-arc seas (A – Alboran, T – Tyrrhenian; Ae – Aegian) and "downfall" seas (B –Black, C – Caspian);
2 – back-arc sedimentary basins (P- Pannonian, Po – Po valley); 3 – late Cenozoic andesite-latite volcanic arcs (in circles): 1 – Alboran, 2 – Cabil-Tell, 3 – Sardinian, 4 – South-Italian, 5 – Drava-Insubrian,
6 – Evganey, 7 – Carpatian, 8 – Balkanian, 9 – Aegian, 10 -12 – Anatolian-Caucasian-Elburssian
(10- Anatolia-Caucasian, 11 – zone of the Modern Caucasus volcanism, 12 – Caucasus-Elburssian);
4 – areas of flood basaltic volcanism (in square): 1 – South Spain and Portugal, 2 – Atlas, 3 – Eastern Spain, 4 – Central France massif, 5 – Rhine graben, 6 – Czech-Silesian, 7 – Pannonian,
8 – Western Turkey, 10 – northern Arabia; 5 – suture zones of major thrust structures

According to geophysical data, lithosphere of the Alpine Belt has very complicated structure and rather different beneath ridges and basins (Hearn, 1999; Artemieva et al., 2006, Kissling et al., 2006). M. Artemiev (1971) firstly showed that two types of basins exist here (Fig. 3). The first type (Tyrrhenian, Aegian, and Alboran seas a well as Pannonian depression) is characterized by large positive isostatic anomalies, which evidence about excess of mass beneath them, and presence of basaltic volcanism. Very likely, they represent occurrence of the present-day extended mantle plume heads, which continues to receive fresh material resulting in onwards displacement of volcanic arcs in time. Judging by the magnitude of the anomalies, the most intense it arrives in the areas beneath the youngest back-arc sea of the Mediterranean - Aegean, as well as under the Pannonian basin.

Propagation of the plume heads, judging on geological data, had diversy-oriented character - under the Carpathians material moved to the east (Royden, 1989; Harangi et al., 2006), under the Tyrrhenian Sea - in the south-east (Rehault et al., 1987; Harangi et al., 2006), under the Aegean Sea - to the south (Harangi et al., 2006), under the Alboranian Sea – to west (Lonengran, White, 1997). At that crustal material above extended plume heads was transported to their front, where involved into descending mantle flows with the formation of subduction zones with the corresponding magmatism; back-arc basins with transitional and oceanic crust were formed in their rear (Bogatikov et al, 2009).

Simultaneously with formation of depressions in the collision zone rift systems formed and basaltic volcanism develops, represented mainly Fe-Ti subalkaline basalts typical for intraplate (plume) magmatism (rifts in Central and Western Europe, Atlas, basalt plateaus of North Africa and Western Arabia, etc.), emerging in front of the mountain ranges (Grachev, 2003; Wilson, Downes, 2006). Based on isotopic and geochemical data, this anorogenic Circum-Mediterranean magmatism has a common source - the so-called Common Mantle Reservoir (Common Mantle Reservoir, Hofmann, 1997; Lustrino, Wilson, 2007). This obviously indicates presence of the modern mantle superplume beneath the whole region; Alpine orogen with its complex system of ridges and basins located over the central part of this superplume. It is in a good agreement with geophysical data: according to (Hearn, 1999; Smewing et al., 1991, etc.), these plumes are merged into a single asthenospheric uplift at a depth of 200-250 km.

The second type of depressions presented by the Eastern Mediterranean, including the Ionian Sea, and the Black and Caspian seas. In contrast to the aforementioned first type of depressions, there are large isostatic minimums occurred beneath them, indicating a deficit of mass, probably due to the downward currents in the mantle ("cold plumes") between extended heads of "hot" plumes. For these seas are typical passive margins, thick (up to 16-20 km) covers of Meso-Cenozoic sediments and oceanic crust; magmatic activity absent in connection with them. One of the main minima located in the eastern Mediterranean (Levantine basin), where approximately 3-3.5 Ma rapid subsidence area below sea level occurred (Emels et al., 1995). Substantial isostatic anomalies are absent in connection with the Black Sea, which, probably, indicates a reduced rate of downward movement.

Basins of the Black and Caspian seas look like a large "downfalls" which cutoff pre-Pliocene structures of the Caucasus and the Kopetdag. The formation of these seas began, apparently, in the early Cretaceous, but a significant deepening of the basins occurred in the late Oligocene-early Miocene, and in Miocene gradual shallowing of the basins took place (Zonenshain, Le Pichon, 1986). New deepening of the Black Sea and South Caspian basins occurred in Pliocene-Quaternary, which happened almost simultaneously with the uplift of the Caucasus and the Crimea, which in the Oligocene-early Miocene were not expressed in the relief (Grachev, 2000).

Judging on geophysical data, along the sides of "downfall" basins (for example, under the northern parts of the Black Sea and Eastern Mediterranean), steeply-dipping seismically active zones occur, which stretching in the mantle to depths of 60-70 km (Zverev, 2002; Shempelev et al, 2001).

An exception to the general rule, a large positive isostatic anomaly is, situated beneath the Trans-Caucasian Transverse Uplift on the Great and Lesser Caucasus (see Fig. 3). It is located between the "downfalls" of the Black and Caspian Seas, and continues to south to eastern Anatolia and north of the Arabian Peninsula. Obviously, this anomaly is also

Features of Caucasian Segment of the Alpine-Himalayan Convergence Zone: Geological, Volcanological, Neotectonical, and Geophysical Data

7

associated with ascending of a mantle plume. There is no depression here, but the front of the Anatolian-Caucasian-Elburssian andesite-latite arc shifted sharply to the north, forming a zone of young volcanism of the Caucasus.

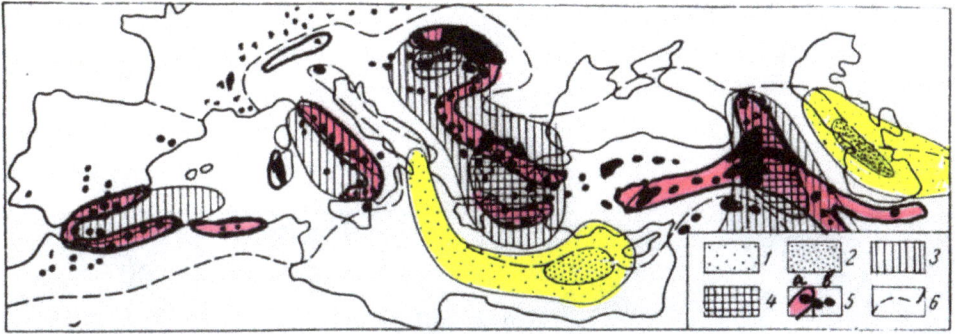

1 - regional lows of average intensity; 2 - of high intensity; 3 - regional highs of average intensity; 4 - intense; 5 - volcanic rocks: a - calc-alkaline series, b - basalt series; 6 – boundaries of the Alpine Belt

Figure 3. Distribution of regional isostatic anomalies and areas of Cenozoic volcanism in Alpine Belt. After M.Artemiev.

3. Features of structure of Caucasian segment

The Caucasus Mountains are located in the eastern part of the actual Alpine zone itselfs in the Arabian-Eurasian syntaxis, between the "downfalls" of Black and Caspian seas (Fig. 1). As it was mentioned above, these seas, very likely, represent small relics of the Neotethys Ocean, which gradually shallowed in the Miocene; their new significant deepening began in the Pliocene-Quaternary, along with the rise of the Caucasus and the Crimea.

Main Caucasian Ridge (Greater Caucasus, GC), in essence, is the southern edge of the Eurasian plate, raised along the Main Caucasus Fault (MCF). The latter is a part of a super-large deep-seated fault, traceable from the Kopetdag across the Caspian Sea, the Caucasus and the Crimea. Very likely, that its further continuation is the Trans-European Suture Zone (Tornquist-Teisseyre Fault Zone), which separates the East European craton from European Variscides and Alpides (Artemieva et al., 2006). In essence, this Kopetdag-Caucasus-Trans-European super-fault divides the Alpine orogen and stable Eurasian plate *sensu stricto* (Fig. 4).

There is a consensus now that formation of the Alpine tectonic structure of the Caucasus occurred under leadership of submeridional horizontal compression. It is generated by a counter movement of two plates: the Arabian ledge (the indenter) and the East European

Craton. Pressure is transmitted from the Arabian to the area of the Greater Caucasus (Arabi-(Arabian-Caucasian syntaxis). The current "invasion" of the Arabian wedge into the Eurasian plate occurs along the Bitlis-Zagros toward the Greater Caucasus, and its displacement relative to Eurasia takes place at a rate of several centimeters per year (Saintot et al., 2006). Reducing the space between Arabian indenter and Eurasian plate in the Late-Alpian time reaches a total of 400 km, but it is unevenly distributed over the area (Leonov, 2007). The bulk of it falls on the territory lying south of the Main Caucasian Fault. To the north of it, within the Greater Caucasus, some reduction also takes place, but it is small and, apparently, does not exceed a few tens of kilometers.

TESZ – Trans-European Suture Zone; MCF – Main Caucasian Fault; KDF – Kopetdag Fault

Figure 4. Kopetdag-Caucasian-Trans-European superfault.

In this context, great importance is the structure of the MCF. According to popular opinion, it is a large overthrust or underthrust Transcaucasian massif under the Great

Features of Caucasian Segment of the Alpine-Himalayan Convergence Zone: Geological, Volcanological, Neotectonical, and Geophysical Data

9

Caucasus (Khain, 1984; Saintot et al., 2006 and references therein). However, the direct and indirect observations of geological data, as well as the most straightforward form of the fault suggests that the leading role play here steep imbricate structures and reverse faults; a steep or vertical inclination of the MCF observed from geophysical data to depths of 70-80 km (Shempelev et al, 2005). From this position the MCF is a reverse fault with a large value of vertical displacement, but without a large horizontal component (Leonov, 2007). In this case, the reduction of space to the south of the GC may be due to lateral "diffluence" of the matter to both sides before the hard "stop" of the East European Craton under the pressure of the Arabian indenter, as it can follows from the results of the study of modern GPS-deformation in the zone of Africa-Arabia-Eurasia continental collision (Reilinger et al., 2006). In other words, judging on geological and geophysical data, convergence of Arabia with Eurasia is substantially accommodated by lateral transport of material within the interior part of the collision zone and lithospheric shortening between the Caucasus and Zagros mountains.

4. Late Cenozoic volcanism of the Caucasus

An important feature of the area of syntaxis is a large belt of Late Cenozoic (up to practically present-day) volcanism, which extends in submeridional (Transcaucasian) direction from the eastern Anatolia via the Lesser to the Greater Caucasus, where large Elbrus and Kazbek volcanoes occur. Volcanics of this belt on their petrological and geochemical features are often close to the suprasubduction calc-alkaline magmas, and represented mostly by basaltic andesites, andesites and dacites under subordinate role of low-Ti basalts and rhyolites (Koronovsky, Demina, 2007; Keskin et al., 2007). Volcanic structures themselves with a lot of calderas and acid pyroclastics are also very close to the volcanoes of island arcs and active continental margins. Along with this type of magmatism, extensive lava plateaus, formed by eruptions (often fissured) of moderately alkaline basalts of within-plate type, such as Javakheti, Geghama, Syunik, Kars, and others, are also observed in the region.

Volcanics of the calc-alkaline series ("suprasubduction type") is of a prime interest. In contrast to the island arcs and active continental margins, Eastanatolian-Caucasus volcanic belt stretches across the overall structure of the Greater Caucasus, following the direction of positive Transcaucasian isostatic anomaly. Just south of the Van Lake, this belt is divided into two branches: one of them can be traced to the west, to Central Anatolia, and the other - to the east, toward the Elburs and the Zagros. Moreover, although the Arabian-Caucasian syntaxis is characterized by high seismicity, the earthquake hypocenters with depth not more than 50-60 km dominated, and deep-focus earthquakes (up to 120 km) are extremely rare and occurred only in the eastern part of the GC at its border with the Caspian Sea (Fig. 5 and 6). From this it follows that there is no any subduction beneath the eastern Anatolia (Sandvol et al., 2003) and Caucasus at present.

Isotopic and geochemical data suggest that origin of magmatism of "suprasubduction type" associated with interaction between a mantle plume head and continental crust material (Lebedev et al, 2006; 2011; Chugaev et al., 2012). Judging by the fact that the orientation of

East-Anatolian-Caucasian zone, where joined Anatolian-Caucasian and Caucasian-Elburssian arcs, practically coincides with the zone of syntaxis, i.e. with the area of maximum stress, we think that melting of crustal material in processes of deformations at high pressures played an important role in generation of these magmas. As shown earlier, crystal lattice of minerals under such conditions is at the stress state, which making them easier to disintegrate, and the conversion to the liquid phase requires much less heat (Sharkov, 2004 and references herein).

An essential role in this process of melting can play frictional heat generated during deformation (Frischbutter, Hanisch, 1991; Molnar, England, 1995) and mantle fluids from degassing due to decompression of the plume head, which penetrated a deformable crust, introducing some warmth and some components, how it is determined by isotopic and geochemical methods. From this standpoint the emergence of the Anatolian-Caucasian and Caucasian-Elburssian arcs may be due to the diffluence of deep-seated crustal material to both sides from the Arabian indenter, fixing appearance of foci of melting related to the tectonic flowage of material. In other words, these volcanic arc trace suture zone, on which outflow of crustal matter from the "stop" of the Eurasian plate occurred in the process of continental collision. Likely, the location of these sutures zones was determined by configuration of the plume head which extends to the north and oncoming "streamlined" from the both sides by tectonized material of shallow lithosphere.

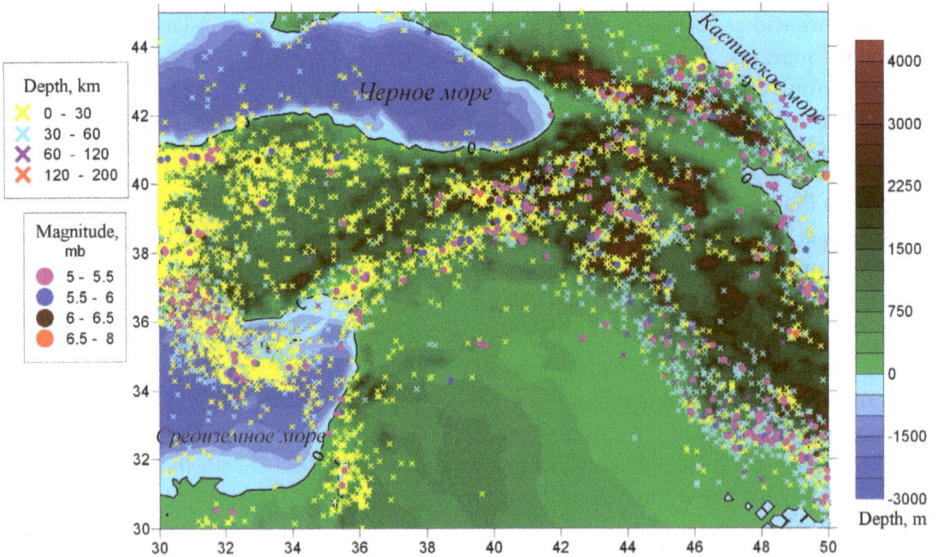

Figure 5. Distribution of the earthquakes in the region. Shallow focuses sharply predominate.

Features of Caucasian Segment of the Alpine-Himalayan Convergence Zone: Geological, Volcanological, Neotectonical, and Geophysical Data

11

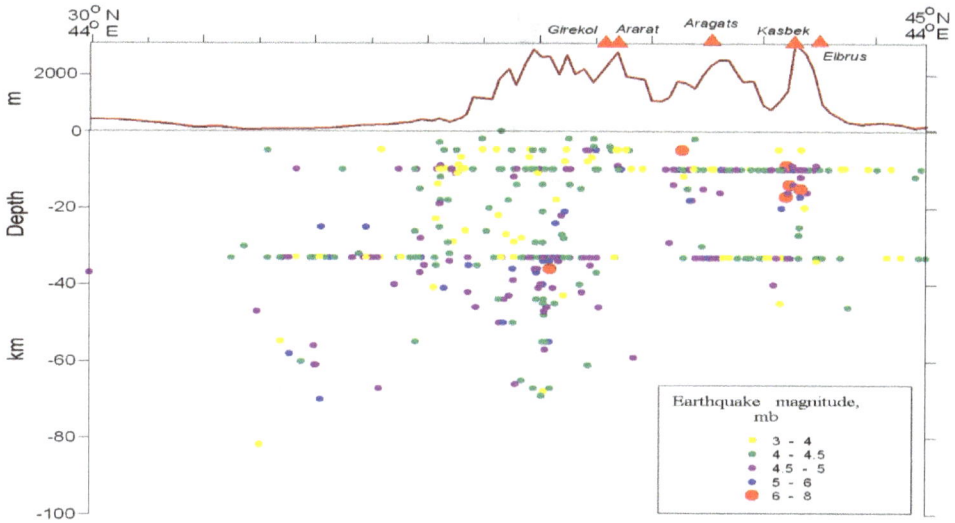

Figure 6. Distribution of earthquake focuses beneath the volcanic area Eastern Turkey – Caucasus

Similar isotopic and geochemical characteristics were established for the late Cenozoic calc-alkaline volcanic rocks in the zone of continental collision in the Alpine-Mediterranean region, where these features of magmas are usually explained by the complex composition of the mantle sources, strongly contaminated by crustal material (Harangi et al., 2006 and references herein).

5. Discordance between deep-seated situation and geological structure of the region

Attention is drawn to discordance of deep-seated processes and geological situation on the surface: the Kopetdag-Caucasian-Trans-European superfault sinking beneath level of the Caspian Sea, where it can be traced only by a strip of earthquakes. This fault is distorted in the north of the Black Sea, which may be indicative of the continued deepening of the sea which disturb the subsurface geological structures (Fig. 7). In addition, geophysical and geological data indicate that the head of the mantle plume from the late Miocene has extended to the north, crossing at the depth the MCF and resulting in the appearance of the modern volcanism including the late Quaternary volcanoes Kazbek and Elbrus with its present-day shallow magmatic chambers (Masurenkov, Sobisevich, 2010; Gurbanov et al., 2011). It is possible that "diving" of the plume head under the edge of the Eurasian plate, which occurred in the Miocene and continued at present, caused a regeneration of an older suture zone, leading to the rise of the Greater Caucasus. Thus, the geological situation in the region continues to develop, mainly due to large-scale deep-seated processes.

Figure 7. Discordances of the Kopetdag-Caucasian-Trans-European superfault in the region

It is pay attention that orientation of the Anatolian-Caucasian volcanic arc does not coincide with the largest neotectonic structures on the Turkish territory – North-Anatolian and East-Anatolian fractures zones: it is located between them. There is no clear correlation also between volcanism and the present-day motions of crustal material which established in the zone of Africa-Arabia-Eurasia continental collision by GPS constraints (Reilinger et al., 2006). All of these also evidence that deep-seated processes in the mantle not always found their reflection on the relatively shallow crustal level.

What could be further scenario? Most likely, this process of propagation of the mantle plume head to the north could lead to "cut open" of the lithosphere of the Eurasian plate, the separation of the Caucasian mountain system into two parts and the formation here continental rift zone like the Rhine Graben.

6. Conclusions

1. The Caucasus is a part of huge Late Cenozoic Alpine-Himalayan convergence zone. The Greater Caucasus is an edge of the Eurasian plate, raised along the large reverse fault - the Main Caucasian Fault. This fault, in turn, is a part of the super-fault, stretching from the Kopetdag to the Trans-European Suture Zone (zone Tornquist-Teisseyre).
2. The Caucasus is limited from both sides by large depressions modern Black and Caspian seas of "downfall" type, which "cut" pre-Pliocene structures both the Caucasus and the Kopetdag; origin of these seas is associated with downward currents in the mantle ("cold plumes").
3. The peculiar structure of the region is north-south Transcaucasian Rise, which is located in the northern part of Caucasian-Arabian syntaxis. Large positive isostatic anomaly is confined with it, apparently indicating presence here of the mantle plume head.
4. Along the zone of syntaxis the belt of Neogene-Quaternary volcanism occurs which begins in Eastern Anatolia and traced through the Lesser and Greater Caucasus. Two

types of volcanic rocks are represented here: (1) prevailing volcanics of calc-alkaline se-series, very close in petrological and geochemical characteristics to suprasubduction type, and (2) extensive lava plateaus formed by basalts of intraplate (plume related) type.

5. However, subduction zone under the Caucasus region, as well as throughout the Caucasian-Arabian syntaxis is absent and shallow earthquakes (50-60 km) are dominated here. We considered that origin of calc-alkaline magmas is associated with interaction between the mantle plume head and crustal material at relatively shallow depths under conditions of deformation at high pressures, leading to melting of the material in the zone of collision. In other words, appearance here of such magmas has not considered to any subduction zone.

6. Reduction of space in the area of Caucasian-Arabian syntaxis, which occurred during the Late Cenozoic, reached 400 km; such shortening in absence of subduction was apparently achieved mainly due to "diffluence" of the crustal material to both sides before hard "stop" of the East European Craton under the pressure of the Arabian indenter.

7. Situation in the region continues to develop now mainly due to deep-seated mantle processes, destroying the structure of the pre-Pliocene collision zone, while the development of the underlying processes occurs independent of the processes in the earth's crust.

Author details

Evgenii Sharkov, V.A. Lebedev and A.V. Chugaev
Institute of Geology of Ore Deposits, Petrography, Mineralogy and Geochemistry RAS, Moscow, Russia

A.G. Rodnikov, N.A. Sergeeva and L.P. Zabarinskaya
Geophysical Center RAS, Moscow, Russia

7. References

Artemiev, M.E., 1971. Some peculiarities of deep-seated structure of depressions of Mediterranean type: evidence from data on isostatic gravity anomalies. *Bull. Soc. of the Nature Investigators of Moscow*. Geol. Dept., 4, 5-10 (in Russian with English abstract)

Artemieva, I.M., Thybo, H. & Kaban, M.K., 2006. Deep Europe today: geophysical synthesis of the upper mantle structure and lithospheric processes over 3.5 Ga. In: *European Lithosphere Dynamics*. D.G. Gee and R.A. Stephenson (eds). Geol. Soc. London Mem. 32: 11-42.

Chugaev, F.V., Chernyshev, I.V., Lebedev, V.A., Eremina, A.V., 2012. Isotopic composition of plumbum and origin of Quaternaly lavas of Elbrus volcano (Greater Caucasus, Russia): data of high-precisious MC-ICP-MS method, *Petrology*, 20, in press.

Emels, K.-Ch., Robertson, A. & Richer, C., 1995. Mediterranean Sea.1. *Science operator report DSDP* Leg.160: 11-20.

Frischbutter A., Hanisch M., 1991. A model of granitic melt formation by frictional heating on shear planes. *Tectonophysics*, 194: 1-11.

Gautheron, G., Moreira, M. & Allegre, C., 2005. He, Ne and Ar composition of the European lithospheric mantle. *Chem. Geology*, 1-2: 97-112.

Gize, P. & Pavlenkova, N.I., 1988. Structural maps of the Earth's crust of Europe. *Physics of the Earth*, N 10, 3-14.

Grachev, A.F. (ed.), 2000. *Neotectonics, geodynamics and seismicity of the Northern Eurasian*. PROBEL, Moscow, 487 p. (in Russian with English abstract).

Grachev, A.F., 2003. Final volcanism of Europe and it's geodynamic nature. *Physics of the Earth*, N 5: 11-46.

Gurbanov, A.G., Bogatikov, O.A., Karamurzov, B.S., Tsukanova, L.E., Lexin, A.V., Gazeev, V.M., Mokhov, A.V., Gornostaeva,T.A., Zharikov, A.V., Shmonov, V.M., Dokuchaev, A.Ya., Gorbacheva, S.A., & Shevchenko, A.V., 2011. Unusual type of degassing from melts of peripheral magmatic chambers of "asleep" Elbrus volcano (Russia): geochemical and mineralogical features. *Volcanology & Seismology*, 4: 3-20.

Harangi, S., Downes, H. & Seghedi, I., 2006. Tertiary-Quaternary subduction processes and related nagmatism. In: *European Lithosphere Dynamics*. D.G. Gee and R.A. Stephenson (eds). Geol. Soc. London Mem. 32: 167-190.

Hearn, T.M., 1999. Uppermost mantle velocities and anisotropy beneath Europe. *Journ. Geophys. Res.*, 104(B7): 15123-15139.

Hofmann A.W., 1997. Mantle geochemistry: the message from oceanic volcanism. *Nature*, 385: 219-229.

Keskin, M., 2007. Eastern Anatolia: A hotspot in a collision zone without a mantle plume. In: Foulger, G.R. and Jurdy, D.M. (Eds.). Plates, plumes, and planetary processes. *Geological Society of America Special Paper* 430, 693-722.

Khain, V.E., 1984. *Regional Geotectonics. Alpian-Mediterranean Belt*. Moscow, Nedra, 344 p. (in Russian with English abstract).

Kissling, E., Schmid, S.M., Lippitsch, R., Ansorge, J. & Fugenschuh, B., 2006. Lithosphere structure and tectonic evolution of the Alpine arc: new evidence from high resolution teleseismic tomography. In: *European Lithosphere Dynamics*. D.G. Gee & R.A. Stephenson (eds). Geol. Soc. London Mem. 32: 129-145.

Lebedev, V.A., Bubnov, S.N., Chernyshev, I.V., and Gol'tsman, Yu.V., 2006. Basic Magmatism in the Geological History of the Elbrus Neovolcanic Area, Greater Caucasus: Evidence from K-Ar and Sr-Nd Isotope Data. *Doklady Earth Sciences*, 2006, Vol. 406, No. 1, pp. 37-40.

Lebedev, V.A., Chernyshev, I.V., Chugaev, A.V., Gol'tsman, Yu.V. & Bairova, E.D., 2010. Geochronology of Eruptions and Parental Magma Sources of Elbrus Volcano, the Greater Caucasus: K-Ar and Sr-Nd-Pb Isotope Data. *Geochemistry International*, Vol. 48, No. 1, pp. 41-67.

Lonergan, L., White, N., 1997. Origin of the Betic-Rif mountains belt. *Tectonics*, 16: 504-522.

Features of Caucasian Segment of the Alpine-Himalayan Convergence Zone: Geological, Volcanological, Neotectonical, and Geophysical Data

15

Louden, K.E., Chian, D., 1999. The deep structure of non-volcanic rifted continental margins. *Roy. Soc. of London Phil. Trans*, 357: 767-804.

Lustrino, M. & Wilson, M., 2007. The circum-Mediterranean Anorogenic Cenozoic Igneous Province. *Earth Sci. Rev.* 81. P. 1-65.

Magmatic Rocks. V. 5. Ultramafic Rocks, 1988. E.E. Laz'ko & E.V. Sharkov (Eds.). Moscow, Nauka Publishers, 508 p. (in Russian)

Masurenkov, Yu.P. and Sobisevich, A.L. Fluid-magmatic system of Pyatigosk volcanic center. Doklady Earth Sci., 425, 815-820.

Rehault, T. G., Moussat, E. & Fabri, A., 1987. Structural evolution of Tyrrhenian back-arc basin. *Marine Geology*, 74, 123-150.

Reilinger, R., McClusky, S., Vernant, P. et al., 2006. GPS constraints on continental deformation in the Africa-Arabia-Eurasia continental collision zone and implications for the dynamics of plate interactions. *J. Geophys. Res.*, 2006. 111, B05411, 1 of 26, doi:10.1029/2005JB004051.

Ricou, L.E., Dercourt, J., Geyssant, J. et al., 1986. Geological constrain on the Alpine evolution of the Mediterranean Tethys. *Tectonophysics*, 123, 83-122.

Royden, L.H., 1989. Late Cenozoic tectonics of the Pannonian Basin System. *Tectonics*, 8, 51-61.

Saintot, A., Brunet, M.-F., Yakovlev, F., Sebrier, M., Stephenson, R., Ershov, A., Chalot-Prat, F. & McCann, T., 2006. The Mesozoic-Cenozoic tectonic evolution of the Great Caucasus. In: *European Lithosphere Dynamics*. D.G. Gee and R.A. Stephenson (eds). Geol. Soc. London Mem. 32: 129-145.

Sharkov, E.V., 2004. Role of the energy of interface formation in the melting and retrograde boiling. *Geochemistry Intern.*, 42: 950-961.

Sharkov E., 2011, Does the Tethys Begin to Open Again? Late Cenozoic Tectonomagmatic Activization of the Eurasia from Petrological and Geomechanical Points of View. *New Frontiers in Tectonic Research - General Problems, Sedimentary Basins and Island Arcs*, E.V. Sharkov (Ed.), Rijeka: InTech, 2011, 4-18.

Sharkov E. & Svalova V., 2011. Geological-Geomechanical Simulation of the Late Cenozoic Geodynamics in the Alpine-Mediterranean Mobile Belt. In: *New Frontiers in Tectonic Research - General Problems, Sedimentary Basins and Island Arcs*, E.V. Sharkov (Ed.), Rijeka: InTech, 19-38.

Shempelev, A.G., Prutsky, N.I., Feldman, I.S., & Kukhmazov, S.U., 2001. Geologo-geophysical model along profile Tuapse-Armavir. In: *Tectonics of the Neogea: general and regional aspects. Proceedings of XXXIV tectonic meeting*. Moscow, GEOS, 316-320 (in Russian).

Shempelev, A.G., Prutsky, N.I., Kukhmazov, S.U. et al,, 2005. Materials of geophysical investigation along Near-Elbrus profile (volcano Elbrus-Caucasian Mineral Waters). In *Tectonics of the Earth's crust and mantle. Proceedings of XXXVIII tectonic meeting*. Moscow, GEOS, 361-365 (in Russian).

Spakman, W., van der Lee, S. & van der Hilst, R., 1993. Travel-time tomography of the European-Mediterranean mantle down to 1400 km. *Phys. Earth Planet Inter.*, 79, 3-74.

Wilson, M. & Downes, H., 2006. Tertiary-Quaternary intra-plate magmatism in Europe and its relationship to mantle dynamics. In: *European Lithosphere Dynamics*. D.G. Gee and R.A. Stephenson (eds). Geol. Soc. London Mem. 32, 147-166.

Ziegler, P.A., Schumacher, M.E., Dezes, P., van Wees, J.-D. & Cloetingh, S., 2006. Post-Variscan evolution on the lithosphere in the area of the European Cenozoic Rift System. In: *European Lithosphere Dynamics*. D.G. Gee and R.A. Stephenson (eds). Geol. Soc. London Mem. 32, 97-112.

Chapter 2

Cyclic Development of Axial Parts of Slow-Spreading Ridges: Evidence from Sierra Leone Area, the Mid-Atlantic Ridge, 5-7°N

E.V. Sharkov

Additional information is available at the end of the chapter

1. Introduction

Slow- and ultraslow-spreading mid-ocean ridges became come to attention of researchers in recent years again because identification of so-called oceanic core complexes (OCCs). These complexes are characterized by tectonized and heterogeneous lithosphere, large yields of altered gabbros and serpentinized mantle at the oceanic bottom and presence of large deep-sea hydrothermal fields and mineralization (Conference…, 2010). For example, OCCs are quite common in the slow-spreading Mid-Atlantic Ridge (MAR) where they make up ~30% of its length (Escartín et al., 2008; Smith et al., 2008; MacLeod et al., 2009 and references therein). OCC form about 50% of ultra-slow South-West Indian Ridge length (Cannat, 2010); the most studied site here is Atlantis Bank (Thy, 2003; Schwartz et al., 2009). OCCs are known in back-arc seas too, for example in the Philippine Sea (Ohara et al., 2001).

The largest of the OCC is the Godzilla Mullion in the Philippine Sea. The second in the world and the largest in the MAR is the St. Peter and St. Paul complex about 90 km long and up to 4000 m in height, located near the axial zone of MAR in the equatorial region, south of the Sierra Leone area. A feature of this OCC is a dissected topography, with its most elevated blocks even reach the ocean surface to form the St. Peter and St. Paul Rocks. They are composed mainly serpentinized often sheared mantle hornblende (metasomatized) peridotites, containing hornblendite schlierens and veins (Roden et al., 1984; Hékinian et al., 2000). Such peridotites are commonly found as xenoliths in intraplate (plume-related) basalts of oceans and continents, representing fragments of the cooled upper parts of mantle plume heads above its melting zone (Magmatic …, 1988); so, that is tectonic block of the upper edge of a mantle plume moved out to the surface here.

The name for OCCs was given by analogy to metamorphic core complexes (metamorphic cores), located in core (inner) parts of orogens on continents. In essence, such complexes are represented by exposed metamorphosed deep-crustal rocks, which underwent by viscous-plastic and brittle deformation. The same situation is typical for OCCs, which are outcrops of tectonized and altered deep-seated crustal and mantle rocks in the axial parts of mid-oceanic ridges. Because of characteristic striated surface (mullion structure), such complexes often referred as megamullions (Tucholke, 1998).

According to the commonly accepted (Penrose) model of plate tectonics, the occurrence of mid-ocean ridge (MOR) is associated with uprising of hot deep mantle material, which, reaching shallow depths, begins to melt due to adiabatic decompression. It is assumed that formation of new oceanic crust occurs here, which symmetrical spreads to both sides of the ridge due to convection currents in underlying mantle. Resulted excess of the crust is absorbed in subduction zones beneath island arcs and active continental margins. The axial part of the MORs, where crust is generated (constructive plate boundaries), considered as centers or zones of oceanic spreading.

From such positions, outcrops of plutonic rocks in the spreading zones do not fit in the traditional model of plate tectonics. According to numerous studies of OCCs, axial parts of ridges are uplifted relative to their average height, and often have asymmetrical structure, where outcrops of plutonic rocks are disposed outside of axial valleys, where neovolcanic hills are located (Ildefonse et al., 2007; Smith et al., 2008; MacLeod et al., 2009, etc.). Modern volcanism is practically absent, however, numerous hydrothermal vents occurred.

In this regard, it was suggested that *oceanic core complex* results from activity of an *oceanic detachment fault* (Conference…, 2010). This fault is a large-offset normal fault formed at or in the vicinity of a mid-ocean ridge axes that accommodates a significant fraction of the plate separation (Fig. 1); offsets range from kilometers to tens kilometers or more. According to this model, oceanic detachment faults may initiate as steep normal faults at depth, and turn into shallow low angle extensional faults through rotation of the footwall. It is suggested that this type of spreading should be recognized as a fundamentally distinct mode of seafloor spreading that does not result in a classical Penrose model of oceanic crustal structure. However, many elements and details of this hypothesis of "one-side spreading" are poorly justified (i.e., unknown fault geometry at depth, structure of magmatic systems, route of hydrothermal currents, etc.), as well as motives of appearance of such detachment faults, which absent in fast-spreading ridges.

Identification of the OCCs set to geologists a number of problems which solution is possible only using the complex of geological, petrological and geochemical studies. Such work was done on example of Sierra Leone area, located in axis of the MAR (5-7°N). It was based on materials dredged during the cruises of R/V "Akademik Ioffe" (10th cruise, 2001-2002) and "Professor Logachev" (22th cruise, 2003) (Sharkov et al., 2005, 2007, 2008; Savelieva et al., 2006; Simonov et al., 2009; Aranovich et al., 2010). Judging on presence here of serpentinites upon mantle peridotites and altered tectonized lower-crustal gabbros, as well as widespread

Cyclic Development of Axial Parts of Slow-Spreading Ridges: Evidence from Sierra Leone Area,
the Mid-Atlantic Ridge, 5-7°N

19

development of extensional structures, including normal faults, the Sierra Leone area can be determined as OCC. However, unlike of typical OCCs, the altered deep-seated rocks were found mainly in the bottom and slopes of deep graben-like depressions, whereas surface of the ridge is covered by the flows of fresh pillow lavas with chilled glassy crusts; i.e. a kind of structural discordance occurred in the area. Marked asymmetry in structure of the ridge is not found here as well as clear evidence of oceanic detachment fault existence. In this context, studied area is of great interest as a possible example of the transition from the typical OCCs to regions of the ridge between them where only basalts developed and spreading is symmetrical.

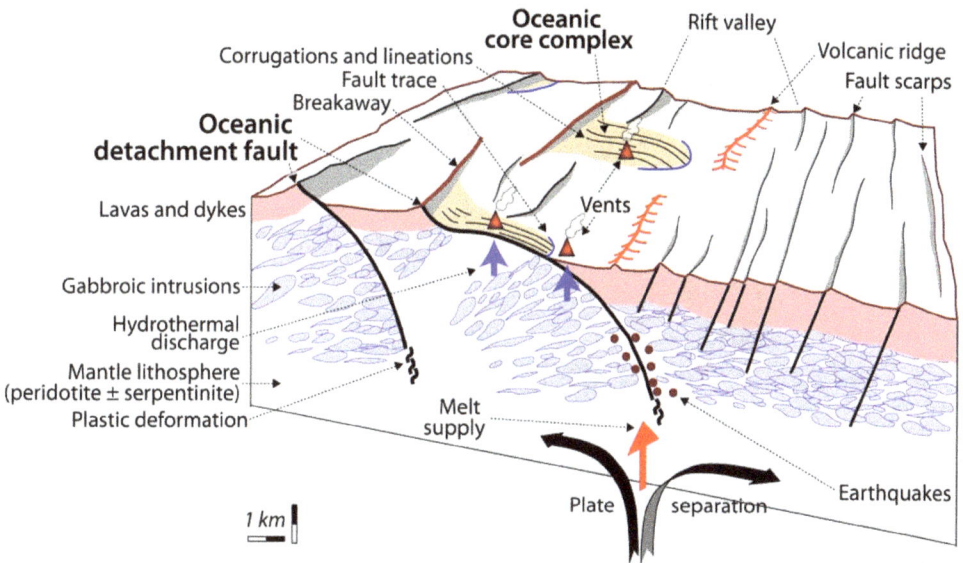

Figure 1. Scheme of oceanic core complex with oceanic detachment fault. After: Conference ..., 2010.

The aim of this paper, based on our data from the Sierra Leone area and published information, to discuss diverse processes, occurred in axes of the modern slow-spreading MORs, and give a new way to interpret of geological, petrological and geodynamical data both in their spreading zones and in underlying mantle.

2. Brief description of geological background

The studied segment of the MAR with strongly dissected relief is located in the vicinity of non-transformed Sierra Leone Fault, between Bogdanov Fracture Zone (7°10' N) and 5°00' N (Fig. 2). South to area, from 5°00' N and to the Strakhov FZ, the MAR represents leveled basaltic plateau, crossed by narrow meridionally-oriented axial rift valley. Geological structure of the area is showed in (Pushcharovsky et al, 2004).

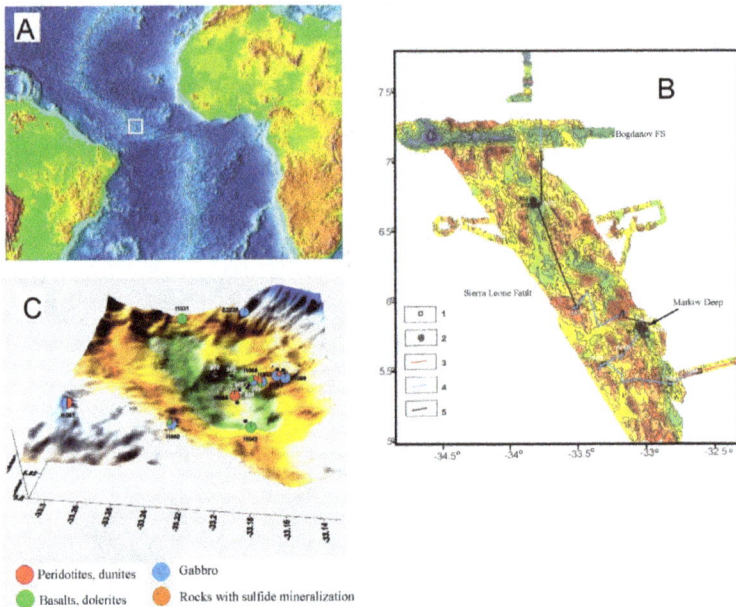

A - location of the Sierra Leone area in the Atlantic; B – Bathymetric map of the Sierra Leone area: 1- sites of dredgins; 2 – sites of Rosetta module; 3-6 – profiles; C – Markov Deep

Figure 2. Sierra Leone area, the Central Atlantic

Feature of morphostructural image of the area is lack of transform faults and the spreading zone is represented here by en echelon system of graben-like depressions (valleys) of 4-5 km depth from ocean surface. As it mentioned above, altered deep-seated rocks found mainly in the sides of rift valleys and on their floor, at that outcrops of plutonic rocks are traced for about 60 km along the MAR axis. Flows of fresh pillow lavas cover top of the ridge and partly fill bottom of some rift valleys. Thickness of these flows is small because within the area of their distribution are found outcrops of altered plutonic rocks. Despite the uneven sampling, we can say with confidence that both sides of the rift valleys formed by the same complex of rocks that characterize the entire section of oceanic crust.

The structure of bedrocks on the eastern slope of the deepest (~5 km) Markov Deep can be seen on Fig. 3, which were finding by marine acoustic complex (sidescan sonar) GBO MAK-1M during the 22th cruise of R/V "Professor Logachev". The crust has a well-defined subhorizontal layered structure, partially masked by sediments, and looks like structure of the Kane OCC (23°30 'N) (Dick et al., 2008). Numerous steep-dipping normal faults are clearly visible here; one of them (at the left), apparently filled with dolerite dike which, probably, represents a lava flow's feeder.

Cyclic Development of Axial Parts of Slow-Spreading Ridges: Evidence from Sierra Leone Area, the Mid-Atlantic Ridge, 5-7°N

21

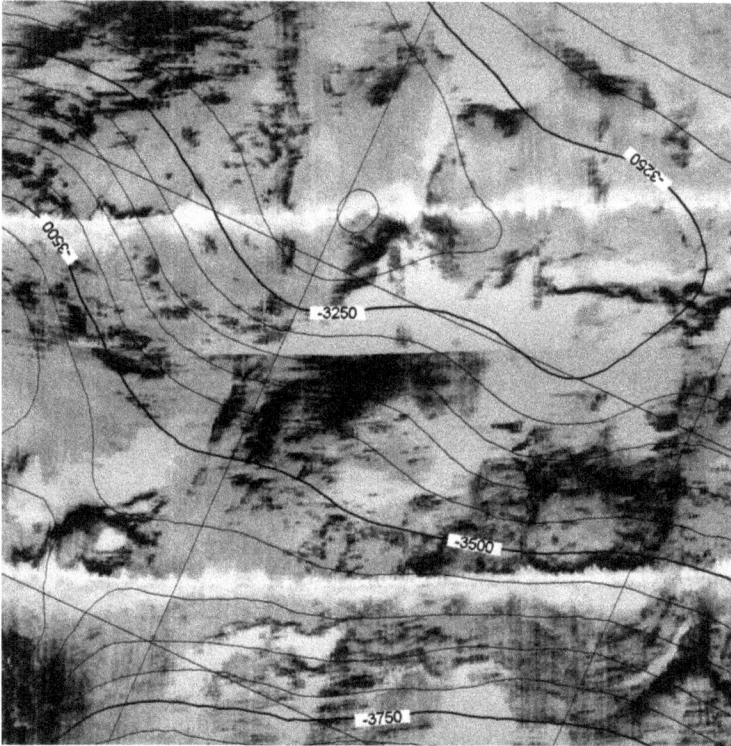

Clearly visible layering of the lithosphere, partially overlapped by sediments (gray). Numerous steeply dipping normal faults are visible; one of them (on the left side of the figure) seems to be filled with dolerite dikes and, probably, was a feeder of lava flow.

Figure 3. Structure of the eastern slope of the Markov Deep by data of remote sensing obtained using marine acoustic complex (sidescan sonar) GBO MAK-1M from board of R/V "Professor Loga-chev" (22th cruise, 2003).

3. Dredged rocks from the Sierra Leone area

The spectrum of dredged rocks at the area is typical for slow-spreading ridges (Sharkov et al., 2005, 2008;. Savelieva et al., 2006):

1. strongly serpentinized ultrabasites (depleted lherzolite and harzburgite, rare dunite); most of them are mantle restites, but in some samples relics of cumulate structures preserved, attesting their intrusive origin;
2. two types of altered tectonized gabbros of lower oceanic crust: (i) primitive magnesian gabbros (troctolite, olivine gabbro, and gabbro), which related to MORB and (ii) ferrogabbros (Fe-Ti-oxides-bearing gabbronorite, hornblende-bearing gabbro and gabbrodiorite), related to specific siliceous Fe-Ti-oxide series (see Section 2.3);
3. small veins and nests of plagiogranites (trondhjemite);
4. dolerite dike and/or sill complexes, including ilmenite- and hornblende-bearing varieties also;
5. basalts – fragments of pillow and massive lavas; very fresh varieties with chilled glassy crusts predominate among them.

Most of the rocks were undergone to secondary alterations. Magmatic minerals (olivine, pyroxene, and plagioclase) often display deformation textures resulted from early high-temperature cataclasis, associated with plastic flow of solidified, but still hot rocks. Judging by the Ti-zircon thermometry, it occurred within a temperature range from 815°C to 710°C (Zinger et al, 2010). During pervasive low-temperature alterations, peridotites underwent strong serpentinization, while gabbros and some basalts – amphibolization with appearance of fibrous actinolite upon pyroxenes and thin veins of prehnite, carbonate and chlorite along the fractures. In some cases rocks were schistozed and brecciated, and underwent by metasomatic processes; the thickest metasomatic zone bear veinlet-disseminated sulfide mineralization (see Section 3).

3.1. Features of the fresh basalts

Most of studied fresh basalts with chilled glassy crusts often have porphyritic structure with phenocrysts of three major types: Ol±Chr, Ol+Pl±Chr and Pl+Cpx, which is typical for MORB (Langmuir et al., 1992). Equilibrium cumulates in transitional magma chambers have to correspond with dunite, troctolite, olivine gabbro and gabbro, typical for many layered mafic-ultramafic intrusions on continents (Sharkov, 2006).

Sometimes partly-melted xenocrysts were found in basalts and volcanic glass: olivine Fo_{88-89}, similar in composition to the olivine of mantle restite, and plagioclase An_{83-86} (Fig. 4), similar in composition to the plagioclases of lower-crustal primitive gabbro. This is evidence that the basaltic melts crossed rocks of the shallow lithosphere on their way to surface.

All studied fresh lavas are commonly oceanic plateau basalts (T-MORB) and more rare close to E-MORB in composition. They are characterized by the same level of REE with typical for MORB flat character of distribution; the Ce/Yb ratio ranging from 0.95 to 1.69. Judging on #mg (56-63) and mineral compositions, they are not primary mantle-derived melts and

Cyclic Development of Axial Parts of Slow-Spreading Ridges: Evidence from Sierra Leone Area, the Mid-Atlantic Ridge, 5-7°N

23

underwent by crystallizing differentiation in transitional (intrusive) magma chambers. It is in a good agreement with small negative Eu-anomaly which reflects the fractionation of plagioclase in intermediate magma chamber (Sharkov et al., 2005).

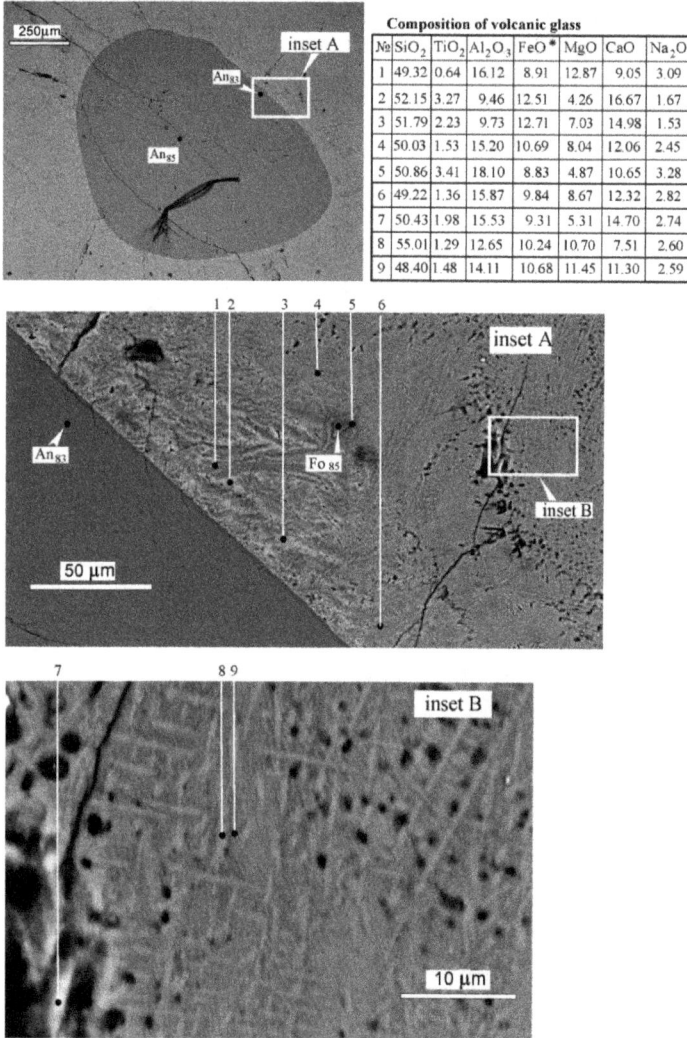

Composition of volcanic glass

№	SiO_2	TiO_2	Al_2O_3	FeO^*	MgO	CaO	Na_2O
1	49.32	0.64	16.12	8.91	12.87	9.05	3.09
2	52.15	3.27	9.46	12.51	4.26	16.67	1.67
3	51.79	2.23	9.73	12.71	7.03	14.98	1.53
4	50.03	1.53	15.20	10.69	8.04	12.06	2.45
5	50.86	3.41	18.10	8.83	4.87	10.65	3.28
6	49.22	1.36	15.87	9.84	8.67	12.32	2.82
7	50.43	1.98	15.53	9.31	5.31	14.70	2.74
8	55.01	1.29	12.65	10.24	10.70	7.51	2.60
9	48.40	1.48	14.11	10.68	11.45	11.30	2.59

Diamonds denote basalts and their glasses (glasses are shown in the inset by open symbols), boxes denote gabbros, and circles are trondhjemites. Dashed lines indicate the position of the data point of metasomatite replacing gabbro. Hypothetical sources: (1) HIMU, (2) EM2. Continuous line outlines the field of the MAR basalts between 3° and 46°S (Fontingie and Schilling, 1996), dashed line denotes MAR basalts between 30° and 50°N (Yu et al., 1997), dotted line outlines basalts of Sao Migel Island, Azores (Widom et al., 1997).

Figure 4. Dissolution of plagioclase xenocrysts and textural–compositional heterogeneity of the chilled glass Sample I1052/38. Image in back-scattered electrons.

Fresh basalts of the Sierra Leone area in terms of Sr-Nd isotopic characteristics fall into central part of the field of modern MORB for the southern hemisphere and occupy an intermediate position between the most depleted basalts of the MAR ($^{87}Sr/^{86}Sr$ <0.7025, εNd >12) and enriched basalts of high latitudes both the northern and southern hemisphere ($^{87}Sr/^{86}Sr$ > 0.7030, εNd <8) (Fig. 5). Variations of the isotopic characteristics within a relatively small (less than 300 km in the meridional direction) studied area is comparable in scale with variations along the 15-20 times more extended segments of the MAR (Sharkov et al., 2008). The points form an elongated box on the diagram which suggests the presence here of two finite member (depleted and enriched) mixing in different proportions.

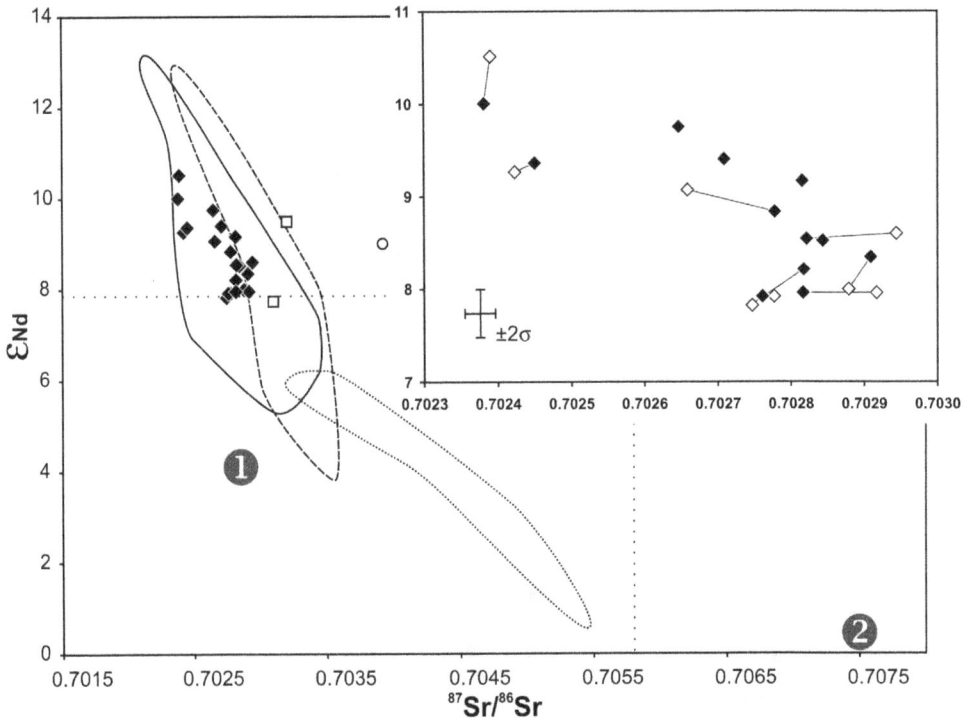

Figure 5. Sr–Nd isotope diagram for the studied basalts and their glasses dredged at the Sierra Leone area, Mid-Atlantic ridge, 5°–7°N.

Significant nonsystematic differences in $^{87}Sr/^{86}Sr$ ratio and less significant differences in εNd value between basalts and their chilled glassy crusts were firstly found in some samples (Sharkov et al., 2008). Higher Sr isotopic ratios can be observed both in the glasses and the basalts at the same lava fragments (Fig. 5, inset), at that isotope and geochemical characteristics of the samples show no essential correlation. So, seawater did not affect to the Sr and Nd isotope system in the chilled crusts of the studied pillow lavas. It is suggested that such isotopic differences are related to a small-scale heterogeneity of the melts which had no time to homogenized during their rapid ascent to the surface. The heterogeneity was

Cyclic Development of Axial Parts of Slow-Spreading Ridges: Evidence from Sierra Leone Area, the Mid-Atlantic Ridge, 5-7°N

25

presumably related to the partial contamination of basaltic melts by older plutonic rocks material (especially, lower-crustal gabbros) (Fig. 3).

3.2. Primitive gabbro

The magnesian primitive gabbros are represented by troctolite, olivine gabbro and gabbro, which dominated among lower-crustal rocks of the area. The same gabbros are widespread in rocks of the lower oceanic crust and traditionally viewed as intrusive equivalents of MORB (Pearce, 2002; Ildefonce et al., 2007; Dick et al., 2008, etc.). Absence of reactionary subsolidus pyroxene-spinel rims between Mg-olivine and Ca-plagioclase, typical for deep-seated gabbros of continents evidence, that crystallization of parental melts occurred under pressure, essential low 5 kbar (Sharkov, 2006).

According to our geochemical data, these gabbros have lower contents of REE, Th, Nb, Ta, Zr and Hf as compared to the fresh basalts of the area (Sharkov et al., 2005). From this follows that the fresh basalts and older primitive gabbros were formed from some different mantle sources.

3.3. Ferrogabbro

Different ferrogabbros play essential role among lower-crustal rocks (about 1/3 of the gabbros' samples). They are represented by melanocratic troctolite, norite, gabbro-norite, and gabbronorite-diorite, enriched in Fe-Ti oxides (ilmenite and magnetite) and often by brown primary-magmatic hornblende (kaersutite) (Sharkov et al., 2005). Subvolcanic analogues of ferrogabbro are represented by hornblende Fe-Ti-oxide dolerites, and very rare basaltic flows with essential amount of Fe-Ti oxides, mainly ilmenite.

Ferrogabbros, like primitive gabbros, are characterized by low concentrations of light REE; however, they are enriched in ore components – Zn, Sn and Mo, have elevated contents of Cu and Pb, and low – Ni and Cr. In contrast to the primitive gabbro, ferrogabbros have positive anomalies of Nb and Ta. Study of melt inclusions in chromites from rocks of this series showed that their composition vary from Fe-Ti basalt to andesite (icelandite) and dacite (Simonov et al., 2009). The ion-microprobe study of the melt inclusions yielded direct evidence for elevated water content (up to 1.24–1.77 wt %) in the melts that produced ferrogabbros; small globules of Fe-Ni sulfides were found in them also. So, these rocks from one hand are saturated and supersaturated by SiO_2 and have increased H_2O content, which typical for subduction-related magmas, and on the other hand have high contents of Ti, Fe, Nb, Ta and P, typical for magmas of plume origin.

Ferrogabbros are obligatory component of the lower-crustal sections of rocks in OCCs, where they play essential role. Many people thought that the ferrogabbros were produced by fractional crystallization of the MORB-type melts (Dick et al., 1992; Thy, 2002 and references therein). However, they often intruded primitive gabbros (Thy, 2002) and their quantity usually exceed possibility of the MORB crystallizing differentiation.

We believe that these rocks belong to magmatic siliceous Fe-Ti-oxide series, specific for the oceanic environment, which origin related to melting of hydrated oceanic lithosphere by action of a new mantle plume (Sharkov et al., 2005). Newly formed mantle-derived melts passed through the upper cooled part of the plume head, accumulated at the mantle-crust boundary and produced a magma chamber, which started to ascend according to the zone refinement mechanism, i.e. by melting the roof and crystallizing at the bottom (Fig. 6). The melt was continuously enriched in components not only melted rocks of the chamber's roof but also from the partly melted rocks at the heated peripheries of the melting zone, where processes of anatexis occurred (see Section 2.4), as well as fluid material from the heated rocks on the distant periphery. Obviously, unusual characteristics of these melts, like their enrichment in SiO_2, H_2O, and some ore components, typical for hydrothermal activity (Pb, Cu, Zn, etc.), can be explained by such features of melting process.

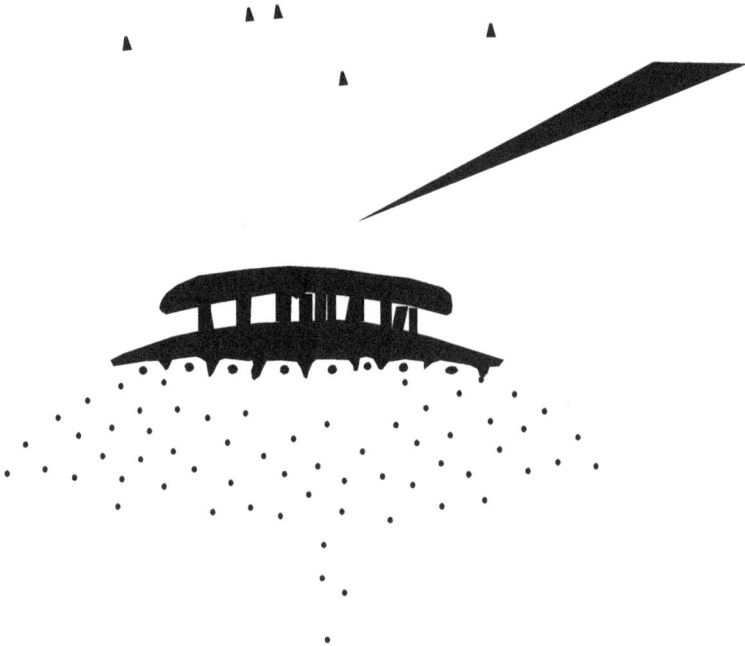

Figure 6. Hypothetical scheme of the melts of the siliceous Fe–Ti-oxide series genesis

Cyclic Development of Axial Parts of Slow-Spreading Ridges: Evidence from Sierra Leone Area, the Mid-Atlantic Ridge, 5-7°N

27

D. Pearce (2002) drawn attention to the fact that, unlike the fast-spreading ridges, in the slow-spreading ridges volcanic equivalents ferrogabbros very rare. Truly, there are only two small fragments of hornblende basalt with ilmenite, as well as a sample of variolite (Krassivskaya et al., 2010) in our collection. Apparently, the main cause of the limited vertical mobility of such melts is their water saturation, which decreases sharply at a pressure about 1 kbar (Fluids..., 1991). This leads to separation and removal of water, following increase of the solidus temperature and, as a result, to rapid solidification of melts at depth; therefore, the volcanic eruption of such magmas are rare, how it is clearly seen in the most OCCs and, particularly, in studied area.

3.4. Oceanic plagiogranites

As in all OCCs, small quantities of plagiogranites (tonalites and trondhjemite) are found on the area. Their origin is usually attributed to later stages of magmatic crystallization. However, according to our data, formation of such melts can be explained by anatexis of hydrated lithospheric rocks near intrusive contacts (Aranovich et al, 2010). Special role in this process belongs to "metamorphosed" sea-water from subfloor hydrosphere, enriched in NaCl due to absorption of pure part of sea-water during formation of the secondary hydrous minerals (serpentine, chlorite, actinolite, etc.) in the bedrocks.

4. Metasomatism and ore mineralization

Hydrothermal metasomatic zone with rich sulphide mineralization was found in the Markov Deep (Sharkov et al., 2007 and references herein). According to results of dredging, at least two zones of intense tectonic deformation and metasomatism (at depth 4400-4600 and 3700 m) occur here, extending in NW direction with gentle angles (30°- 40°) dip to east (Fig. 7). These zones are formed by brecciated and schistosed ferrogabbros to thin-foliated cataclasites upon them of chlorite-amphibole-epidote-clinozoisite composition. The presence of chaotic plication, striation, grooves, slickensides and slip-scratches on the surface of the clasts as well as fragments of small folds with distinct axis, which are oriented along lineation, point to the fact that tectonic movements have evolved under shear conditions.

Sulfide mineralization represented by quartz-sulfide and prehnite-sulfide veins, sulfide dissemination and massive ore deposits. Mineral composition of ores is represented by pyrite, chalcopyrite, sphalerite, pyrrhotite, bornite, and atacomite as well as native Cu, Pb, Zn, Au and Sn and intermetallides (isoferroplatinum, tetraferroplatinum, and brass).

According to our data, hydrothermal-metasomatic processes occurred under low pressure (0.5-1 kbar) and started at temperature ~750°C; however, the major ore-forming metasomatic processes occurred in range 400-160°C. Sm-Nd isotopic data and $\delta^{34}S$ value evidence that ore-bearing fluid initially had magmatic origin and then were progressively diluted with sea-water of oceanic subfloor hydrosphere.

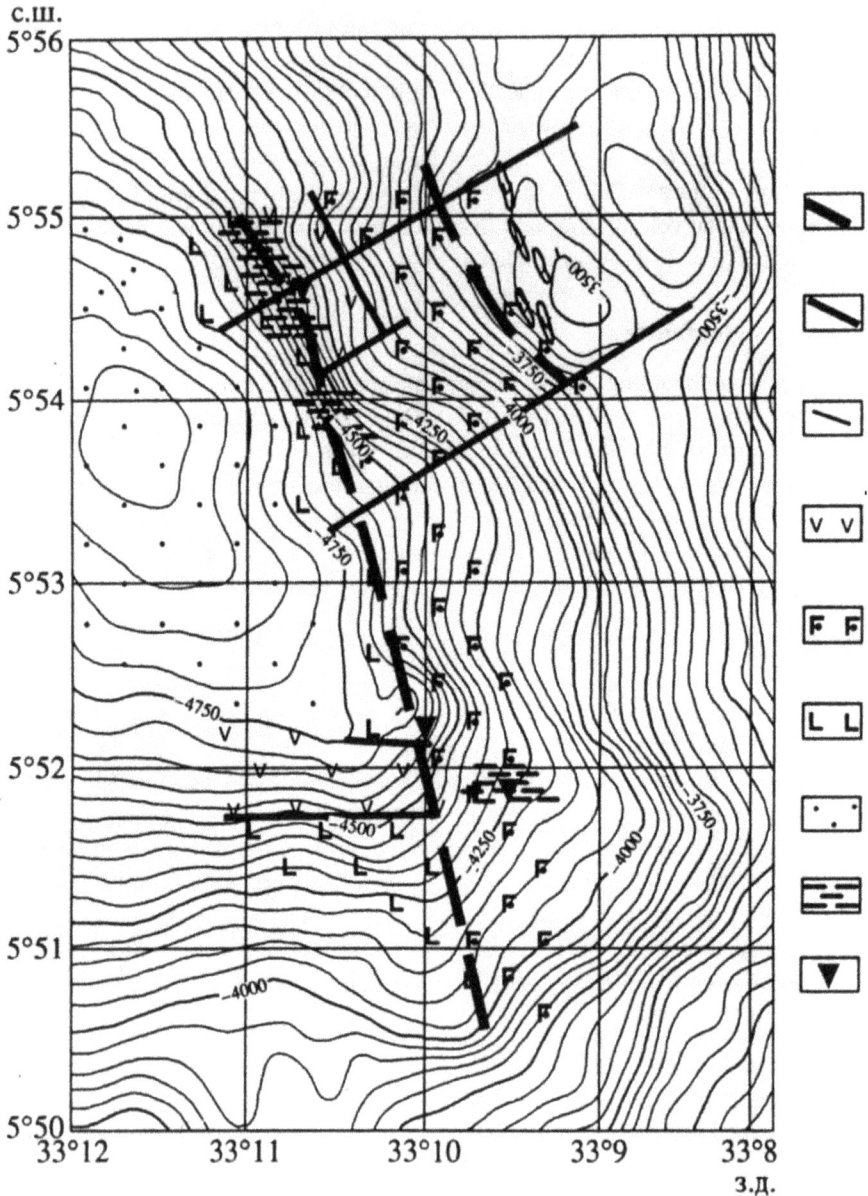

1 - zones of tectonic deformation and metasomatism, 2 - faults, 3 - geological boundaries; 4-7 – fields of preferential distribution: 4 - basalts, 5 - gabbro, 6 - ultrabasic, 7 - sedimentary cover; 8 - zone of hydrothermally altered rocks; 9 - sulfide mineralization in the bedrock.

Figure 7. Scheme of the geological structure of the eastern edge of the Markov depression. According to Beltenev et al. (2004).

The main source of ore-bearing fluids could be an intrusion (Fig. 8), formed by water-saturated melt of siliceous Fe-Ti-oxide series, which contains sulfur and ore components (see Section 2.3). Addition of the ore material can be leached from the host gabbros on the way up, and part of the sulfur been introduced by sea water. Separation of fluids from such magmas commonly occurs at a pressure of about 1 kbar, when the solubility of water in them decreases sharply (see Section 2.3), i.e. at the depth of 3-4 km below the seabed, where, apparently, was located solidifying intrusion.

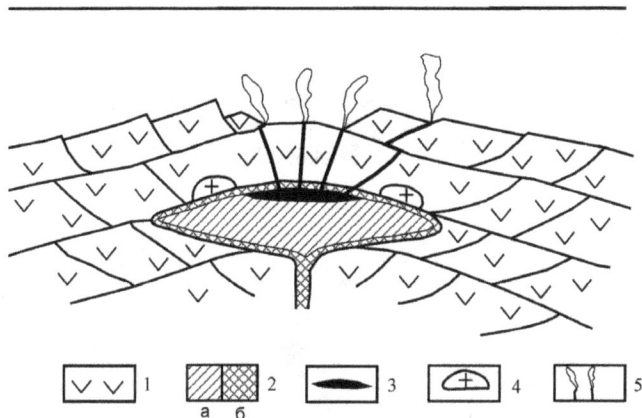

1 – vents of hydrothermal systems at the ocean bottom o ("black smokers"), 2 – solidifying intrusion of siliceous Fe-Ti-oxide series with lenses of the residual melt; 3 – place of formation of plagiogranite melts; 4 – chilled zone of the intrusion, 5 - fractured rocks of hydrated oceanic lithosphere.

Figure 8. Scheme of the ore-bearing hydrothermal system structure

The appearance of relatively high-temperature metasomatites on the oceanic floor indicates that these rocks were moved to the surface after their formation at the depth. Judging by presence of atakamite and weak oxidation of sulfides, it happened very recently, apparently in the process of ongoing formation of the present Markov Deep. We suggest that this ore-bearing zone is fragment of feeding system of an extinct black smoker.

5. Results of U-Pb dating of zircon from gabbros by SHRIMP-II

The U-Pb SHRIMP-II dating of zircons from the gabbros of the area showed that these rocks, which are in the present-day oceanic spreading axis of the MAR, were formed earlier, in the Holocene-Pleistocene, 0.7-2.3 Ma; above, presence of zircon grains with age up to Mesoarchean (older than ~ 87 Ma up to 3117 Ma) were established (Bortnikov et al., 2008) (Fig. 9). The magmatic nature of young zircon with thin oscillatory zoning and sectorial structure suggests that its age defines the crystallization age of the host magmatic rocks; "old" zircon are defined as xenocrysts. Later the presence of uneven-aged zircon grains were found not only in the gabbros of the Sierra-Leone area, but in gabbros as well as ultramafics, plagiogranites, diorites, and even basalts of other parts of the MAR, i.e., presence of ancient xenogenic zircon grains in oceanic bedrocks is widespread (Skolotnev et al, 2010 and references herein)

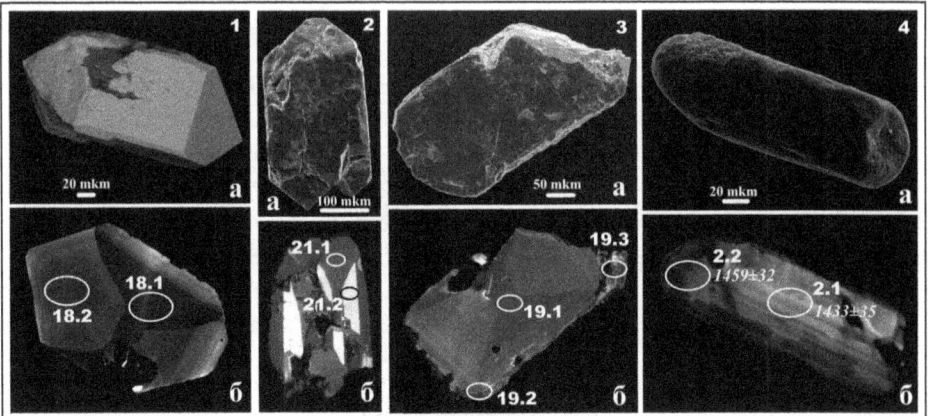

Site I 1028, sample I 1028-1

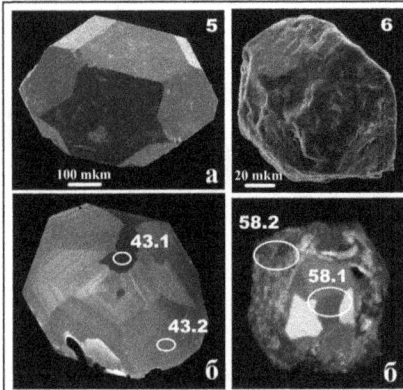

Site I 1028, sample I 1028-5

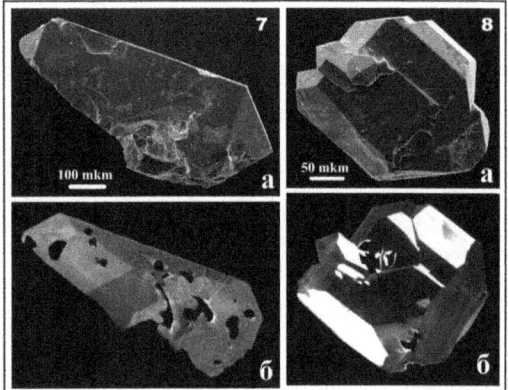

Site L 1153, sample L 1153-49

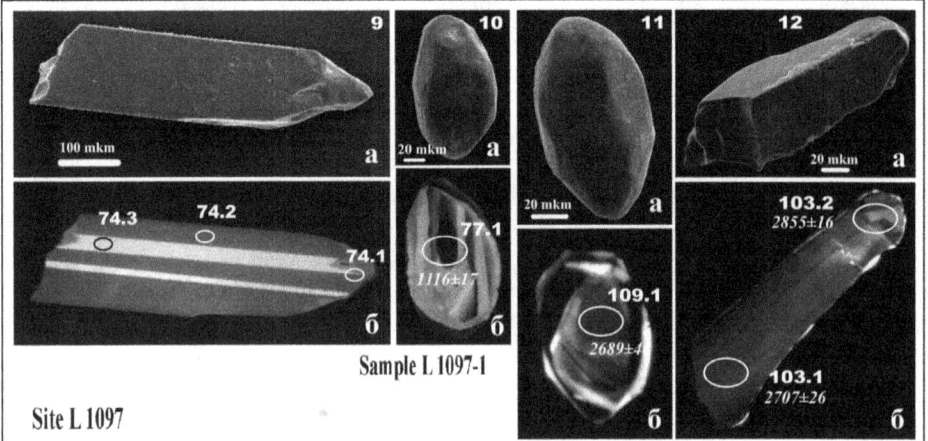

Site L 1097

Sample L 1097-1

Sample L 1097-3

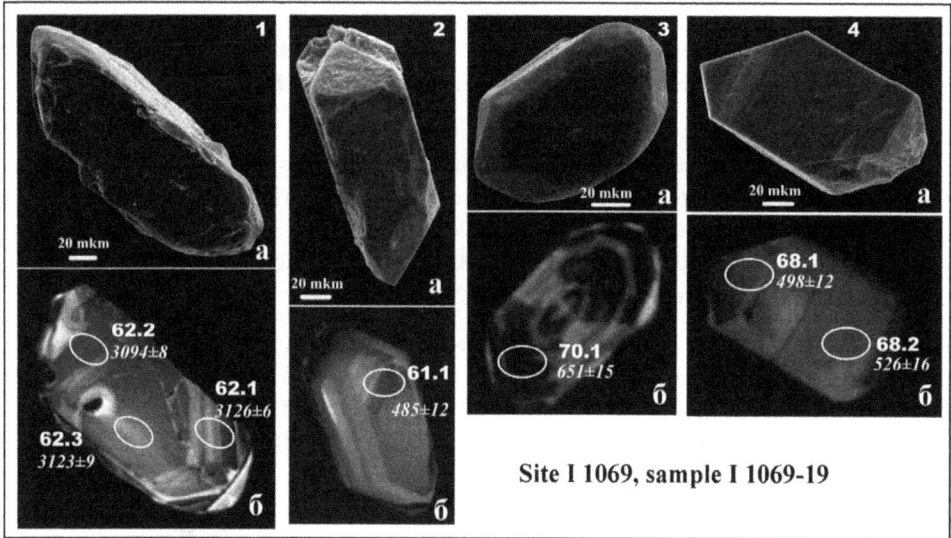

A. Microimages of zircons from gabbronorites of the Markov Deep (dredge sites I1028, L1153, and L1097). Hereinafter: (a) natural appearance, (b) cathodoluminescence image (CL); index of dredge site: (I) R/V *Akademik Ioffe* and (L) R/V *Professor Logatchev*. Spot numbers of U–Pb age determination are as in Tables 1 and 2. Age is shown for xenogenic zircons. (1) Euhedral zircon with weakly corroded prism surface and diopside inclusion (dark); sectorial and oscillatory zonings are seen; (2) bipyramidal zircon with corroded prism face; oriented light fragments of bands (reflection of high-temperature cataclasis) are seen in the CL images; (3) zircon with corroded prism and pyramid faces; the CL images demonstrate sectorial zoning, resorption of prism and pyramid faces, and formation of colloform shell; (4) prismatic zircon with weakly corroded surface; the Cl image shows coarse zoning, weak resorption of prism and pyramid faces, and thin shell; (5) euhedral grain with well-expressed oscillatory zoning; (6) zircon with corroded surface; the CL image shows core with fragments of light bands (reflection of high-temperature cataclasis) and colloform shell; (7) subhedral zircon with corroded prism surface; the CL image demonstrates fragments of sectorial zoning and small inclusions of plagioclase (dark) of irregular shape (poikilitic structure); (8) growth of secondary small pyramidal zircons due to redeposition of matter on the other side of the crystal; oriented light bands produced by high temperature cataclasis are seen clearly in the CL; (9) fragment of long-prismatic zircon with corroded pyramidal termination; deformation-related light bands are observed in the CL image; (10) rounded zircon with coarse zoning in the core and thin shell; (11) analogue of zircon 10, but with a wider shell; (12) fragment of prismatic zircon lacking internal structure in the CL.

B. Microimages of zircons from troctolite, Site I1069-19. (1) Prismatic grain with corroded surface; fragments of coarse zoning and shell are seen in the CL; (2) analogue of 1, with wider shell; (3) subhedral zircon with coarse concentric zoning, elements of sectorial zoning, and fragments of thin shell; (4) subhedral grain with weak corrosion of one pyramid; no internal structure was identified in the CL.

Figure 9. Zircons from gabbros of the Markov Deep. After N.S. Bortnikov et al. (2008).

We believe that such xenogenic zircon could initially belong to fragments of material from the "slab graveyards" in the deep mantle, captured by mantle plume, which moved from the core-mantle boundary (Bortnikov et al, 2008). Such "graveyards" may contain rocks of different ages and backgrounds, including the Precambrian gneisses and sedimentary rocks involved in subduction zones. A detailed study of exhumed slabs presented by ultrahigh-pressure complexes of Kazakhstan, China, Norway and others, which were formed at $P > 2.8\text{-}4$ GPa (and possibly up to 8.5 GPa) and $T = 600\text{-}900°C$, showed that zircon could persist even under these conditions (Ernst, 2001 and references herein). Apparently, these PT-

conditions preserved in rocks "graveyard slabs" also, because, according to seismic tomography (Karason, van der Hilst, 2000), they form a great bodies of hundreds kilometers thick; billions years are required to warm up them by conductive heat.

Possibility of existence of buried subducted material beneath the Central Atlantic also evidence from results of study of lithium isotopy in basaltic glasses at 12-16°N (Casey et al., 2010). Like in Sierra Leone area, these basalts in composition are intermediate between E- and T-MORB and are characterized by positive Ta- and Nb-, as well as Ti-, Sr- and Eu- anomalies. They set the lowest of recorded value of δ^7Li, indicating presence in magma components of refractory rutile-bearing eclogite.

According to geophysical data, asthenosphere beneath the MAR is represented by lens-like body about 200-300 km thick (Fig, 10) (Anderson et al., 1992; Ritsema, Allen, 2003). So, our finding of ancient zircons in gabbros and Li-isotopic data support the idea about existence of colder mantle beneath the axial part of the MAR, which has penetrated by mantle plumes.

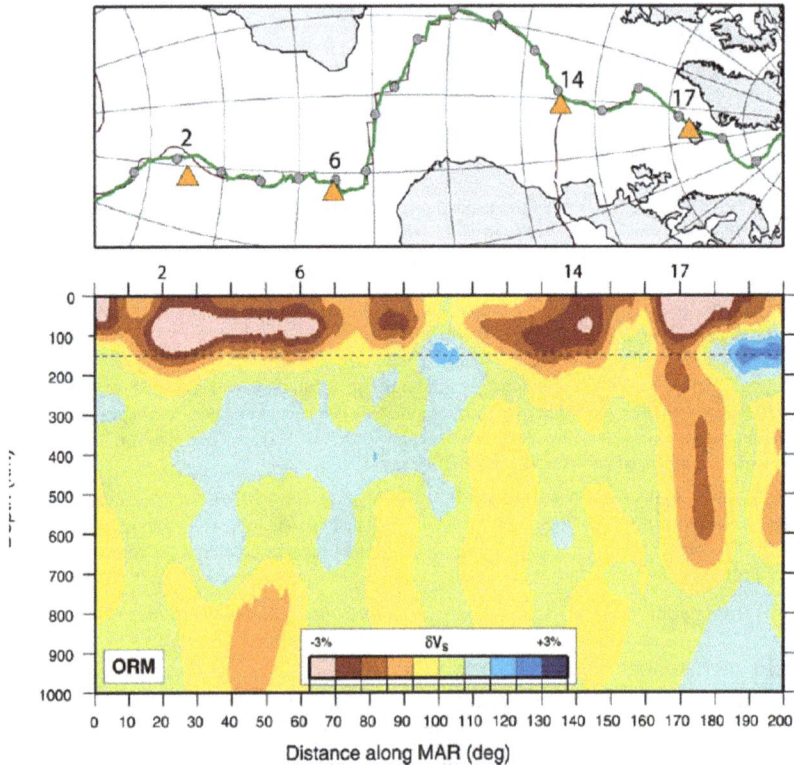

Figure 10. Tomographic profile along the axis of the Mid-Atlantic Ridge, showing that the highest speed anomalies of transverse waves are localized under the "hot spots" (triangles): 2 - Tristan, 6 - Ascension, 14 - Azores and 17 – the Iceland;; the latter can be traced to the lower mantle and possible to the mantle-liquid core boundary (after Ritsema & Allen, 2003).

Cyclic Development of Axial Parts of Slow-Spreading Ridges: Evidence from Sierra Leone Area, the Mid-Atlantic Ridge, 5-7°N

33

6. Processes of formation of shallow lithosphere (oceanic crust and lithospheric mantle) in the Sierra Leone area

According to the classical model, the occurrence of mid-ocean ridges associated with localized upwelling of deep hot mantle material, which melts due to adiabatic decompression producing specific MORB volcanism. However, it was found that asthenosphere beneath the MAR has lens-like shape up to 200-300 km thick, which is located between colder and dense material of shallow lithosphere and underlied mantle (see section 5).).

As it was shown above, oceanic crust of the Sierra Leone area was formed at least three independent episodes of magmatic activity: the modern, attributed with the eruptions of fresh pillow-lavas, and two previous ones, which led to formation of lower-crustal gabbros (altered primitive gabbros and ferrogabbros consequently). Accordingly, the fresh basalts are not genetically related to the altered lower crust and question arises about it's origin.

6.1. Origin of lower oceanic crust: Evidence from Sierra Leone area

There is a consensus that it exists between the mantle and the upper oceanic crust and composed of various gabbros, often alternated with peridotites. It implies existence between mantle and upper oceanic crust transitional magmatic chambers (intrusions), which solidification provide formation of the lower crust. At the same time mechanism of this crust formation is open to question because its geological study in present-day oceans is concerned with serious technical difficulties. About its composition and structure we can judge only by random samples, gave by dredges, inhabited submarine devices, or "pinpricks" of deep-water drilling wells. According to Pearce (2002), two main points of view on the origin of the oceanic lower crust dominate now: (1) its formation during crystallization in a single melt lens followed by the flow of crystal mush down and away from the ridge (model "gabbro glacier", Quick, Denlinger, 1993) and (2) through grows from series of sill-like bodies throughout the crust (the "Christmas tree" model). However, the situation is stayed uncertain.

From this point of view essential help for understanding processes of formation and development of lower crust of the modern oceans can provide gabbro complexes of ophiolites – fragments of ancient oceanic or back-arc seas lithosphere, find in orogens (Knipper et al, 2001; Dilek, Furnes, 2011 and references therein). In contrast to modern oceanic floor, about which structure we can judge only by fragmental data, they are available for direct geological studies.

Of particular interest in this regard is well-preserved Voikar (Voikar-Syninsky) ophiolite assemblage in the Polar Urals (Russia). Its gabbro complex consists on two major megarhythms (Fig. 11), generally similar in structure to large layered mafic-ultramafic intrusions, formed in calm tectonic settings, were found there above the mantle restite complex (Sharkov et al, 2001). At that for the lower megarhythm are typical primitive gabbro and olivine gabbro, and for the upper – mainly gabbro-norite, sometimes hornblende-bearing, often with increased concentrations of ilmenite and titanomagnetite, which resemble the rocks of the siliceous Fe-Ti-oxide series. All rocks of the assemblage are cut by diabase dikes. Thus, as is the case of the Sierra Leone area, there are two independent sets of intrusive rocks recorded here, successively formed by different magmas.

In contrast to majority of continental large layered mafic-ultramafic intrusions, almost all plutonic rocks of the Voikar's ophiolites were undergone shearing. Like the site of Sierra Leone, it started with ductile flow of rocks at high temperatures and changed by the brittle-plastic and brittle deformation under conditions of the greenschist facies during cooling. This led to a strong serpentinization of ultramafic rocks and amphibolization of gabbros with extensive development of fibrous amphibole upon pyroxenes. Cumulative structures remain rare, although the overall shape of rocks indicates their intrusive origin.

The absence of cryptic layering in cross-section of the megarhythms suggests that their formation occurred by crystallization of the transitional magma chambers accompanied by replenishment of fresh melts under conditions of open magmatic system (Sharkov et al., 2001). At such circumstances, solidifying from the bottom up magma chamber could be a lens, gradually moving up and leaving a "tail" of hardened hot material. In other words, although these intrusions were not necessarily initially large, but could gradually grew with the arrival of new portions of melt.

According to Sm-Nd and Re-Os isotopy data, significant differences between material of the lower and upper megarhythms as well as mantle section occur in the Voikar ophiolite assemblage. Thus, presence of ancient material determined in the rocks of the upper megarhythm, where the $^{187}Os/^{188}Os$ ratio is 6.5-7.1, which is much higher than in the mantle rocks of the assemblage and two times higher then in diabases of the sheeted complex dikes (Sharma et al., 1995, 1998). These data indicate that: (1) formation of the gabbro complex was happened at two stages, i.e., a two-stage build-up of the lower crust occurred here; (2) judging from the relatively well-preserved sections of the complex, formation of each of them happened during the relative calm of tectonic processes; (3) still hot rocks, soon after their solidification, were involved in processes of plastic flow, gradually changed by plastic-brittle and brittle deformations; and (4) there are marked differences in isotopic characteristics between major constituents of the assemblage: mantle rocks, dikes sheeted complex, as well as two megarhythms of the gabbro complex.

The data available on the lower crust of the Sierra Leone area and others OCCs (see above) are in good agreement with data on the Voikar gabbro complex. The presence in the area's lower-crustal gabbros relic cumulative structures and elements of the primary magmatic layering (Fig. 3) suggests that this crust is formed by large layered mafic-ultramafic intrusive bodies of different age and origin. Very likely, that its formation happened mainly through underplating, i.e. building up from below, through accumulation of newly formed basaltic melts at crust-mantle boundary how it established on the continents (Rudnick, 2000).

Presence in the lower crust of the area both primitive gabbro, derived from MORB-type melts, and ferrogabbros of siliceous Fe-Ti-oxide series, shows that, like in Voikar, at least two different types of layered intrusions occurred here.

Appearance not numerous relatively fresh gabbros, olivine gabbros and troctolites among dominated altered gabbro can be considered them to the recent cycle of activity. In any case, judging on the phenocrysts in the fresh basalts (see Section 2.2), those rocks were probably formed in transitional chambers of young magmatic systems.

Strongly serpentinized mantle peridotites, like in the most ophiolites including Voikar's, represented here by typical mantle restites – harzburgites and subordinate depleted lherzolite and dunites (Savelieva et al., 2006). Some of these peridotites, judging on rare good preserved samples, have cataclastic structure (Fig. 11) evidenced about their involving in deformation processes.

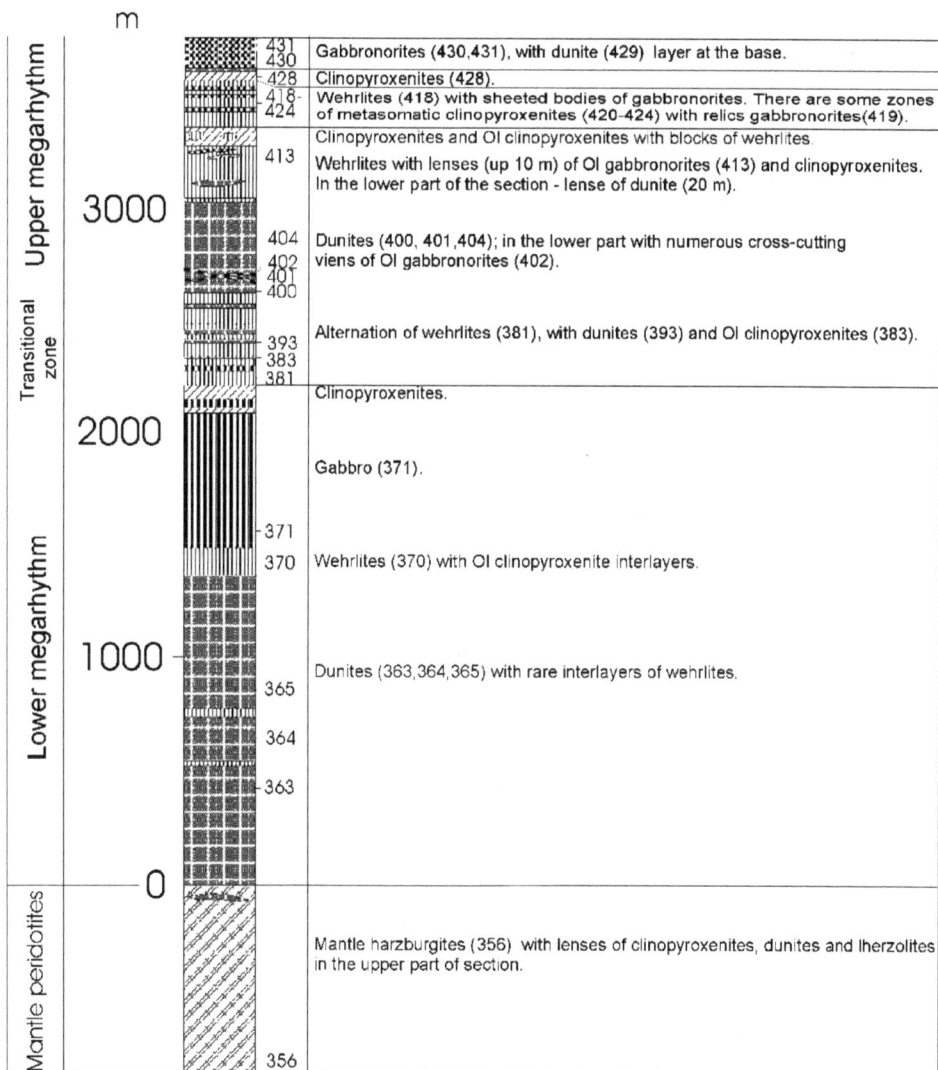

Figure 11. Geological section of gabbro (layered) complex Voykar ophiolite assemblage (Polar Urals), by (after Sharkov et al, 2001).

Thus, according to data available, the most OCCs (Escartín et al., 2008; Smith et al., 2008; MacLeod et al., 2009: Silantyev et al., 2011 and references therein), as well as ophiolite assemblages (Knipper et al., 2002; Dilek and Furnes, 2011), have the similar structure and composition of lower crust and lithospheric mantle. So, the structure of the studied area represents common type of the shallow oceanic lithosphere and we can discuss some general problems of origin and functioning of slow-spreading ridges on its example.

7. Discussion

7.1. Origin of oceanic core complexes: Evidence from Sierra Leone area

As it follows from study of typical OCCs (see Introduction), they are parts of slow-spreading ridges with asymmetric structure with one-way sliding of crustal material; it is suggested that their origin is attributed to activity of hypothetical oceanic detachment faults (Fig. 1). It assumes that these faults develop as a result of strain focusing around rheologically strong gabbro plutons hosted in weaker serpentinized lithospheric mantle; hence it deduced that OCCs were formed during periods of relatively enhanced magma supply. However, as mentioned above, even sticklers of this hypothesis of "one-sided spreading" recognize that many of its elements and details are still poorly substantiated (Conference ..., 2010). In fact, it is determined only asymmetry of the structure of these parts of mid-ocean ridges with exposed altered deep-seated rocks and presence there gently sloping and normal faults.

There are two main hypotheses of the OCCs origin exist now. Predominant model of their appearance is considered with activity of oceanic detachment faults during periods of reduced magmatic activity or its absence ("dry spreading») (Tucholke, Lin, 1994; Tucholke et al., 2008; Dick et al., 2008; Escartin et al, 2008, etc.). According to another view, based on widespread gabbro in such structures, the OCCs were formed at period of relatively depressed (but not reduced) magmatism, realized as large plutons from overlapping access of magma to the surface by oceanic detachment faults (Ildefonse et al., 2007). Sort of these conceptions is model of "life cycle of OCC" (MacLeod et al., 2009). According to this model, spreading becomes markedly asymmetric when the core complex is active, and volcanism is suppressed or absent; when the asymmetry is such that the detachments accommodate more than half the total plate separation, the active faults migrate across the axial valley. As a consequence magma is emplaced into and captured by the footwall of the detachment fault rather than being injected into the hanging wall, explaining the frequent presence of gabbro bodies and other melt relicts at oceanic OCC. Core complexes are ultimately terminated when sufficient magma is emplaced to overwhelm the detachment fault.

However, a numbers of problems remain unsolved in context of these models: motives of ascending of older altered lithospheric rocks at high hypsometric levels, lack of genetic interrelations between fresh basalts and older lithospheric rocks, presence essential quantity of rocks of the siliceous Fe-Ti-oxide series (ferrogabbros) among them, etc. Above all, OCCs, in essence, represent outcrops of shallow oceanic lithosphere, which formation has not considered with hypothetical oceanic detachment faults: the latter can only expose them, but not create, especially because this lithosphere was formed much earlier and at greater depths. So, the main problems are origin of this lithosphere, reason for its local ascending in axial parts of slow- and ultraslow ridges, and how these ridges functionate under condition of lens-like asthenosphere beneath them, i.e. what the reasons for the oceanic spreading there?

7.1.1. What has occurred in slow-spreading ridges in geomechanics terms?

Since H. Hess (1962) times, the most researches believe that process of oceanic spreading associated with ascending of hot mantle material to the axial part of mid-oceanic ridges, its adiabatic melting accompanied by formation of oceanic crust, growth of the plate, and their motion to both sides of ridges under influence of deep convection. The most complete geomechanical aspects of the model was considered by D. Turcotte and J. Schubert (2002 and references herein).

It is known that shallow lithosphere in the MAR axis is in a position of mechanical instability, how it evidence from presence of constant shallow earthquakes, caused by processes of stretching, discontinuity and delamination of bedrocks which indicate uplift and spreading of its axial part. According to (Turcotte, Schubert, 2002), upwelling of mantle rocks is accompanied by heat loss, occurred through molecular heat conduction; as a result, they attached to the base of separating plates, becoming part of them. Because material of heated plastic asthenospheric material can flow like a liquid in geological time scale under influence of external forces, increase of load promote flow of this material to the ridge axis, ensuring its stable triangular shape over time.

In accordance with Turcotte-Schubert model, the triangular shape of the ridge should lead to gravitational instability of the system, causing sliding (slumping) of material along its slopes. Mathematical simulation of such process, performed on example of the MAR, revealed that the force of the ridge push are sufficient to implement such a mechanism (Scheidegger, 1987). As a result, the crustal material should slide from uprising dome-shaped part of the ridge axis (tectonic erosion), exhuming of deep-seated rocks on the oceanic floor (Fig. 12).

Such structure of mid-oceanic ridges in terms of geomechanics is typical for piercment structures which formation determined by introduction of plastic less dense and less viscous layer into overlying more dense layer under gravity influence (Scheidegger, 1987 and references herein). According to the theory, penetrating masses, being less dense than overlying rocks will tend moving upwards, regardless of else tectonic forces. Though classical theory of piercing structure formation developed on example of salt domes, we have a close situation in the MAR: heated plastic asthenospheric material and overlied it cold dense shallow lithosphere. In this case, due to tectonic erosion, pressure above growing asthenospheric crest falls and as a result of adiabatic decompression (decrease in the solidus temperature with decreasing pressure) it led to melting of the material (Fig. 13). According to calculations (Girnis, 2003), smelting of MORB begins at pressure ~15 kbar, however, the mass-melting occurred at pressures 8-10 kbar, i.e. at the depths 28-35 km, where major melting zone has to situated.

Figure 12. Micrograph cataclased harzburgite restite; large grains (porfiroclast) deformed orthopyroxene surrounded by small neoblasts of olivine and pyroxene. Sample 1063-39, polarized light (collection of E.V. Sharkov).

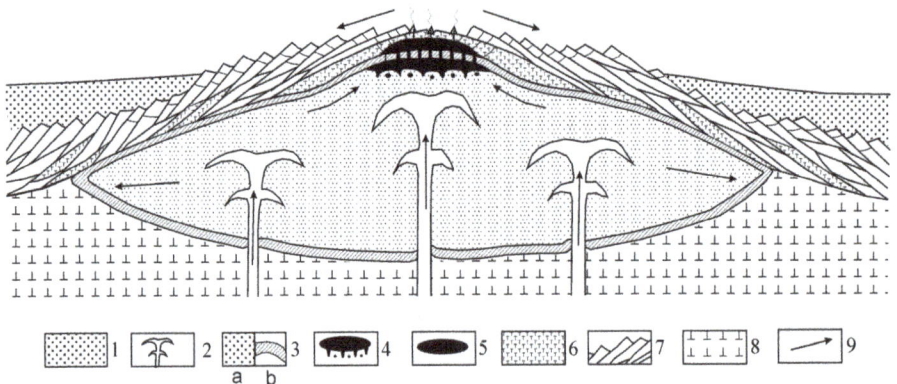

1 – sediments; 2 - mantle plumes penetrating into the asthenospheric mantle, and partly or completely mixed with it; 3 - asthenospheric lens under the MAR, bordered by cooling zone in contact with lithospheric mantle; 4 - melting zone in the upper part of the asthenospheric lens; 5 - transitional magma chamber; 6 - depleted lithospheric mantle (restite from a previous episode of melting), transformed into the lithospheric mantle; 7 - oceanic crust formed by gabbros and basalts; 8 - oceanic lithospheric mantle; 9 - direction of movement of material.

Figure 13. The proposed scheme of the deep structure of the MAR

Appearance of the melting zone brings further contribution in ascending of the ridge's axial part because melting of silicate rocks leads to decrease of density of material in the magma generation zone by 11-13% (Handbook ..., 1969). This led to development of fractures in the overlying lithospheric peridotites, promoted to their serpentinization under influence of subfloor sea-water which reduced their density by 35-38%. All of this stimulate further growth of the dome and strengthen the tectonic erosion of its axis.

In essence, geological sense of oceanic spreading as well as formation of new lithosphere plates lies in this complex of processes. From such standpoint, constructive plate boundaries, at least in the slow-spreading ridges, should be formed by a collage of tectonic slices of shallow oceanic lithosphere and basaltic covers, i.e. these lithospheric plates in geomechanical sense are not monolithic as suggested by Hess (1962). It is in a good agreement with results of study of ophiolite assemblages which are formed by packages of tectonic slices of similar crustal rocks and mantle restites.

Thus, in contrast to the generally accepted views, we do not attribute oceanic spreading with hypothetical convection currents in asthenosphere, but with processes of gravitational instability at the lithosphere-asthenosphere boundary in axes of the slow-spreading ridges, resulting in sliping of rocks on their slopes. Gabbros and restite ultrabasites as well as basalts, genetically related to this tectonomagmatic episode, are involved in this process. Sliding of the rocks is accompanied by their delamination, formation of different faults, tectonic slices, etc., that creates a characteristic "seismic noise".

7.1.2. Processes of the OCCs formation: Evidence from the Sierra Leone area

How it evidence from study of the Sierra Leone area, located in spreading zone of the MAR, formation of its structure occurred at least in two stages. The first stage attributed to formation of shallow oceanic lithosphere (lower crust and restite mantle peridotites) and the second, modern – unconformably overlying them flows of fresh basalts.

Between these stages there was occurred rise of the lithosphere dome, accompanied by the sliding of material (tectonic erosion), exhumed deep-seated rocks to the oceanic floor and the appearance of numerous extensional structures, ensures the existence of pallets (subfloor) hydrosphere and the ways for hydrothermal fluids ascent; remains of a former hydrothermal systems were found in the Markov Deep (see section 3). This stage of the area development can be defined as a formation of oceanic core complex (OCC).

The second (current) stage of the ridge development on the studied area is also characterized by extensive development of extensional structures up to appearance of the rift graben-like structures and a fairly powerful basaltic volcanism. These melts come from intermediate magma chambers, where they were subjected to fractional crystallization, and, before reaching the oceanic floor, passed through the ancient lithospheric rocks, partially assimilating its material.

Inasmuch as situation at the lithosphere-asthenosphere boundary in the slow-spreading MAR before the OCCs formation was in a state of unstable equilibrium (see Section 6.1.1), such development of events demanded of a trigger to start ascent of the asthenospheric crest in studied area. Most likely this role played a mantle plume, which reached the boundary and lifted it, thereby disturbed the unstable equilibrium and initiated rise of the dome (Fig. 14). Its recent existence here follows from isotopic data (Schilling et al., 1994), as well as the general uplifting of the territory and composition of fresh pillow lavas (mainly T-MORB (oceanic plateau basalts) to E-MORB); such characteristic of basalts typical for sites on the ridges next to manifestations of intraplate (plume) magmatism (Basaltic ..., 1981).

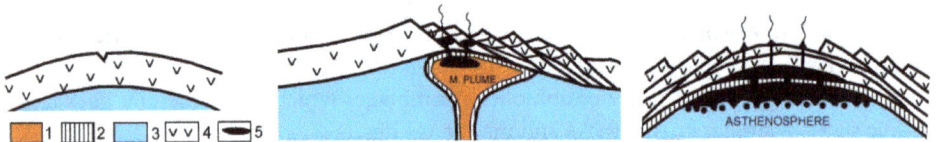

1 – mantle plume; 2 – its cooling borders; 3 – asthenosphere; 4 – shallow lithosphere; 5 – basaltic melt

Figure 14. Scheme of cyclic evolution of tectonomagmatic processes in axial part of the MAR

Most likely, the typical for OCCs hydrothermal fields are also associated with magmatic systems, generated by mantle plumes. This is particularly true for water-saturated melts of siliceous Fe-Ti-oxide series, which have limited mobility in the vertical, and usually do not reach the surface in the thickness of the crust hardens in the form of intrusions (see Section 2.3, Fig. 8). It is in a good agreement with our data on the Sierra Leone area, where we found recently extinct ore-bearing hydrothermal-metasomatic system, attributed to such magma (see Section 3). Very likely, that typical for OOCs phenomenon of wide development of hydrothermal fields under conditions of "dry spreading" (i.e. by practically absence of volcanic eruptions) can be successfully explained by this circumstance.

From this view, appearance of OCC reflects the first stage of the crest uplifting, followed "squeezing-out" by plume head of cold rigid lithosphere as a dome at relatively high hypsometric levels and starting process of tectonic erosion on its axial part, which leads to the appearance (exhumation) of deep-seated rocks on the ocean floor. The head of the plume was, in general, asymmetrical and often provided outside of the ridge axial zone that led to the emergence of on-side spreading. Fragments of the plume heads, as it shown on example of the St. Peter and St. Paul complex (see Introduction), can be found sometimes. Perhaps, they are encountered more often, but it is difficult diagnostics because of strong serpentinization.

Widely represented in various OCCs surfaces with corrugations and striations (mullion structures) are usually interpreted as evidence for the existence of oceanic detachment faults, namely its footwall (see Introduction). However, mullion structures are a common pattern under joint flowage of very different on viscosity tectonic plates (Allaby & Allaby, 1999), in this case – the serpentinite and gabbro, and does not carry specific information about existence of oceanic detachment fault here.

Cyclic Development of Axial Parts of Slow-Spreading Ridges: Evidence from Sierra Leone Area, the Mid-Atlantic Ridge, 5-7°N

41

From all this it follows that formation of the OCCs is likely represent the first stages of dome growth due to appearance a new mantle plume that disturbs the unstable equilibrium at the lithosphere-asthenosphere boundary. This led to uplifting of the area, to beginning of tectonic erosion and provided specific magmatism of siliceous Fe-Ti-oxide series and related hydrothermal activity.

7.1.3. Cyclic character of tectonomagmatic processes in the axial zone of the slow-spreading ridges

One of important points for understanding situation in the ridges axes are processes of adiabatic melting. As a result of the OCC appearance and beginning of sliding of material from the ridge axis, asthenosheric material began to inflow there and the ridge, because of mechanical reasons, gradually got a symmetrical structure. Due to concomitant reduction of pressure in the frontal part of the crest, process of melting had to strengthened. Since position of the solidus isotherm in this case is determined by lithostatic pressure, a melting zone should has form of a flattened lens at the top of the asthenospheric crest. It led to mixing of asthenosphere's and plume's materials and appearance melts of T- and E-MORB composition. Judging on the Sierra Leone area, this change of the melting regime accompanied by temporal interruption of tectonomagmatic processes, after that flows of newly formed basalts began to overlap altered rocks of the former OCC. The next stage of the dome ascending should be already vast eruptions under conditions of a "normal" bilateral spreading, how it occurred to the south and north of the studied area.

As a result of melting, density of asthenospheric lherzolite would gradually decrease, mainly due to removing of Ca and Fe with basaltic melt. Simple calculations show that decrease in the density of material in this case could reach 8-10%, because strongly depleted mantle material (mainly harzburgite) consists mostly of relatively light magnesian olivine and orthopyroxene. Because of this, restite material will accumulate in upper part of the melting zone, forming a separate layer, which cannot be involve in processes of the asthenospheric convection. Formation of such layer of light refractory material should lead to the cessation of melting of the mantle beneath the ridge axis, and, as a result, it become part of shallow lithosphere and situation returned to state of unstable equilibrium.

Thus, there are three main stages of the cyclic development of spreading zones: (i) initial - OCC (often one-sided spreading) → (ii) intermediate, such as Sierra Leone (the transition to a bilateral spreading) → (iii) normal (bilateral spreading). Each of these three types of spreading are observed in different segments of the present-day MAR, suggesting that these sites are various stages of development. In general, once started, the processes in the axial zone of the ridge are mutually self-sustaining conditioned, resulting in almost continuous growth of the oceanic lithosphere in the slow-spreading ridges axes.

8. Processes of slow-spreading ridges formation and development: Evidence from the MAR

8.1. Interaction of asthenosphere and mantle plumes

In contrast to MORB, derived from moderate-depleted mantle material, magmatism, related to mantle plumes, is presented by geochemically-enriched Fe-Ti picrites and basalts, evidence about rather different melting source. According to Anderson et al. (1992), the most centers of intraplate (plume) magmatism in the Atlantic are localized within the MAR. From this it follows that slow-spreading mid-ocean ridges are an areas of joint manifestations of asthenospheric and plume activity, and relationships between them is a key for understanding the functioning and development of slow-spreading mid-oceanic ridges.

Continuous smelting of basalts from the asthenospheric material had to inevitably affect to its composition in terms of increasing degree of depletion. However, this has not happened, and composition of the melting substrate as a whole remains practically the same during at least the latest 140 Myr in case of the Central Atlantic. Because asthenosphere beneath the MAR is a lens-shape body of 200-300 km thick (see Section 4), it requires a constant feed of it by geochemically-enriched material. Evidently, it can be the material of mantle plumes, constantly rejuvenating the composition of the asthenosphere after removing from it the low-melting components (basalts).

How in particular an interaction of the asthenosphere and mantle plumes could be occurred? It is known that under conditions of rigid continental lithosphere plume-related magmatic systems form isolated localities. However, in the case under consideration the situation is quite different: both asthenospheric and plume material have close visco-plastic consistency. Accordingly, only the largest and most stable in time plumes like Iceland, Azores, Tristan, etc., can cross such thick lens. They lose much of their material, which mixes with the asthenosphere matter, leading to the appearance of T-MORB (oceanic plateau basalts) and E-MORB in the adjacent parts of the spreading zone (Basaltic ..., 1981). Only such plumes can pass through the asthenosphere lens and products of their melting reach the surface, forming oceanic islands and seamounts. The existence of less powerful plumes may indicate the appearance of mantle-crustal magmas of the siliceous Fe-Ti-oxide series (see Section 2.3); still weaker plumes "damped" in the thick asthenospheric lens. In this connection, attention is drawn to that this lens itself is not a single uniform body, and is subdivided into several segments (Fig. 10). It is also confirmed by the results of geochemical studies of basalts throughout the MAR length (Fig. 15) (Dmitriev et al, 1999; Silantyev, Sokolov, 2010).

Mantle plumes, penetrating the asthenospheric lens, should contribute to forced convective mixing of its material and lead to practical levelling of its composition. Obviously, for this reason, geochemical and isotopic-geochemical characteristics of MORB, both in the Sierra Leone area and all over the MAR, are close to each other. From this it follows that material of the asthenosphere is a mixture of moderately depleted lherzolites and geochemically-enriched material of mantle plumes as finite compositions. Asthenospheric plastic material, in contrast to the melt, is mixed substantially worse, which are evidence from variations of isotopic and geochemical characteristics of the fresh basalts (see Section 2.1).

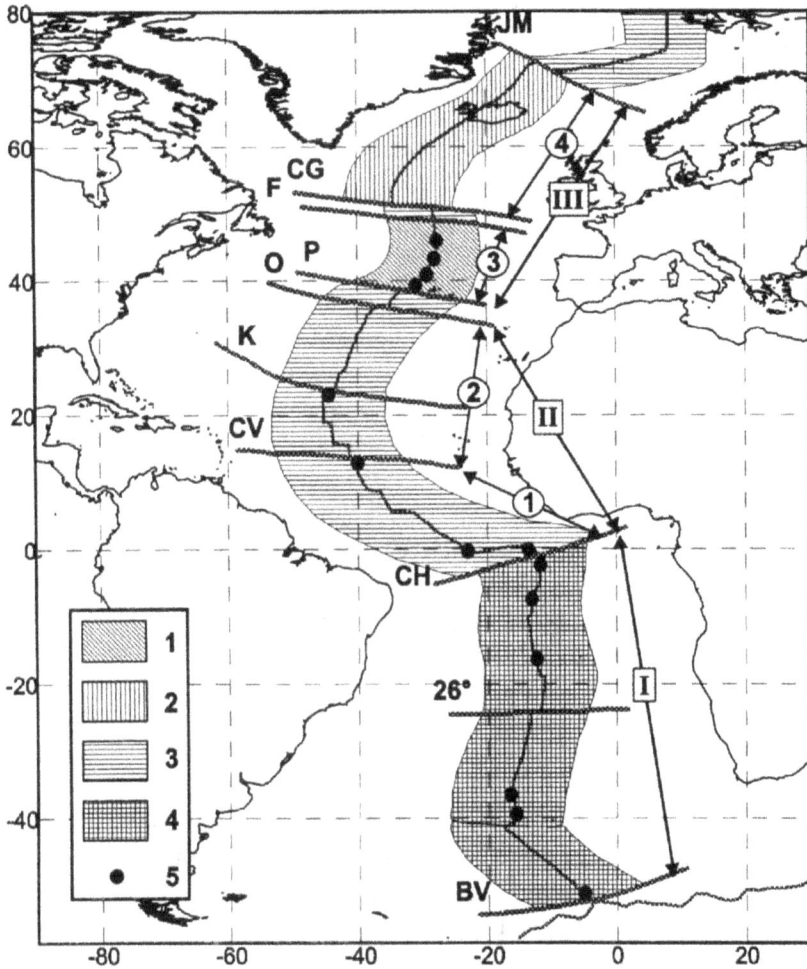

1 – TOR-1; 2 – TOR-1 + TOR-Fe; 3 – TOR-2; 4 – TOR-1 + TOR-2; 5 - TOR-K. I, II and III - segments of the 1st order: I - South Atlantic, II - Southern Region of the North Atlantic, III - Northern Region of the North Atlantic (TOR – varieties of tholeiites of ocean ridges), 1 - 4 – segments of the 2nd order: 1 – Equatorial ; 2 – Central; 3 – Azores; 4 – Icelandic. Fractures zones: JM – Jan Mayen, CG – Charlie Gibbs, P – Pico, D – Oceanographer, K – Kane, CV - Cape Verde, CH – Cheyne, 26°- 26° S, BV – Bouvet. TOR –tholeiite of oceanic ridges..

Figure 15. Scheme of tectonomagmatic segmentation of the Mid-Atlantic Ridge. After (Dmitriev et al, 1999).

As it was mentioned above, according to geophysical data, a colder mantle occurs beneath asthenospheric lens (Fig. 10). It is supported by finds of ancient xenogenic zircon grains, which, very likely, were trapped by rising plumes from the " slab graveyards" in the mantle beneath the ridge (see Section 4). Apparently, the impurities, trapped by plumes from deep mantle, as well as material of the shallow lithosphere, trapped by basalts on their way to the surface (see Section 2.1), play essential role in scatter of points on the Sr-Nd diagrams and the appearance of various "mantle reservoirs," in particular, HIMU.

Thus, mantle plumes bring to the geochemically-depleted oceanic asthenosphere considerable amount of fresh hot geochemically-enriched material, providing forced convection of the material. This leads to more or less effective mixing of the two types of finite materials, as well as to general leveling of the asthenosphere composition and temperature, thereby support their sustainable dynamic equilibrium in considerable time, at least ~140 Myr in case of the Central Atlantic.

8.2. How formation and development of the oceanic asthenospheric lens occurred?

Obviously, constant addition of a new (plume) material had to cause to increasing of the asthenospheric lens size, leading to its extention in both directions: from the ridge axis (spreading itself) and along it (propagation of ridge). As a consequence of the lens extending, it becomes a "trap" for the plumes, rising in the neighborhood, which became parts of the asthenosphere's supply system and contribute to further widening of the ocean floor in width and length.

However, it remains unclear how the asthenospheric lens, which promoted oceanic spreading, was initially formed. Perhaps, its occurrence was attributed to an elongated area of concentrated manifestation of mantle plumes activity. Apparently, in this case the extended heads of neighboring large plumes came in contact with each other and merged (coalescence) into a single body. This body will grow due to involvement of plumes in the neighborhood, gradually increasing in size and can gradually developed into a zone of oceanic spreading. Possible examples of initial stages of the process are ultraslow-spreading ridges (Knipovich Gakkel, Monze, Lena Trough), which develope in the North Atlantic and the Arctic Ocean, where the MAR propagates (Snow, Edmonds, 2007).

Modern example of an elongated area of mantle plumes activity may be the Trans-Eurasian Belt of tectonomagmatic activation, which stretches out along the whole of Eurasia from the Atlantic to the Pacific and appeared after closure of the Mesozoic Tethys Ocean (Sharkov, 2011). If the plumes are distributed uniformly, a large igneous province, like the Permian-Triassic Siberian Traps, formed instead of an ocean.

Thus, data available on the Sierra Leone area and other OCCs allow to complement the existing models of the structure and development of slow-spreading ridges and liberalize present-day views on processes, occurred in their axes. Besides, they provide opportunity to discuss problem of structure of the mantle beneath the slow-spreading ridges. As shown above, the shallow oceanic lithosphere is composed mainly by plutonic rocks and high depleted mantle and qualitatively different from the underlying asthenospheric lens, formed by moderately depleted material. Located beneath the lens deep mantle also differs significantly from the asthenospheric material. From this it follows that all three components of the mantle under the slow-spreading MAR have an independent origin, and whole-mantle convection is absent here.

The latest data on the geology, petrology, geochemistry, isotopy and geophysics of the oceanic bedrocks takes into account in the proposed model. These data were not known a half century ago, when the basic conceptions of plate tectonics were elaborated. Gradually it

Cyclic Development of Axial Parts of Slow-Spreading Ridges: Evidence from Sierra Leone Area, the Mid-Atlantic Ridge, 5-7°N

45

has revealed details that allow to take a new look at the nature and mechanisms of oceanic spreading. In general, our data suggest that real tectonomagmatic processes in the axis of slow-spreading MAR are essential different from existing views on processes and mechanisms of oceanic spreading. Obviously, there is still much uncertainty, but it is also clear that new approaches to the study of geology and petrology of ocean are necessary.

9. Conclusions

1. Sierra Leone area, located in the axial part of the MAR (5°-7°N), is characterized by outcrops of extensive deformed and metamorphosed plutonic rocks of the shallow oceanic lithosphere. The area are characterized by wide development of extensional structures, including rift valleys and variably oriented faults. These features of the structure and composition of rocks can define the area as a kind of oceanic core complex (OCC). However, unlike typical OCCs, outcrops of altered gabbros and serpentinites occur only on valleys' slopes and floors, while surface of the ridge is overlapped by flows of fresh pillow lavas with chilled glassy crusts, i.e., a kind of "structural unconformity" occurs here.

2. Fresh basalts are close in composition to E- and T-MORB (oceanic plateau basalts); judging by #mg and composition of phenocrysts, they are not primary mantle-derived melts, and underwent by crystallization differentiation in the intermediate (intrusive) chambers. On the way to the surface, they crossed the mantle peridotite and lower-crustal gabbros, and were partly contaminated by their material.

3. The lower crust in the area is composed by gabbros of two types: (1) primitive, magnesian, derived from MORB, and (2) often hornblende ferrogabbros derived from melts of siliceous Fe-Ti-oxide series. These melts, on the one hand, were saturated and supersaturated in silica and characterized by elevated water content, which is typical for suprasubduction magmas, and on the other hand – have a high content of Ti, Fe, Nb, Ta and P, which are typical for magmas of intraplate (plume) origin. It suggests that formation of such specific melts was attributed to melting of hydrated oceanic shallow lithosphere under influence of new mantle plumes. Minor trondhjemite occurrences are observed in form of veins and small bodies; their origin is considered to anatexis of hydrated rocks of the lithosphere by influence of mafic intrusions.

4. Sulfide mineralization, found in the Markov Deep, is confined to a zone of hydrothermal-metasomatic processing in cataclasites upon ferrogabbros and, apparently, was attributed to fluids of magmatic origin, gradually diluted by sea-water from the subfloor hydrosphere. The source of these fluids could be shallow intrusions of siliceous Fe-Ti-oxide series. This mineralized zone is probably a piece of a Extinct39 feeder system of former "black smoker".

5. SHRIMP-II U-Pb dating of zircon grains, extracted from gabbros of the area, revealed the two groups: (i) "young", primary magma, with the age of 0.7-2.3 Ma, and (ii) "ancient", xenogenic, with age from 87 to 3117 Ma; at that zircons of different ages may be found in the same samples. It is assumed that the grains of the "ancient" group belong

to material of " slab graveyards", which fragments were captured by ascended mantle plumes in the deep mantle.

6.   It is shown that structure of the MAR in geomechanical terms is an example of a piercing structure, appeared due to the introduction of plastic less dense and less viscous layer (asthenosphere) in a more dense layer (shallow lithosphere) under gravity. The system is in unstable equilibrium state until appearance of mantle plume, which lifted the dense lithosphere at higher hypsometric level in shape of a dome. This is causing sliding of material on slopes of the dome (tectonic erosion), often one-sided, resulting in exhumation of the deep-seated rocks on ocean floor and forming OCC. The interaction of the plume head with the hydrated lithosphere, led to appearance of melts of siliceous Fe-Ti-oxide series which providing hydrothermal activity.

7.   As a result of tectonic erosion, pressure at the ridge axis decreased, which led to starting of asthenospheric crest ascent; because of decompression, a zone of adiabatic melting appeared at the top of the crest and role of basaltic volcanism gradually increasing. The ridge axis gradually got a stable triangular shape and slumping of material becomes bilateral. As a result, intermediate structures, such as Sierra Leone area, with magmatism of E- and T-MORB are formed, and then the stage of vast eruptions of MORB comes. The process goes to end when the melting zone was overfill by light restite material, which is not involved in convection; the restites became a part of the lithosphere, and the system returns to a state of unstable equilibrium.

8.   Geological sense of the oceanic spreading, evidently, is a combination of thermal and geomechanical processes at the lithosphere-asthenosphere boundary, starting with the formation of domes and slumping newly formed material (new lithospheric plates) on their slopes. From this standpoint constructive plate boundaries, at least in slow-spreading ridges, should be represented by a collage of tectonic slices of shallow oceanic lithosphere and basaltic sheets. So, process of spreading in the MAR has a cyclic character. It begins from appearance of OCCs, often characterized by "one-side spreading" and numerous hydrothermal fields, and via structure type Sierra Leone area pass to normal bilateral spreading with vast basaltic eruptions.

9.   Based on these data and taking into account the published materials of seismic tomography, it is developed a new model of oceanic spreading in the MAR. It is shown that the long-term existence of the MAR's oceanic spreading (at least 140 Myr) and stability of composition of basalts can be explained by dynamic equilibrium between permanent removal from asthenosphere of newly formed basalt and replenishment of new geochemically-enriched material of mantle plumes. The constant injections of a hot plume material in asthenospheric lens provide forced convection its material and prevents it from freezing; moreover, it also promote expansion of the asthenosphere both across a ridge axis (oceanic spreading) and along its axis (propagation of the ridge).

10. How it is evidence from the MAR, three independent components in structure of its mantle occur: (i) shallow lithosphere (including basaltic upper crust), (ii) asthenosheric lens beneath the ridge, and (iii) deep mantle with "graveyards of slabs". Each of them, how it was shown above, has own origin and composition. From this evidently follows that the total convection in the oceanic mantle is absent.

11. Thus, the processes in the slow-spreading ridge axes are mutually conditioned self-sustaining character, resulting in almost continuous growth of the oceanic lithosphere in its different parts, and supported the process of spreading, as evidenced by the presence of symmetrical magnetic anomalies.

12. Slow- and ultraslow-spreading ridges are a special class of oceanic spreading, characterized by widespread development of oceanic core complexes and absence of subduction on periphery of the oceans, where developed passive margins. Appearance of such ridges is associated with the elongated areas of concentrated manifestation of sustainable mantle plume activity. Apparently, extended heads of large plume came into contact with each other, merging (coalescenced) in almost single body asthenospheric lenses.

Author details

E.V. Sharkov
Institute of Geology of Ore Deposits, Petrography,
Mineralogy and Geochemistry (IGEM), Russian Academy of Sciences, Moscow, Russia

Acknowledgement

I am very thanks to N.S. Bortnikov, S.A. Silantyev and A.V. Girnis for stimulated discussions and comments.

10. References

Allaby, A. & Allaby, M., 2011. A Dictionary of Earth Sciences. 1999. *Encyclopedia.com*. 9 Sep., pdf.

Anderson, D.L., Tanimoto, T. & Shang, Y., 1992. Plate tectonics and hot spots: the third dimension. *Science,* 256: 1645-1650.

Aranovich, L.Ya., Bortnikov, N.S., Serebryakov, N.S., & Sharkov, E.V., 2010. Conditions of the Formation of Plagiogranite from the Markov Trough, Mid-Atlantic Ridge, 5°52'–6°02'N. *Doklady Earth Sciences,* 434 (1): 1257–1262.

Basaltic Volcanism Study Project 1981. Basaltic volcanism on the terrestrial planets, 1981. New York: Pergamon Press, 1286 p.

Beltenev, V., Ivanov, V., Rozhdestvenskaya, I. et al., 2009. New data about hydrothermal fields on the Mid-Atlantic Ridge between 11°-14°N: 32th Cruise of R/V «Professor Logachev». *InterRidge News,* 18: 14-18.

Bortnikov, N. S., Sharkov, E. V. , Bogatikov, O. A., Zinger, T.F., Lepekhina, E.N., Antonov, A.V. & Sergeev, S.A., 2008. Founding of young an anicient zircons in gabbros of Markov Deep, Mid-Atlantic Ridge, 5°30.4' – 5°32.4' N (results of SHRIMP-II U-Pb dating): meaning for understanding of deep-seated geodynamics of the modern oceans. *Dokl. Earth Sci.* 421(5): 859–868.

Cannat, M., Escartín, J., Lavier, L. et al., 2010. Lateral and Temporal Variations in the Degree of mechanical weakening in the footwall of oceanic detachment faults. *AGU Chapman Conference "Detachments in Oceanic Lithosphere: Deformation, Magmatism, Fluid Flow, and Ecosystems" Agros, Cyprus 8-15 May 2010. Conference Report*: 38-39.

Casey, J.F., Gao, Y., Benavidez, R. & Dragoi, C., 2010. The lowest δ⁷Li yet recorded in MORB glasses: the connection with oceanic core complex formation, refractory rutile-bearing eclogitic mantle sources and melt supply. *2010 AGU Fall Meeting*. 13-17 December 2010. San Francisco, California. Abstract V11A-2245, pdf.

Condie, K.C., 2005. High field strength element ration in Archean basalts: a window to evolving sources of mantle plumes? *Lithos*, 79: 491-504.

Conference Outline. *AGU Chapman Conference "Detachments in Oceanic Lithosphere: Deformation, Magmatism, Fluid Flow, and Ecosystems" Agros, Cyprus 8-15 May 2010. Conference Report:* 20-21.

Dick, H.J.B., Meyer, P.S., Bloomes S.H. et al., 1991. *Lithostratigraphic Evolution of an in Situ Section of Oceanic Layer 3* . Eds. R.P. Von Herzen, P.T. Robinson et al. *Proc. ODP, Sci. Results 118:* 439–538.

Dick, H.J.B., Robinson, P.T. & Meyers, P.S., 1992. The plutonic foundation of a low-spreading ridge. *Amer. Geophys. Union monograph 70:* 1-39.

Dick, H.J.B., Tivey, M.A. & Tucholke, B.E., 2008. Plutonic foundation of a slow-spread ridge segment: Oceanic core complex at Kane Megamullion, 23°30'N, 45°20'W. *Geochem., Geophys., Geosyst.*, 9, Q05014, doi:05010.01029/02007GC001645.

Dilek Y. & Furnes H., 2011.Ophiolite genesis and global tectonics: Geochemical and tectonic fingerprinting of ancient oceanic lithosphere. *GSA Bull.*, 123(¾): 387-411.

Dmitriev, L.V., Sokolov, S.Yu., Melson, V.G. & O'Hirn, T., 1999. Plume and spreading association of basalts and their reflection in petrological and geochemical parameters in northern part of the Mid-Atlantic ridge. *Russian J. Earth Sci.*, 1(6): 457-476.

Escartín, J., Smith, D.K., Cann, J. et al., 2008.Central role of detachment faults in accretion of slow-speading oceanic lithosphere. *Nature*, 4559: 790-795.

Ernst, W.G., 2001. Subduction, ultrahigh-pressure metamorphism and regurgitation of byont crustal slices – implications for arcs and continental growth. *Phys. Earth and Planet. Inter.* 127: 253-275.

Fluids and redox reactions in magmatic systems. A.A. Kadik (ed.). 1991. *Moscow, Nauka Publ.*, 256 p.

Fontingie, D. & Schilling ,J.-G., 1996. Mantle heterogeneities beneath the South Atlantic: a Nd-Sr-Pb isotope study along Mid-Atlantic Ridge (3°S-46°S). *Earth Planet. Sci. Lett.*, 142(1/2): 209-221.

Gillis, K.M. & Coogan, L.A., 2002. Anatectic migmatites from the roof of an ocean ridge magma chamber. *J. Petrol.*, 43: 2075–2095.

Girnis, A. V., 2003. Olivine-Orthopyroxene-Melt Equilibrium as a Thermobarometer for Mantle-Derived Magmas. *Petrology*, 11(2): 101-113.

Handbook of physical constants of rocks. S. Clark (ed.)., 1969. Moscow, Mir: 520 p. (Russian translation)

Hauri, E.H., Whitehead, J.A. & Hart, S.R., 1994. Fluid dynamic and geochemical aspects of entrainment in mantle plumes. *J. Geophys. Res.*, 99(B12): 24275-24300.

Hékinian, R., Juteau, T., Gracia, E. et al., 2000. Submersible observations of Equatorial Atlantic Mantle: The St. Paul Fracture Zone region. *Marine Geophysical Research*, 21: 529-560.

Hess, H.H., 1962. The history of the ocean basins. *Geol. Soc. Am. Buddington Vol.*, 599-620.

Ildefonse, B., Blackman, D.K., John, B.E. et al., 2007. Oceanic core complexes and crustal accretion at slow-spreading ridges. *Geology*, 35(7): 623-626.

Karason, H. & van der Hilst, R.D., 2000.Constraints on mantle convection from seismic tomography. *The History and Dynamics of Global Plate Motions.* Eds. M. Richards, R. G. Gordon, and R.D. van der Hilst. *Geophys. Monogr.* 121: 277–289.

Knipper, A.L., Savelieva, A.L., Sharaskin, A.Ya., 2001. Problems of ophiolite classification. *Fundamental problems of general geotectonics.* Yu.M. Puscharovsky (ed.). Moscow, Nauchny mir: 250-281 (in Russian).

Krassivskaya I.S., Sharkov E.V., Bortnikov N.S., Chistyakov A.V., Trubkin N.V., Golovanova T.I., 2010. Variolitic lavas in the axial rift of the Mid-Atlantic rift and its origin. *Petrology,* 18: P.263-277.

Langmuir, C.H., Klein, E.M. & Plank, T., 1992. Petrological systematics of mid-ocean ridge basalts: constraints on melt generation beneath ocean ridges. *Mantle flow and melt generation at mid-ocean ridges.* Eds.: Morgan J.P., Blackman D.K., Sinton J.M. Geophys. Monogr. Am. Geophys. Union, 71: 183-280.

MacLeod, C.J., Searle, R.C., Murton, B.J. et al., 2009. Life cycle of oceanic core complexes. *Earth Planet. Sci. Lett.,* 287(3-4): 333-344.

Ohara, Y., Yoshida, T., Kasuga, S., 2001. Giant megamullion in the Parece Vela Backarc basin. *Mar. Geophys. Res.,* 22: 47-61.

Pearce, J., 2002. The oceanic lithosphere. *Achievements and Opportunities of Scientific Ocean Drilling.* Spec. Issue of the JOIDES Journal, 28(1): 61-66.

Puscharovsky, Yu.M., Skolotnev, S.G., Peyve, A.A., Bortnikov, N.S., Bazilevskaya, E.S. & Mazarovich, A.O., 2004. *Geology and metallogeny of the Mid-Atlantic ridge: 5-7°N.* Moscow, GEOS, 151 p. (in Russian)

Quick, J.E. & Denlinger, R.P., 1993. Ductile deformation and the origin of layered gabbro in ophiolites. *J. Geophys. Res.,* 98: 14015–14027

Ritsema, J. & Allen, R.M., 2003. The elusive mantle plume. *Earth Planet. Sci. Lett.,* 207: 1-12.

Roden M.F., Hart S.R., Frey F.A., Melson W.G., 1984. Sr, Nd and Pb isotopic and REE geochemistry of St. Paul's Rocks: its metamorphic and metasomatic development of an alkali basalt mantle source. *Contrib. Mineral. Petrol.* 85(8):

Rudnick, R., 1990. Growing from below. *Nature,* 347(6295): 711-712.

Savelieva, G.N. 1987. *Gabbro-ultrabasite complexes of the Urals ophiolites and their analogs in the modern oceanic crust.* Moscow, Nauka Publ., 246 p. (in Russian)

Savelieva, G.N., Bortnikov, N.S., Peyve, A.A. & Skolotnev, S.G., 2006. Utrabasite rocks of the Mrkov Deep, rift valley of the Mid-Atlantic Ridge. *Geochem. Intern.* 44(11): 1192-1208.

Scheidegger, A.E., 1982. *Principle of Geodynamics.* 3rd edition. Berlin-Heidelberg-Bew York,.

Schilling, J.-G., Hanan, B.B., McCully, B. & Kingsley, R.H., 1994. Influence of the Sierra Leone mantle plume on the equatorial MAR: a Nd–Sr–Pb isotopic study. *J. Geophys. Res.,* 99: 12005–12028.

Sharkov, E.V., 2006. *Formation of layered intrusions and their ore mineralization.* Moscow: Scientific World, 364 p. (in Russian),

Sharkov, E.V., Abramov, S.S., Simonov, V.A., Krinov, D.I., Skolotnev, S.G., Bel'tenev, V.E. & Bortnikov, N.S., 2007. Hydrothermal alteration and sulfide mineraliztion in gabbroids of the Markov Deep (Mid-Atlantic Ridge, 6° N). *Geology of Ore Deposits,* 49(6): 467-486.

Sharkov, E.V., Bortnikov, N.S., Bogatikov, O.A., Zinger, T.F., Bel'tenev V.E., & Chistyakov, A.V., 2005.,Third layer of the oceanic crust in the axial part of the Mid-Atlantic Ridge (Sierra Leone MAR segment, 6°N). *Petrology,* 13(6): 540-570.

Sharkov, E.V., Chistyakov, A.V., & Laz'ko, E.E., 2001. The Structure of the Layered Complex of the Voikar Ophiolite Association (Polar Urals) as an Indicator of Mantle Processes beneath a Back-Arc Sea . *Geochemistry Intern.*, 39(9): 831-847.

Sharkov, E.V., Shatagin, K.N., Krassivskaya, I.S., et al., 2008. Pillow Lavas of the Sierra Leone Test Site, Mid-Atlantic Ridge, 5°-7° N: Sr-Nd Isotope Systematics, Geochemistry, and Petrology. *Petrology*, 16(4): 335-352.

Sharma, M., Hofmann, A.W. & Wasserburg, G.J., 1998. Melt generation beneath ocean ridges: Re-Os isotopic evidence from the Polar Ural ophiolite. *Miner. Mag., V.M. Goldschmidt Conference. Toulouse Abstracts.* 62A: 1375-1376.

Sharma, M., Wasserburg, G.J., Papanastassiou, D.A., Sharkov, E.V. & Laz'ko, E.E., 1995. High $^{143}Nd/^{144}Nd$ in extremely depleted mantle rocks. *Earth and Planet Sci. Letters*, 35: 101-114.

Silantyev, S.A., Dmitriev, L.V., Bazylev, B.A. et al., 1995. An examination of genetic conformity between co-existing basalts, gabbro, and residual peridotites from 15°20' Fracture Zone, Central Atlantic: evidence from isotope composition of Sr, Nd, and Pb. *InterRidge News*, 4: 18-21.

Silantyev, S.A., Krasnova, E.V., Cannat, M. et al. 2011. Peridotite-gabbro-trondhjemite association of rocks of Mid-Atlantic Ridge in 12°58' – 14°45'N: hydrothermal fields Ashadze and Logachev. *Geochem. Intern.*, 49(4): 339-372.

Simonov, V.A., Sharkov, E.V. & Kovyasin, S.V. 2009. Petrogenesis of the Fe-Ti intrusive complexes in the Sierra-Leone test site, Central Atlantic. *Petrology*, 17(5): 488-502.

Skolotnev S.G., Beltenev V.E., Lepekhina E.N. & Ipatiev I.S., 2010. Young and ancient zircons from rocks of oceanic lithosphere of the Central Atlantic, geotectonical consequences. *Geotectonics*, 6: 24-59.

Smith, D.K., Escartín, J., Schouten, H. & Cann, J.R., 2008. Fault rotation and core complex formation: Significant processes in seafloor formation at low-spreading mid-ocean ridges (Mid-Atlantic Ridge, 13°-15°N). *Geochem, Geophys, Geosystems*, 9(3): 1525-2027.

Snow, J.E. & Edmonds, H.N., 2007. Ultraslow-spreading ridges. Rapid paradigm changes. *Oceanography*, 20(1): 90-101.

Thy, P., 2003. Igneous petrology of gabbros from Hole 1105A: oceanic magma chamber processes. Eds. J.F. Casey and D.J. Miller. *Proc. ODP, Sci. Results*, 179: 1-76.

Tucholke, B.E., 1998. Discovery of "Megamullions" Reveals Gateways Into the Ocean Crust and Upper Mantle. *OCEANUS*, 41(1): 15-19.

Tucholke, B.E., Behn, M.D., Buck, W.R. & Lin, J., 2008. Role of melt supply in oceanic detachment faulting and formation of megamullions. *Geology*, 36: 455-458.

Tucholke, B.E. & Lin J., 1994.,A geological model for the structure of ridge segments in slow spreading oceanic crust. *J. Geophys. Res.*, 99: 11937-11958.

Turcotte, D.L. & Schubert, G., 2002. *Geodynamics*. 2nd edition. Cambridge: Cambridge Univ. Press, 847 p.

Widom, E., Carlson, R.W., Gill, J.B. et al., 1997. Th–Sr–Nd–Pb isotope and trace element evidence for the origin of the Saõ Miguel, Azores, enriched mantle source. *Chem. Geol.*, 140(1): 49-68.

Sedimentary Basins

Seismic Paleo-Geomorphic System of the Extensional Province of the Niger Delta: An Example of the Okari Field

Muslim B. Aminu and Moses O. Olorunniwo

Additional information is available at the end of the chapter

1. Introduction

Hydrocarbon exploration and exploitation requires that the spatial and depth distribution and interplay of factors favorable to commercial hydrocarbon accumulation are thoroughly appreciated. These factors include the distribution of source rock, reservoir rock, and migration pathways, the fidelity of sealing mechanisms, and timing, the temporal relatioinship between reservoir rock formation and the expulsion of hydrocarbons from the source rock. The distribution of these elements of the petroleum system is a result of the tectonic history and fill processes occuring in a basin.

As hydrocarbon exploration moves into geologically and economically more challenging environments, such as deeper subsurface locations, deepwater regions, sub-ice in the Artic, and into geologically and stratigraphyically more complex environments, the costs of exploration is bound to be on the rise and the risks associated with field development greater. Continued success in the hunt for Oil and Gas reserves therefore, depends upon a thorough understanding of the subsurface geology of exploration fields, the ability to accurately predict and delineate the spatial and depth distribution of subsurface geologic facies (source rock, reservoir rock and seal) and the ability to discriminate the fluids saturating reservoirs (oil, gas or brine) and possibly quantifying such.

Seismic geomorphology which is the integration of three-dimensional (3-D) seismic data with analytical techniques typically used in the study of Earth land forms, is growing in importance as a tool in understanding the stratigraphy of both mature fields and underdeveloped basins (Dunlap et al, 2010). Seismic geomorphologic analysis allows geoscientists to correctly identify, interpret, and predict the distribution of possible hydrocarbon-bearing deposits. It helps to unravel the major factors which have contributed to and or continue to control the evolution of a basin or field.

In this study, we implement a series of conventional interpretation procedures and advanced imaging techniques to unravel the paleo-geomorphology, tectonic history and fill architecture of a hydrocarbon target within the Okari Oil Field.

2. The Niger delta

The Niger delta is situated in the Gulf of Guinea on the west coast of Central Africa (Figure. 1). It is located in the southern part of Nigeria between latitudes 4^0 00'N and 6^0 00'N and longitudes 3^0 00'N and 9^0 00'N. It is bounded in the south by the Gulf of Guinea (or the 4000 m bathymetric contour) and in the North by older (Cretaceous) tectonic elements which include the Anambra Basin, Abakaliki uplift and the Afikpo syncline. In the east and west respectivily, the Cameroon volcanic line and the Dahomey Basin mark the bounds of the Delta, Figure 1. The Cenozoic Niger Delta is situated at the intersection of the Benue trough and the South Atlantic Ocean where a triple junction developed during the separation of South America from Africa (Burke, 1972; Whiteman, 1982). The delta is considered one of the most prolific hydrocarbon provinces in the world, and recent giant oil discoveries in the deep-water areas suggest that this region will remain a focus of exploration activities (Corredor et al, 2005).

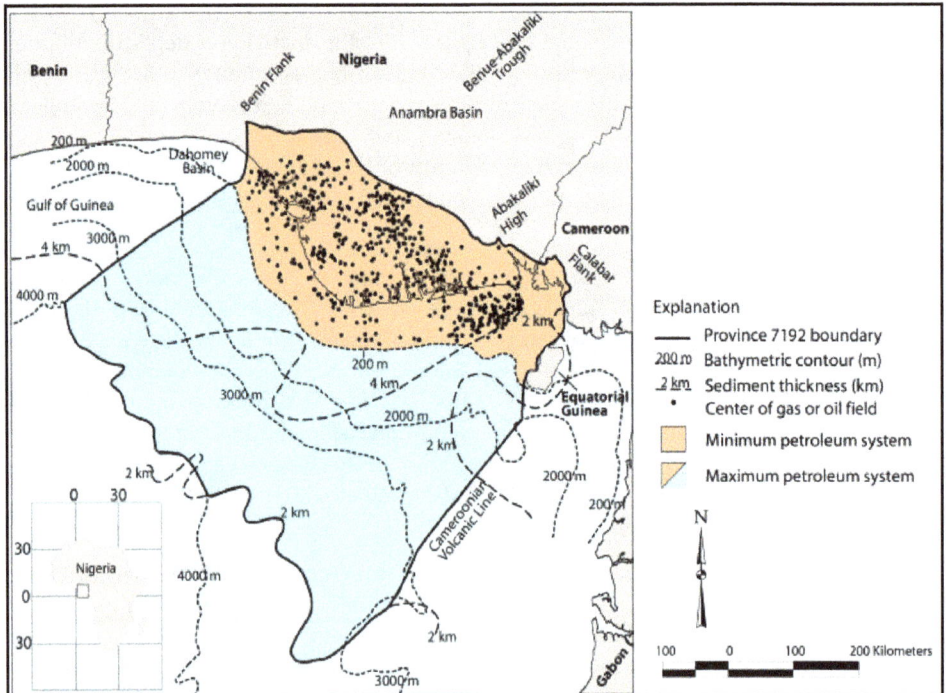

Figure 1. Map of Niger Delta showing province outline (maximum petroleum system); bounding structural features; minimum petroleum system as defined by oil and gas field center points (Modified from Petroconsultants, 1996, as cited in Tuttle et al, 1999).

Furthermore, the Niger delta is one of the world's largest deltas, with a sub-aerial acreage of about 75,000 km². It is composed of an overall regressive clastic sequence, which reaches a maximum thickness of 30,000 to 40,000 ft (9000 to 12000 m). The development of the delta has been dependent on the balance between the rate of sedimentation and the rate of subsidence (Doust and Omatsola, 1990). This balance and the resulting sedimentary patterns were constrained by the structural configuration and tectonics of the basement (Evamy et al, 1978). Important influences on sedimentary rates have included eustatic sea-level changes and climatic variations in the hinterlands. Subsidence has been controlled largely by initial basement morphology and differential sediment loading of unstable shales (Doust and Omatsola, 1990).

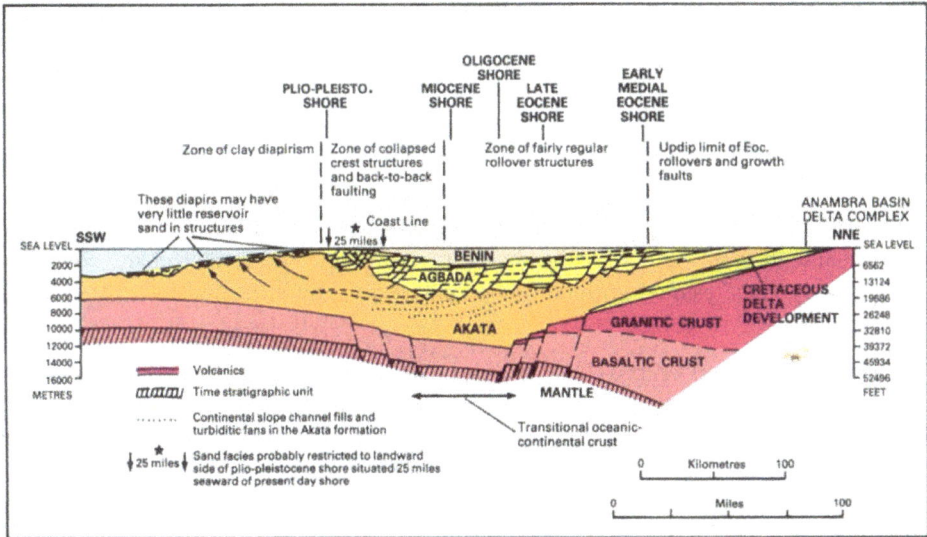

Figure 2. Generalized dip section of the Niger Delta showing the structural provinves of the Delta. The slips resulting from the collapse of shelf sediments in the extentional province is diverted onto thrust ramps through detachemnt structures in the basal *Akata* formation and comsumed by contractional folds in the deep-water fold and thrust belts (Adapted from Whiteman, 1982).

The Delta has built out over the collapsed continental margin, and its core is located above the collapsed continental margin at the site of the triple junction formed during the middle Cretaceous. The main sediment supply has been provided by an extensive drainage system, which in its lower reaches follows two failed rift arms, the Benue and Bida basins. Sediment input generally has been continuous since the Late Cretaceous, but the regressive record has been interrupted by episodic transgressions, some of considerable extent (Weber and Daukoru, 1975). The Niger delta has prograded into the Gulf of Guinea at a steadily increasing rate in response to the evolving drainage area, basement subsidence and eustatic sea level changes. Initially, the delta prograded over extensionally thinned and collapsed continental crust of the West African margin (Figure 2), as far as the triple junction, filling in the basement graben-and-horst topography (Murat, 1970). During the middle and late

Eocene, sedimentary rocks became increasingly sandy, marking the onset of a general regression of the deltaic deposition.

Of dominant importance in the development of the delta, as in similar settings elsewhere in the world has been the influence of synsedimentary listric normal faults. These have been forming at least since the Paleocene and define sites of locally increased subsidence and sedimentation. They lie sub-parallel to the paleo-coastline and are presumed to sole out at relatively shallow depths within marine shale sequences. A number of major basin-wide growth fault zones define depositional realms at succeeding periods of delta history (Doust and Omatsola, 1990). In the paralic interval, growth fault associated rollover structures trapped hydrocarbons. Faults in general play an important role in the hydrocarbon distribution. Growth faults may even function as hydrocarbon migration paths from the overpressured marine clays. There is an intimate relationship between structure and stratigraphy. They both depend on the interplay between rates of sediment supply and subsidence (Evamy et al., 1978).

2.1. Structure

The Niger Delta is regarded as a classical shale tectonics province (Wu and Bally, 2000). The instability and constant motion of the shales in response to the weight of the advancing sediment wedge has resulted in the development of different structural styles in different belts of the Delta.

The Delta has thus been subdivided into five zones. Corredor et al. (2005) identified these structural zones (Figure 3) as [1] an extensional province beneath the continental shelf that is characterized by both basinward-dipping (Roho-type) and counter-regional growth normal faults and associated rollovers and depocenters; [2] a mud diapir zone located beneath the upper continental slope, which is characterized by passive, active, and reactive mud diapirs, including shale ridges and massifs, shale overhangs, vertical mud diapirs that form mud volcanoes at the seafloor, and interdiapir depocenters; [(3] the inner fold and thrust belt, which is characterized by basinward verging thrust faults (typically imbricated) and associated folds, including some detachment folds; [4] a transitional detachment fold zone beneath the lower continental slope that is characterized by large areas of little or no deformation interspersed with large, broad detachment folds above structurally thickened *Akata* Formation; and [5] the outer fold and thrust belt characterized by both basinward- and hinterland-verging thrust faults and associated folds. The inner and outer fold and thrust belts are most evident in the bathymetry, where ridges represent the crests of fault-related folds, and low regions correspond to piggyback basins formed above the backlimbs of fault imbricates. The inner fold and thrust belt extends in an arcuate path across the center of the offshore delta, whereas the outer fold and thrust belt consists of northern and southern sections that define two outboard lobes of the delta (Corredor et al, 2005). These two lobes, and their associated fold belts, are separated by a major rise in the basement topography that corresponds to the northern culmination of the Charcot fracture zone (Figure 3). The break between the northern and southern sections of the outer fold and

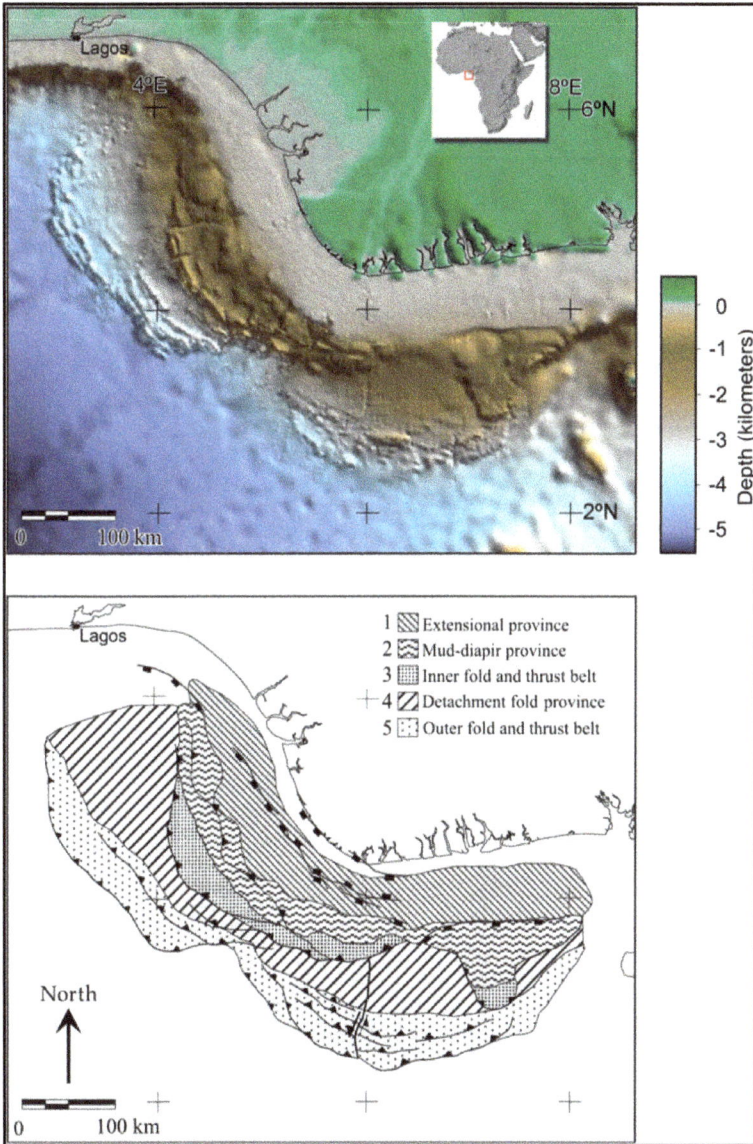

Figure 3. Bathymetric Sea-floor image of the Niger Delta obtained from a dense grid of two-dimensional seismic reflection profiles and the global bathymetric database (Smith and Sandwell, 1997, as cited in Corredor et al, 2005), showing the main structural provinces.

thrust belt are the results of thrust sheets being stacked in a narrow zone above and behind this major basement uplift (Connors et al., 1998; Wu and Bally, 2000). Deformation across these structural provinces is active today, resulting in pronounced bathymetry expressions of structures that are not buried by recent sediments, as illustrated in Figure 3 (Corredor et al, 2005).

2.2. Stratigraphy

The lithostratigraphy of the Niger Delta Basin (Figure 4) consists of three main rock-stratigraphic units of Cretaceous to Holocene origin (Short and Stauble, 1967; Frankl and Cordry, 1967; Avbovbo, 1978). These units represent the prograding depositional environments (Corredor et al, 2005).

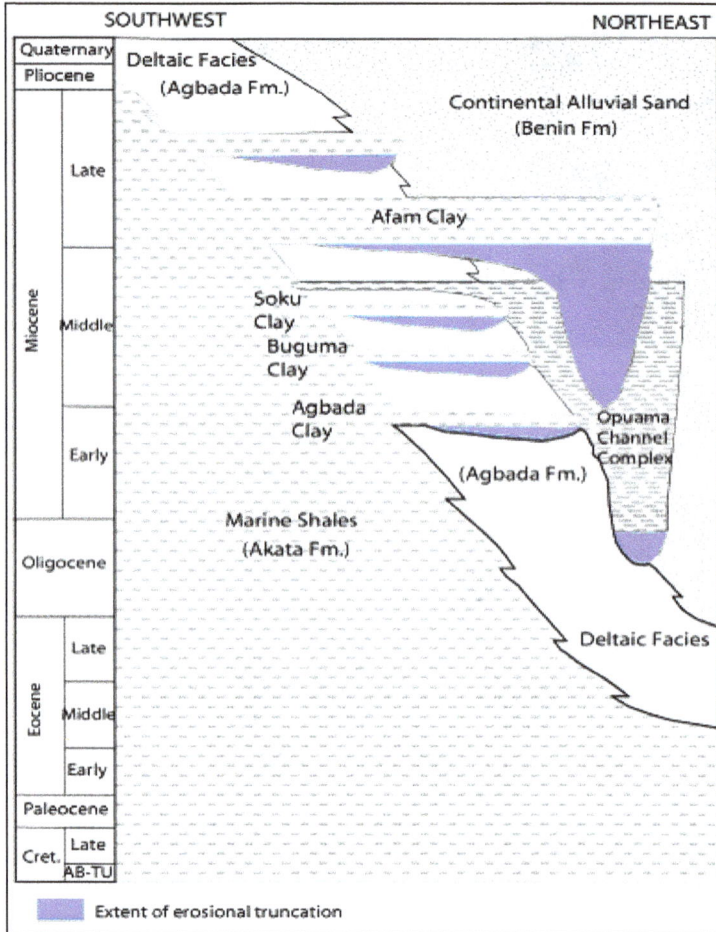

Figure 4. Stratigraphic column showing the three Formations of the Niger Delta, the Marine *Akata* shale, the paralic *Agbada* formation and the continental *Benin* sandstone. (Modified from Doust and Omatsola, 1990).

At the base of the system is the *Akata* Formation, a sequence of planktonic foraminifera rich undercompacted transgressive Paleocene-to-Holocene marine shales, clays, and silt. The *Akata* ranges in thickness from 2000 m (6600 ft) at the most distal part of the delta to 7000 m (23,000 ft.) thick beneath the continental shelf (Doust and Omatsola, 1990). In the outer fold and thrust belts, the *Akata* Formation reaches 5000 m (16,400 ft.) thick as a result of structural repetitions by thrust ramps and in the core of large detachment anticlines (Bilotti and Shaw, 2005). The *Akata* shales are typically undercompacted and frequently move either downward or laterally along the continental shelf, or in an upward diapiric motion along growth faults, in response to the lithostatic pressure of overlying sediment (Ekweozor and Daukoru, 1984). Lateral movements along the continental shelf result in thrusting in more outer bound regions of the Delta. The *Akata* Formation is conformably overlain by a paralic sequence of alternating Lower Eocene to Pleistocene sandstones and sand bodies with shale intercalations, which is known as the *Agbada* Formation (Doust and Omatsola, 1990). The *Agbada* Formation is more than 3500 m (11,500 ft.) thick and represents the actual deltaic portion of the sequence. This clastic sequence was accumulated in delta-front, delta-topset, and fluvio-deltaic environments. The *Agbada* formation is highly faulted with assays of roll-over extension induced growth faults, compensation listric faults and high angle thrust faults depending on which belt of the Niger Delta you are in. Channel and basin-floor fan deposits in the *Agbada* Formation form the primary reservoirs in the Niger Delta. The *Agbada* Formation is in turn covered by the *Benin* Formation which consists of late Eocene to Holocene massive, porous, and unconsolidated freshwater bearing continental deposits, including alluvial and upper coastal-plain deposits that are up to 2000 m (6600 ft) thick (Avbovbo, 1978).

In the outer thrust belt, the *Akata* Formation lies upon oceanic crust and is diachronously overlain by deep-water channel complexes, debris flows, and shales of the *Agbada* Formation, which represent the sedimentary overburden of the deltaic succession (Maloney et al., 2010). The *Benin* Formation is absent in this belt (Maloney et. al., 2010). The *Akata* Formation and the shaly intercalations of the *Agbada* Formation are believed to be the source rocks of the Niger Delta while the paralic *Agbada* Formation constitutes the reservoir rock. It has also been suggested (Corredor et al, 2005) that the *Akata* Formation could contain some turbidite sands which could represent potential reservoirs in deep-water environments.

2.3. The Extensional Province

The development of the Niger Delta was initiated by rifting on the onshore portion of the West African Shield (Merki, 1972). The rifting is associated with the opening of the South Atlantic Ocean which began in the Late Jurassic and spanned through the Middle Cretaceous (Lehner and De Ruiter, 1977, as cited in Tuttle et al, 1999). Within the Delta, rifting had ceased by the Late Cretaceous. Thereafter, gravity induced tectonics became the primary deformational process and represent the key factor determining the structural style of the Extensional Province of the Niger Delta. Shale mobility induced internal deformation and occurred in response to two processes (Kulke, 1995, as cited in Tuttle et al, 1999). First, shale diapirs formed from loading of poorly compacted, over-pressured, prodelta and delta-slope shales of the basal *Akata* Formation by the advancing higher density delta-front paralic wedge of the *Agbada* Formation, and second, slope instability resulting from a lack of lateral, basinward, support for the under-compacted delta-slope *Akata* Formation. These movements initiated syn-sedimentary growth faults which compartmentalized the belts into fault blocks. As faulting advanced fault blocks rotated in fault planes and subsurface horizons formed roll-over anticlines. The resulting seaward extension of subsurface strata was often accommodated by landward dipping compensational listric faults. Other structures which formed include shale diapirs, collapsed growth fault crests, back-to-back features, and steeply dipping, closely spaced flank faults (Evamy et al, 1978). The *Benin* Formation was deposited only after the cessation of gravity induced tectonics and is unaffected by faulting.

3. The Okari Field example: A gravitational shale tectonics field

At the time of this study, the Okari Oil Field had six wells, four of the wells are vertical and the other two deviated. Though the wells had varying suites of logs, they generally possessed lithologic logs (Gamma Ray), porosity logs (sonic, density and neutron), resistivity logs (LLD and MSFL) and checkshot surveys. Available seismic data consisted of 401 inlines and 221 crosslines and a total of 88,621 post-stack seismic traces with a record length of 5 seconds, sampled at 4 ms interval and covering an area of about 56.5 sq km.

Three of the wells, *B1*, *B4* and *B5* had encountered hydrocarbon in a stack of three horizons in a single roll-over anticlinal structure created by the rotation of an east-west trending fault block into the plane of a major structure building fault. The other three wells had turned out dry at the target horizon encountering water saturated sands and shales. It was understood that the variability of subsurface facies was more intricate than initially presumed. This is a known situation in the Niger Delta, the fast rate of sedimentation along the shelf edge and strong wave action are believed to have ensured rapid spacial variation in subsurface facies distribution patterns (Aminu and Olorunniwo, 2011). There was therefore a need to fully image the distribution of reservoir facies within the field to assist with further development and production.

3.1. Methodology

Our study of the field involved reservoir evaluation using well-logs, hydrocarbon saturated zones were identified and their respective petrophysical parameters were computed. Three horizons, *H1*, *H2* and *H3*, were found to be petroliferous. However, only the deepest horizon, *H3*, was considered prospective. It is a fairly shaly sand with average porosity of 27% and hydroarbon saturation of 77%. Next we created a tectono-stratigraphic model of the field by mapping all three horizons and the field's cascade of intersecting faults (both growth and compensational listric faults). In this way we mapped a faulted anticlinal closure against a major structure building fault in the middle of the survey area. Thereafter, we used tri-variate crossplots to study rock-physics relations within the field and to provide a basis for calibration and interpretation of subsurface facies prediction. We then computed several seismic attributes, some of which include the instantanous amplitude, frequency and phase, 1st and 2nd derivative of the seismic trace, and various geometric seismic attributes. We experienced considerable difficulty deriving correlation between computed attributes and well information, and at obtaining geologically meaningful maps of subsurface property distribution. This may be due to one of two factors, (1) the seismic attribute - log relationships are often non-linear, and (2) the Niger Delta is a soft clastic basin with under-compacted subsurface formations, the seismic amplitude response in such cases are often weak and not well correlated to geology (Chopra and Marfurt, 2007). This necessitated that we impliment multi-attributes transforms to predict subsurface facies from the seismic response. We elected to use the multi-linear resgression transform and an artifcial neural network. The multi-linear transform attempts to derive a multi-variate linear relationship between seismic attributes and log derived petrophysical properties. If for a single attribute, the relationship between the target log L and the attribute A is written as

$$L = a + bA \tag{1}$$

Where the coefficients a and b are obtained by minimizing the mean-squared prediction error:

$$E^2 = \frac{1}{N}\sum_{i=1}^{N}\left(L_i - a - bA_i\right)^2 \tag{2}$$

According to Hampson et al, 2001, the multivariate case for a single attribute is written as:

$$L(t) = w_0 + w_1 A_1(t) + w_2 A_2(t) + ... + w_m A_m(t) \tag{3}$$

Where $L(t)$ represents the log sample, $A(t)$ are the corresponding seismic attributes and w_0 ... w_m are the coefficients of the linear transform. If the log consists of n samples then the equation is written as:

$$L_1 = w_o + w_1 A_{11} + w_2 A_{21} + ... + w_m A_{m1}$$

$$L_2 = w_o + w_1 A_{12} + w_2 A_{22} + ... + w_m A_{m2}$$

$$L_n = w_o + w_1 A_{1n} + w_2 A_{2n} + ... + w_m A_{mn} \tag{4}$$

Where A_{ij} is the jth sample of the ith attribute (Hampson et al, 2001).

The neural network on the other hand is able to process multi-parameter and multi-dimensional data sets in a non-linear fashion via a learning process. The task of the neural network is to perform a non-linear mapping k of a set of variables onto a target property,

$$k : A_{mn} \rightarrow L_m \tag{5}$$

In the case of log prediction from seismic data, the network reads a set of input predicting variables, a matrix A_{mn} of seismic attributes and a set of desired output results, a vector L_m of log responses, computes its own output a vector L_m' of pseudo-log responses, computes the error vector $E_m(L_m - L_m')$ and adjust its *weights*, a matrix W_{ij} using a gradient descent algorithm in a direction as to minimize the disparity between the network's output and the desired target output. The neural network approximates a mapping function such that:

$$L_m = \hat{k}(W_{ij}, A_{mn}) + E_m \tag{6}$$

where $\hat{k} \cong k$. The error given as:

$$E(W_{ij}) = \frac{1}{2}\sum_m (L_m - L'_m)^2 \tag{7}$$

and can be minimized using the back-propagation algorithm (Callan 1999, Liu and Liu, 1998). In practice a momentum factor is usually added to the gradient decent algorithm to speed up the search for an optimal solution and to prevent the network from getting caught-up in a local minimal. This process proceeds iteratively until the error bound is within pre-set acceptable limits. At this point the network is presumed to have trained adequately. Training is achieved with seismic attribute data and well information at well locations. The network is next implemented to predict pseudo-well-log responses throughout the field by being presented with seismic attribute data from beyond the wells.

3.2. Discussion

3.2.1. Structure

A total of 12 normal faults were mapped in the study area and are annotated *F1* to *F12* (Figure 5). Seven of the faults dip southward towards the sea while five are listric compensations for extentions within the fault blocks and dip landwards. These fault are gravity in-duced and are the result of the downwad and lateral response of the over-pressured mobile shales of the basal *Akata* Formation. Two of the faults *F1* and *F2* are major structure building growth faults which cut across the entire survey area and compartmentalize it into three east-west trending fault blocks which are downthrown and dip in the south-southwestward seaward direction (Figure 6). These faults appear to sole out as presumed (Corredor et al, 2005) at relatively shallow depths near the top of basal shale sequences. Fault blocks were created synchronously with sedimentation, this is seen from the relative thickening of sub-surface formations on the downthrown side of the faults. Basal shales displaced by these fault blocks are believed to flow laterally along the continental shelve and the slips thereof tranferred through detachment structures in the basal *Akata* Formation onto thrust ramps and consumed by contractional folds in the deep-water fold and thrust belts (Bilotti and Shaw, 2005).

The back-to-back alternations of seaward and landward dipping normal faults resulted in the formation of horst and graben structures partcularly in the western end of the survey area (Figure 6). These horst and graben structures possibly resulted in the creation of mini depo-centres and are probably key factors controlling drainage and depositional patterns in the field.

Contour closures against the major structure building *F1* fault represent a faulted anticlinal structure juxtaposed against the fault. These structures form the majority of structural traps in the Niger Delta. Exploration wells in this field targeted this closure. Fault throw on the eastern arm of the anticlinal structure is much greater than on the western arm (Figure 7). This could indicate that there has been a relative clockwise rotation of the horizon along a crestal hinge line over the anticline lying orthogonal to the *F1* fault. This rotation appears to have added a sense of south easterly dip to the orientation of the fault block. This sense of dip could play a key role in determining drainaige and sediment deposition within the field.

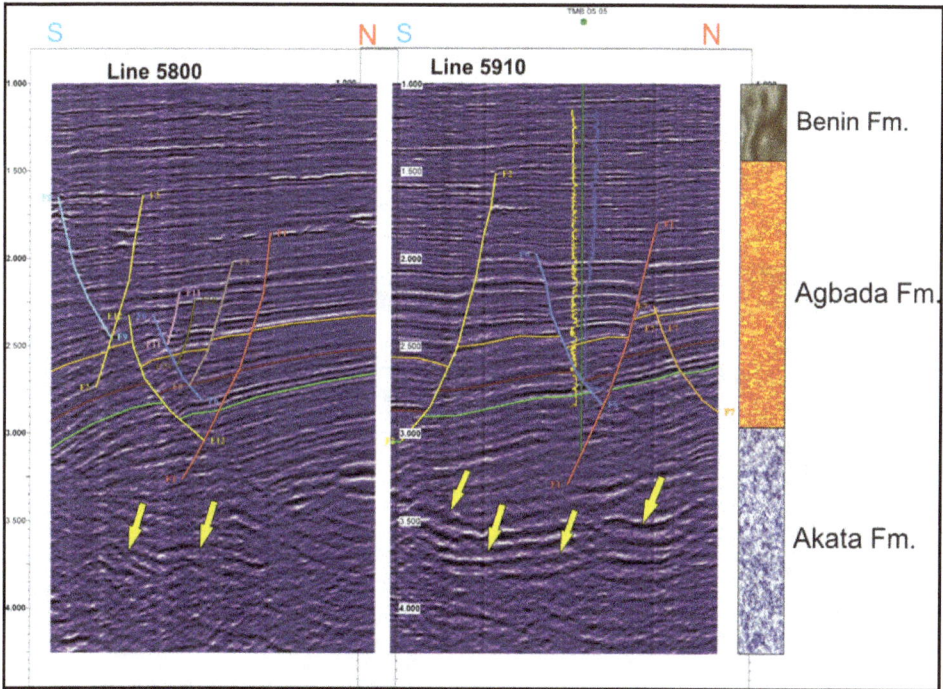

Figure 5. Interpreted seismic lines from the Okari field along with the litho-section. The target horizon in this study the *H3* is indicated in green. Growth faults affect only the paralic Agbada Fm. and compartmentalize the field into east-west trending blocks downthrown to the south. Horizon thickening on the downthrown blocks indicates that faulting was synchronous with sediment deposition. The Akata shales respond (as indicated by yellow arrow) by downward and seaward flowage due to the lithostatic pressure of the advancing paralic wedge. The Benin Formation were deposited after the ceasation of faulting and thus are not affected by faulting.

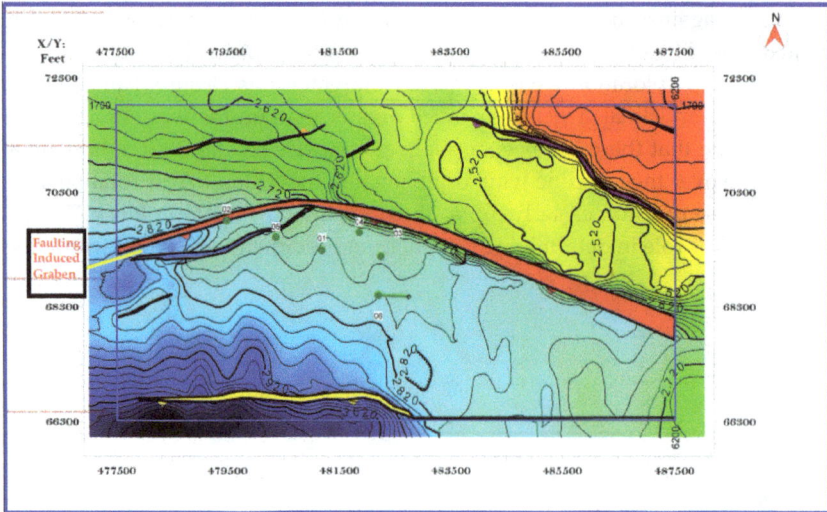

Figure 6. Horizon time map of the target *H3* horizon. Overall dip is in the south westerly direction with fault blocks trending roughly in the east-west direction. Seaward tensional stress is accomodated by landward dipping listric faulting to the west of the field thereby creating a graben structure which apparently played a key role in defining the paleo-drainage route on the horizon. Exploration wells targeted the contour closure against the major *F1* fault (red).

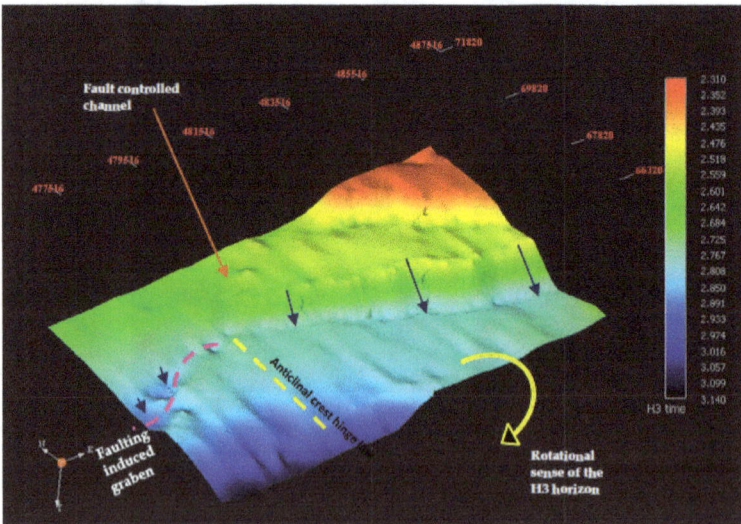

Figure 7. Annotated 3D view of the *H3* horizon. Fault throw (indicated in blue arrows) on the eastern arm of the major *F1* fault is much greater than on the western arm. This appears to have been due to possible rotation of the fault block along a crestal line (yellow dashed line) over the anticlinal structure thereby giving the fault block a south easterly dip. A fault controlled channel is seen to trend in the south westerly direction and terminates against the *F1* fault. The fault induced graben structure has a fairly sinuous configuration.

On the upthrown block of the *F1* fault, a channel-like depression trending in the northeast-southwest direction appears to terminate against the *F1* fault. This depression could have been a major path through which continentally derived detrites was supplied to the region of the anticlinal structure on the downthrown block of the *F1* fault.

3.2.2. Cross Plots

In cross-plots of *p*-wave impedance vs. porosity (Figure 8), we observed that *p*-wave impedance decreases with increase in porosity (θ), with lower-porosity sands generally having higher impedance values, likely a reflection of tighter grain packing. Porosity also increases with increase in shale content (V_{sh}), with clean sands being associated with lower porosities than shaly units of corresponding densities. The shales of the Niger Delta are known to be under-compacted and over-pressured, and the platy nature of the shale minerals ensures they have high though ineffective porosities. The litho-facies are almost separable in two (along the blue line). Furthermore, high hydrocarbon saturation appears associated with lower-porosity dirty sands, with porosity in the range 23 to 27%. The shales and water-bearing sands of the Okari Field appear, generally, to have higher porosities than hydrocarbon-saturated sands.

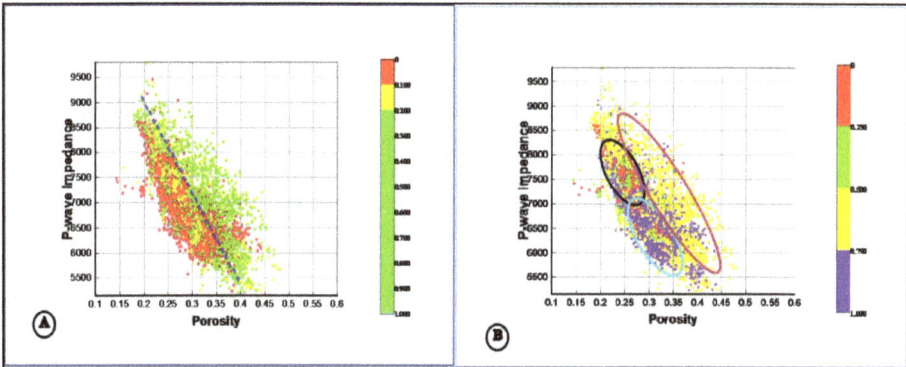

Figure 8. *P*-wave impedance vs. Porosity crossplots, (A) color coded in shale volume. Impedance decreases with porosity increase and shaly units generally possess high porosities. Between shales and sand facies, separation is almost possible along the dashed blue line, and (B) color coded in water saturation. Polygons represent subsuface facies; black for hydrocaron sands, light blue for water saturated sands and brown for shales. Hydrocarbon saturation is associated with lower porosity shaly sand facies.

These lower-porosity shaly reservoir sand units possibly benefitted from reduced hydrocarbon mobility due to the shale infilling of their pore spaces, while cleaner sands in the field may have lost their hydrocarbon saturations to invading brine as a result of their higher porosities and permeabilities, and consequent greater hydrocarbon mobility. Three distinct litho-facies are fairly well discriminated in this field; hydrocarbon sands associated with low relative porosities (23 - 27%) and high *p*-wave impedance, water-saturated sands associated with medium to high porosities (25 - 35%) and low impedance values, and shales associated with generally higher relative though ineffective porosities (30 - 42%) and the entire spread of possible p-wave impedance values (from low through medium to high) in the field.

3.2.3. Predicted litho-facies maps

3.2.3.1. Predicted Shale Volume Maps:

Six seismic attributes; original trace, integral of the trace, average of the trace, average of instantaneous amplitude, and 1st derivative and 2nd derivative of the trace were multi-linearly regressed against the target well logs (shale volume and porosity) in the vicinity of wells penetrating the H3 horizon. These attributes were selected for two major reasons; firstly, forward entry statistical regression of attributes against target well logs (porosity and shale volume) had turned up mainly amplitude-related attributes. Secondly, reviewed literature suggested that amplitude related seismic attributes have robust physical relationships to lithology and porosity (Banchs and Michelena, 2002; Dorrington and Link, 2004; Calderon and Castagna, 2005). The resulting multi-linear transforms were used to predict shale volume and porosity distributions for H3 horizon. Employing the same set of seismic attributes used for the multi-linear regression, we implemented a total of 50 simulations for shale volume prediction over the H3 horizon using a Multi-Layered Feed-forward Back-propagation Neural Network (MFLN). Thereafter, we selected the network which produced the most geologically valid map and strictly honored well information.

The multi-linear regression of well log derived shale volume against the earlier mentioned six seismic attributes (statistics not shown here) turned up a correlation coefficient R of 0.23. The low statistical correlation underscores the difficulty in predicting lithology using linear regression algorithms. Mathematically, the chances of correctly imaging lithologic distribution over the horizon using multi-linear regression of these attributes will result only in a 23% correlation or similarity with actual shale volume distribution at the target horizon. Figure 9, is the result of this multi-linear attribute transform for the H3 horizon. Our optimal neural network achieved 0.59 and 0.67 as validation and training correlation coefficients respectively. The validation coefficient is a significant improvement over the multivariate statistical regression coefficients (0.23). As neural networks are able to implement non-linear mapping, they are better suited to codifying seismic-well-log relationships. The neural network predicted lithology map of the H3 horizon is shown in Figure 10.

The two predicted shale volume maps have stricking semblance in the horizon lithologic distribution and reservoir geometry and internal architecture they reveal. This engenders confidence in the robustness of the distribution so imaged. However, the neural network predicted shale volume distribution possesses greater detail compared to the multi-linear regression map. The maps show intricate details of lithologic variation both in the region of well control (reservoir region) and beyond. They reveal the subsurface paleo-drainage system and geologic situation of the field, one typical of the Niger Delta with several meandering sand-filled channels and sand bars especially in the southwest, north, and west of the field. The paleo-channels generally cross the study area in an east-west direction and possess shale volumes in the range of 15–20%. This directional trend is most possibly

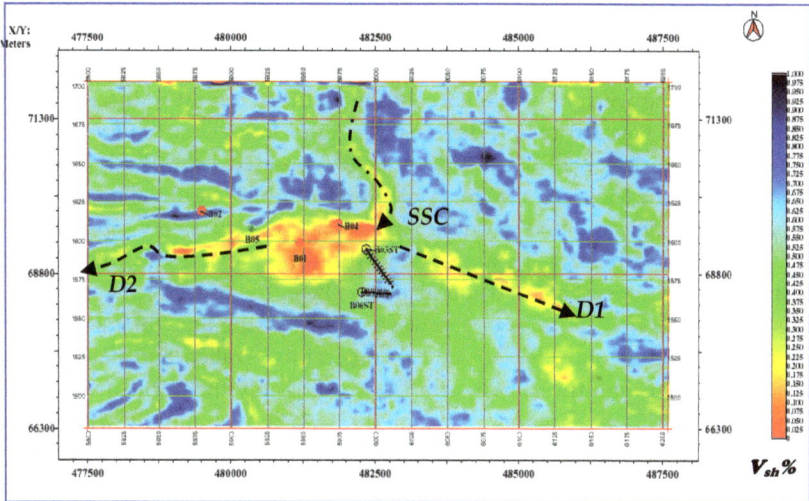

Figure 9. Multi-linear regression predicted shale volume map over the top of *H3* horizon. The sediment supply channel is a North-South trending sinuous channel *SSC* which terminates against the *F1* fault.In crossing the fault the sediments fan out while plunging into the rapidly emerging accomadation. *D1* was possibly the initial drainage channel for the reservoir region but ceased altogether and the sand bodies became detached due to the onset of faulting induced channelization which created another path *D2* in the western region.

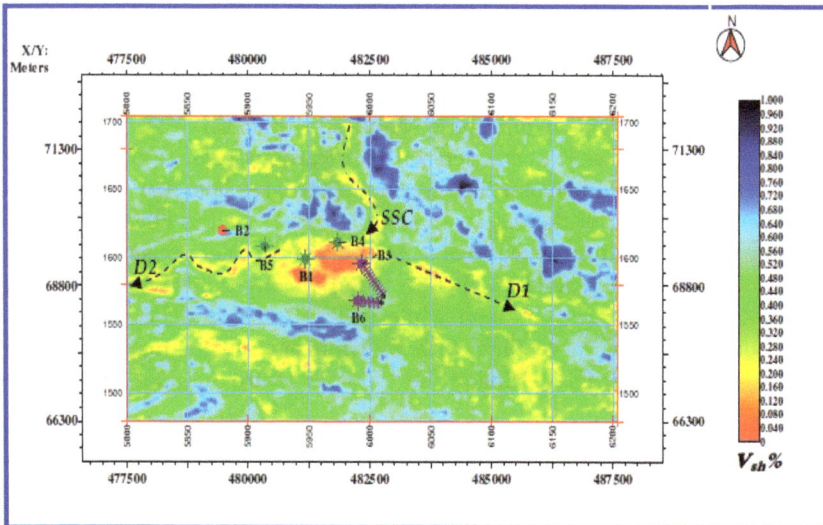

Figure 10. Predicted shale volume map over the top of *H3* horizon. The neural network prediction has high semblance with the result from regression but it provides greater detail in horizon property distribution and the reservoir geometry and internal architecture it delineates. The western *D2* channel is seen to possess greater sinuosity than on the regression map. Lateral shale seals are predicted to the south of the reservoir.

in response to dips on either sides of the anticlinal structure, that on the eastern arm resulting from possible clockwise rotation of the horizon in the plane of the *F1* fault around an axis along the crest of the anticline and roughly orthogonal to the fault plane, while the accentuated dip on the western arm is largely the result of a fault induced graben structure. Channels in the western region demonstrate good lateral continuity while those in the eastern region are less continuous and appear separated from the reservoir. This is due in part to structural controls in form of grabens which guide channel flow in the western region. Channel sand bodies on the eastern arm were possibly separated from the reservoir as a result of horizon flexing during the formation of the roll-over anticline which hosts the known reservoir. It could also have been the result of the creation of considerably steeper dips on a portion of the western arm of the anticlinal structure due to the initiation of faulting controlled structural constraints (graben and half graben structures). The channel through the eastern arm *D1* was possibly the initial drainage path for the reservoir region, this was temporal and might have ceased entirely with the onset of extension induced listric faulting on the western arm of the anticline. The faulting in the western region apparently defined an alternative drainage *D2* path for this anticline. A high-quality sand-filled paleo-channel *SSC*, trending roughly north-south on the upthrown block of the *F1* fault and possibly turning northwest above the reservoir region is imaged on both maps. This channel appears to have been the source of sediment fill for the reservoir region. In crossing the growth fault from north to south, the sediment load transported by the channel fanned out while plunging into the transient depo-center created by the active faulting.

The delineated reservoir is an oblate fan which coincides with the position of the structural closure on Figure 6. The sand distribution within the reservoir region indicates that clean reservoir sands are interspersed with shaly sand units. Initial deposits were possibly reowrked by strong wave acton. These shaly-sand units serve to compartmentalize the reservoir and may act as flow barriers to hydrocarbons present and preclude communication between the segments of the reservior. Extensive lateral shales are predicted to the south and southeast of the reservoir. Coupled with the growth fault north of the reservoir, this possibly indicates reasonable lateral sealing mechanisms. Hydrocarbon-bearing wells (*B1, B4* and *B5*) penetrated clean to fairly shaly sands with shale volume in the range 15–20% at the target horizon while wells *B2, B3,* and *B6,* penetrated shaly sands/shales. This result is in keeping with well information for the *H3* horizon. It is nearly possible to trace out the outline of the major structure building *F1* fault from the maps.

3.2.3.2. Predicted Porosity Maps

The multi-linear regression of well log derived porosity against the same seismic attributes set used for shale volume prediction (statistics also not shown) turned up a correlation coefficient R of 0.43. The correlation coefficient is considerably higher than that obtained for shale volume. This implies that it is easier to predict porosity from amplitude based seismic attributes using multi-linear regression than predicting lithology. Lithology index parameters such as grain shape index, minerology and platiness, to mention but a few, have more intricate relationships to the seismic amplitude response than coresponding determinants of porosity such as cementation and compaction. The predicted porosity map is shown in Figure 11. A neural network with the same network architecture and configuration as that used for predicting shale volume was trained using the same set of attributes *viz.* sample number, original trace, integral of trace, average of trace, average instantaneous amplitude, 1st derivative and 2nd derivative to predict porosity over *H3* horizon. Twenty-five (25) simulations of the neural network were run and the networks satisfying the two earlier mentioned criteria were selected. The most geologically valid map, which honored well information and correlated well with log information, was selected.

Figure 12 is the predicted porosity map over the *H3* horizon. The validation and training correlations for this network were 0.60 and 0.78 respectively. Though the training correlation in much higher than that for shale volume prediction, the validation correlation is hardly better. The validation correlation coefficient indicates the ability of a trained network to generalize on fresh data sets and is the true measure of the robustness of the prediction accuracy of a neural network. The minimal difference between the validation correlations for shale volume prediction and porosity prediction indicates that unlike the multi-linear regression transform, neural networks have almost no difficulty codifying the relationship between seismic attributes and either of well derived shale volume or porosity. This capability is no doubt due to their ability to perform non-linear multi-variate mapping in complex data spaces.

As with the case of shale volume prediction, there is striking semblance in the imaged reservoir architecture and horizon facies distribution revealed by the porosity maps predicted by both the multi-linear regression and the neural network. However, the neural network map provides higher resolution and images greater details of poro-facies variation both within the reservoir and beyond. The reservoir appears to be a fragmented lobe system with lower porosity facies (likely hydrocarbon bearing) interspaced with medium porosity facies (likely water bearing sands and or shales as indicated rock-physics cross-plots). This compartmentalization could be due to the re-working of the initial deposits by strong wave action. Key stratigraphic features imaged on the predicted shale volume maps such as the north-south trending paleo-channel and the eastern and western channels which drained the reservoir region are indicated as sinuous low porosity facies. The maps also indicate that the hydrocarbon-bearing wells penetrate lower porosity (20 - 27% porosity) sands in the reservoir region. Rock-physics studies earlier discussed showed that hydrocarbon sands in the study area are associated with lower porosities compared to water saturated sands and shales. The maps also locate the non-hydrocarbon-wells in regions of relatively higher porosities (porosity greater than 32%) indicating that they penetrate either water saturated sands or shales at the target horizon. The reservoir is bounded to the south by high porosity lithologies which represent shales, indicating again that the reservoir is surrounded by sufficient lateral shale seals to the south and the *F1* fault to the north.

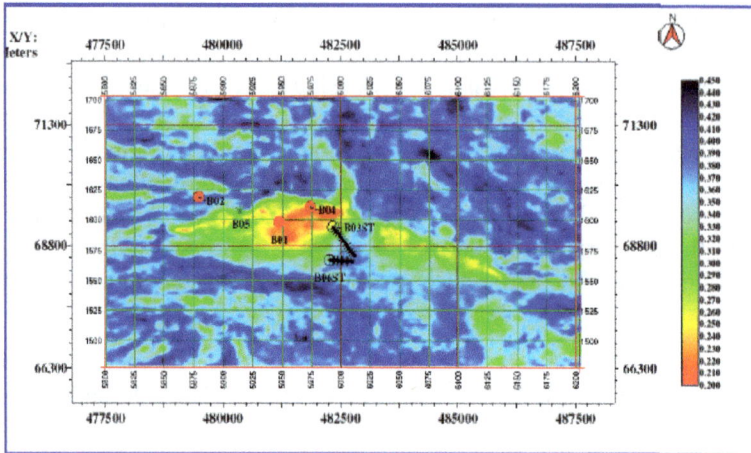

Figure 11. Multi-linear regression predicted porosity map over the top of *H3* horizon. Hydrocarbon wells are indicated to have penetrated lower porosity facies while dry wells encountered higher porosity facies (shales) at the target horizon. Earlier imaged channel features (*SSC, D1* and *D2*) possess low to medium porosities in keeping with earlier crossplot results. The reservoir appears as an oblate fan.

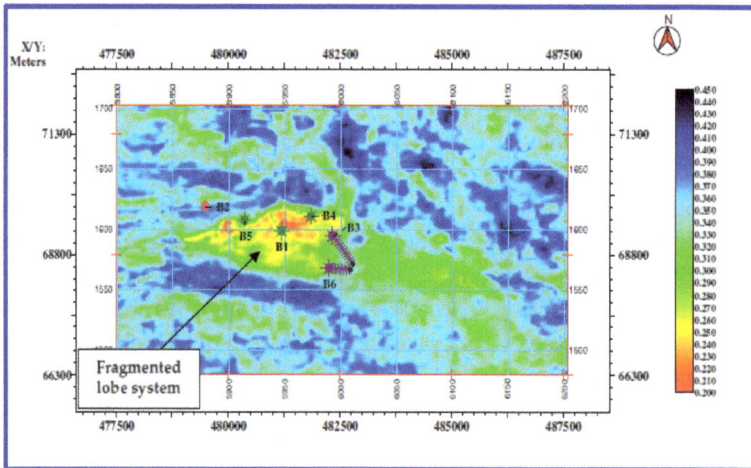

Figure 12. Neural network predicted porosity map over the top of *H3* horizon. The reservoir is a fragmented lobe system with low porosity facies (hydrocarbon sands) interspaced with medium porosity facies (water saturated sands), possibly a result of later re-working of sediments by strong wave action.

4. Conclusion

In this chapter, we have presented our attempt at unravelling the paleo-geomorphic settings of the Okari Oil field, a classical example from the extentional province of the Niger Delta. We employed conventional seismic interpretation techniques to build a tectono-stratigraphic model of the Field and to determine key factors which have contributed to and continue to define the development of this province of the Delta. We have further implemented advanced characterization techniques to image details of facies distribution within the field and further appreciate the effects of major factors on field evolution. Key factors affecting field evolution appear to have been the onset of regional seaward dipping syn-depositional growth faults and the accomodation of resultant extensions by landward dipping compensational listric faults. This resulted in rapid creation of accommodation in the form of mini depo-belts into which continentally derived detritus plunged. In traversing these growth faults the deposits fanned out as mini-deltaic fans which were possibly latter reworked and fragmented by strong wave action. Fault block rotation possibly played an important part in defining drainage paths but appear to have been overridden by the subsequent creation of graben structures by compensational faults. Paleo-geomorphic elements in the field include meandering channels, detached sand bars and a fragmented lobe system. Favorable reservoir facies are to be found in low-to-medium porosity shaly sands units.

Author details

Muslim B. Aminu
Adekunle Ajasin University, Akungba-Akoko, Nigeria

Moses O. Olorunniwo
Obafemi Awolowo University, Ile-Ife, Nigeria

5. References

Aminu, M. B., and Olorunniwo, M. A. (2011). Reservoir characterization and paleo-stratigraphic imaging over the Okari Field, Niger Delta, using neural networks. Leading Edge, v.30, no. 6, pp. 650-655.

Avbovbo, A.A. (1978). Tertiary lithostratigraphy of Niger delta. American Association of Petroleum Geologist Bulletin, v. 62, pp. 295-306.

Banchs, R. E. and Michelena, R, J. (2002). From 3D seismic attributes to pseudo well-log volumes using neural networks: Practical considerations. The Leading Edge, v.21, no. 10, pp. 996-1001.

Bilotti, F., and . Shaw, J. H. (2005). Deep-water Niger Delta fold and thrust belt modeled as a critical-taper wedge: The influence of elevated basal fluid pressure on structural styles. AAPG Bulletin, v. 89, no. 11, pp. 1475–1491.

Burke, K. (1972). Longshore drift, submarine canyons and submarine fans in development of Niger delta: American Association of Petroleum Geologist Bulletin, v.56, pp. 1975- 1983.

Calderon, J. E. and Castagna, J. (2005). Porosity and Lithologic Estimation Using Rock Physics and Multiattribute Transforms In the Balcon Field, Columbia-South America. Expanded Abstracts, 75th SEG Annual International Meeting, pp. 444-447.

Callan, R. (1999). The Essence of Neural Networks. Prentice Hall Europe. P. 232.

Chopra, S., and Marfurt, K. (2007). Curvature attribute applications to 3D surface seismic data. The Leading Edge, v. 26, no. 4, pp. 404-414.

Corredor, F., Shaw, J. H., and Bilotti, F., (2005). Structural styles in the deepwater fold and thrust belts of the Niger Delta: American Association of Petroleum Geologist Bulletin, v. 89, no. 6, pp. 753–780.

Dorrington, K. P. and Link, C. A. (2004). Genetic-algorithm/ neural network approach to seismic attribute selection for well-log prediction. Geophysics, v. 69, no. 1, pp. 212- 221.

Doust, H. and Omatsola, E. (1990). Niger Delta, in J. D. Edwards, and P.A. Santogrossi, eds. Divergent / passive margins basins: American Association of Petroleum Geologist Memoir 48, pp. 201-238.

Dunlap, D. B., Wood, L. J., Weisenberger, C., and Jabour, H. (2010). Seismic geomorphology of offshore Morocco's east margin, Safi Haute Mer area. American Association of Petroleum Geologist Bulletin, v. 94, no. 5, pp. 615–642.

Ekweozor, C. M. and Daukoru, E. M. (1984). Petroleum source bed evaluation of Tertiary Niger Delta. American Association of Petroleum Geologists Bulletin, v. 70, pp.48-55.

Evamy, B.D., Harem boure, J., Knaap, P., Molloy, F. A. and Rowlands, P.H. (1978). Hydrocarbon Habitat of Tertiary Niger Delta, American Association of Petroleum Geologist Bulletin, v. 62, pp. 1-39.

Frankl, E. J. and Cordry, E. A. (1967). The Niger Delta oil Province: Recent development, onshore and offshore. Mexico City. Seventh World Petroleum Congress Proceedings v. 2, pp. 195-209.

Hampson, D. P., Schuelke, J. S. and Quirein, J. A. (2001). Use of multiattribute transforms to predict log properties from seismic data. Geophysics, v. 66, no. 1, pp. 220-236.

Liu, Z. and Liu, J., (1998): Seismic-controlled nonlinear extrapolation of well parameters using neural networks. Geophysics, v. 63, no. 6, pp. 2035-2041.

Maloney, D., Davies, R., Imber, J., Higgins, S., and King, S. (2010). New insights into deformation mechanisms in the gravitationally driven Niger Delta deep-water fold and thrust belt. American Association of Petroleum Geologist Bulletin, v. 94, no. 9. pp. 1401–1424.

Merki, P. J., (1972). Structural Geology of the Cenozoic Niger Delta. In: Dessauvagie, T. F. J. and Whiteman, A. J. (eds.), African Geology, University of Ibadan Press, Nigeria, pp. 635-646.

Murat, R.C., (1970). Stratigraphy and Paleogeography of the Cretaceous and lower Tertiary in southern Nigeria, in 1st Conference on African Geology Proceedings, Ibadan University Press, pp. 251-266.

Short, K. C. and Stauble, A. J. (1967). Outline of geology of Niger Delta. American Association of Petroleum Geologist Bulletin, v.51, pp. 761-799.

Tuttle, M. L. W., Charpentier, R. R., and Brownfield, M. E. (1999). The Niger Delta Petroleum System: Niger Delta Province, Nigeria, Cameroon, and Equatorial Guinea, Africa. United States Geological Survey, Open-File Report 99-50-H, P. 65.

Weber, K. J. and Daukoru, E. (1975). Petroleum Geology of the Niger Delta, Tokyo. 9th World Petroleum Congress Proceedings, v.2, pp. 209-211.

Whiteman, A. (1982). Nigeria – Its petroleum geology, resources and potential, London, Graham and Trotman, P. 394.

Wu, S., and A. W. Bally. (2000). Slope tectonics— Comparisons and contrasts of structural styles of salt and shale tectonics of the northern Gulf of Mexico with shale tectonics of offshore Nigeria in Gulf of Guinea, in W. Mohriak and M. Talwani, eds., Atlantic rifts and continental margins: Washington, D.C., American Geophysical Union, pp. 151–172.

Chapter 4

3D Modelling and Basement Tectonics of the Niger Delta Basin from Aeromagnetic Data

A.A. Okiwelu and I.A. Ude

Additional information is available at the end of the chapter

1. Introduction

Basement structure is crucial in determining the origin, deformation and evolution of basin as well as the influence of the basement in the overlying Phanerozoic rocks and deposition and migration of hydrocarbon within a basin [1]. In petroleum exploration the structural surface interpreted from magnetic depth estimates is often the best available approximation to the true crystalline (metamorphic/igneous) basement configuration and estimate of basement depth (sedimentary thickness) is a primary exploration risks parameter [2]. Specifically, the magnetic basement is very relevant in the application of magnetics to petroleum exploration. Magnetic basement is the upper surface of igneous or metamorphic rocks whose magnetization is so much larger than that of sedimentary rocks. Magnetic basement may or may not coincide with geologic or acoustic basement. In the application of magnetic method, the source depth is one of the most important parameters. Others are the geometry of the source and contrast in magnetization. Basement structure determined from magnetic depth estimates provides insight into the evolution of more recent sedimentary features (subbasin, localization of reservoir bearing structures) in areas where the inherited basement fabrics or architecture has affected either continuously or episodically basin evolution and development [2]. Depths to the magnetic basement are very useful in basin modeling such as determination of source rock volume and source rock burial depth. The identification and mapping of geometry, scale and nature of basement structures is critical in understanding the influence of basement during rift development, basin evolution and subsequent basin inversion [3], [4]. From regional aeromagnetic data sets, information such as tectonic frame of the upper crust can be obtained. The patterns and amplitude of anomalies reflect the depth and magnetic character of crystalline basement, the distribution and volume of intrusive and extrusive volcanic rocks and the nature of boundaries between magnetic terrains [5].

Magnetic anomalies are a result of two things: a lateral contrast in rock composition (lithology) or a lateral contrast in rock structure [6], [7]. Where there is no contrast in magnetization no anomaly is produced. The magnetization could be due to normal induction in the Earth's field or due to remanent magnetization. For accurate modeling and interpretation of magnetic data it is important to recognize and incorporate the remanent component where they exist.

Magnetic anomaly transformation/enhancement provides the opportunity to unravel the basement structure and lithology. Such information is not readily available from the total intensity data sets especially if they are of low resolution. Our objective in this study is to demonstrate the relationship between basement framework, magnetic expression and hydrocarbon prospect in the Niger Delta basin using 3-D modelling and enhancement data sets. In the Tertiary Niger Delta basin exploration (seismic) for hydrocarbon is confined to the sedimentary section despite the fact that basement structure analysis has been used in locating hydrocarbon targets in other sedimentary basins of the world. We show that the geodynamics of the deep basement are important phenomena to the explorationist and could be an important factor that can directly lead to the risk assessment of specific prospect sites in hydrocarbon exploration. Specifically, we demonstrate that basement structure in the offshore Niger Delta have control on oil and gas discoveries even though the basement is known to be beyond drillable depths. It is not possible to prove basement control neither with subsurface mapping, as few wells penetrate basement, nor with seismic, as the basement reflector is not always mappable; residual aeromagnetics is the principal technique used in mapping basement and it is generally applied only to outline the basement fault block pattern [8]. [9] used aeromagnetic data to show that axis of hydrocarbon pool in Alberta basin is coincident with the strike of the basement sourced magnetic signals. [10] reported the relationship between tectonic evolution and hydrocarbon in the foreland of the Longmen Mountains and showed that superimposed orogenic movement and related migration of sedimentary basins controlled the generation, migration, accumulation and disappearance of hydrocarbons. [11] reported three-sets of traps from geophysical and geological data in offshore United Arab Emirate of which one is basement related.

2. Location, geologic and tectonic setting

The study area, fig.1 (Niger Delta) is situated in the Gulf of Guinea. From the Eocene to the present, the delta has prograded southwestward, forming depobelts that represent the most active portion of the delta at each stage or its development [12]. The ideas expressed on location, geologic and tectonic setting is from [13]. The depobelts in this basin form one of the largest regressive deltas in the world with an area of some $300,000km^2$, sediment volume of $500,000km^3$ [14] and a sediment thickness of over 10km in the basin depocenter. The onshore portion of the Niger Delta Province is delineated by the geology of Southern Nigeria and Southwestern Cameroon. The northern boundary is the Benin Flank – an east-northeast trending hinge line south of the West Africa basement massif. The northeastern boundary is defined by outcrops of the Cretaceous on the Abakaliki High and further east-south-east by the Calabar Flank – a hinge line bordering the adjacent Precambrian. The off-shore boundary of the province is defined by the Cameroon volcanic line to the east, the eastern boundary of the Dahomey basin (the eastern-most West African transformed-fault passive margin) to the west [13]

The Niger Delta Province contains only one identified petroleum system [15]. This system is referred to as the Tertiary Niger Delta (Akata-Agbada) petroleum system. The maximum extent of the petroleum system coincides with the boundaries of the province. Most of the petroleum is in the fields that are onshore or on the continental shelf in waters less than 200 meters deep and occurs primarily in large, relatively simple structures. The Tertiary section of the Niger Delta is divided into three formations (fig.2), representing prograding depositional facies that are distinguished mostly on the basis of sand-shale ratio. The type sections of these formations are described by [16] and [17]. The Akata Formation at the base of the delta is of marine origin and is composed of thick shale sequences (potential source rock), turbidite sand (potential reservoirs in deep water) and minor amounts of clay and silt. Beginning from the Paleocene and through the Recent, the Akata formation formed during lowstands when terrestrial organic matter and clays were transported to deep water areas characterized by low energy conditions and oxygen deficiency. Turbidity currents likely deposited deep sea fan sands within the upper Akata Formation during development of the delta. Deposition of the overlying Agbada Formation, the major petroleum-bearing unit, began in the Eocene and continues into the Recent. The formation consists of paralic siliciclastics over 3700 meters thick and represents the actual deltaic portion of the sequence. The clastics accumulated in delta front, delta-topset and fluvio-deltaic environments. In the lower Agbada Formation, shale and sandstone beds were deposited in equal proportions but the upper portion is mostly sand with only minor shale interbeds. The Agbada Formation is overlain by the third formation, the Benin Formation, a Continental latest Eocene to Recent deposit of alluvial and upper coastal plain sands that are up to 2000m thick [16].

The tectonic framework of the continental margin along the West Coast of equatorial Africa is controlled by Cretaceous fractures zones expressed as trenches and ridges in the deep Atlantic. The trough represents a failed arm of a rift triple junction associated with the opening of the south Atlantic [13].

In the Delta, rifting diminished altogether in the Late Cretaceous. After rifting ceased, gravi-ty tectonics became the primary deformational process. Shale mobility induced internal deformation occurred in response to two processes. First, shale diapirs formed from loading of poorly compacted, over-pressured prodelta and delta-slope clays (Akata Formation) by the higher density delta-front sand (Agbada Formation). For any given depobelt, gravity tectonics were completed before deposition of the Benin Formation and are expressed in complex structures, including shale diapirs, roll-over anticlines, collapsed growth fault crests (fig.3), back-to-back features and steeply dipping closed spaced flank faults [18]. Deposition of the three formations occurred in each of the five off-lapping Siliciclastic Sedimentation Cycle that comprises the Niger Delta. The cycles (depobelts) are defined by synsedimentary faulting that occurred in response to variable rates of subsidence and

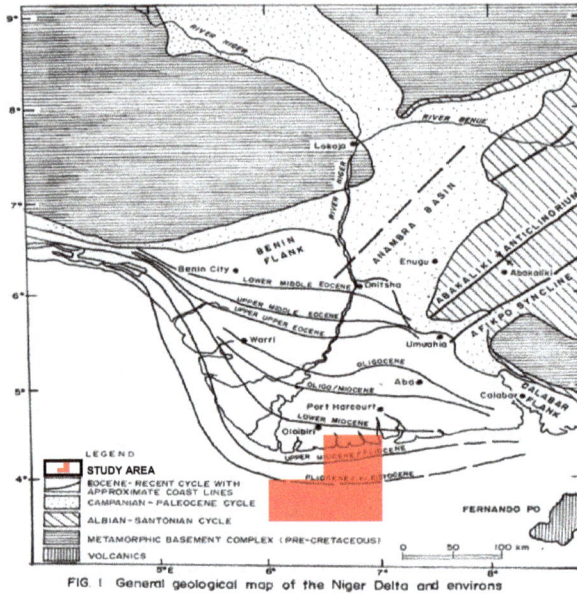

FIG. I General geological map of the Niger Delta and environs

Figure 1. Generalized geological map of Niger Delta Basin

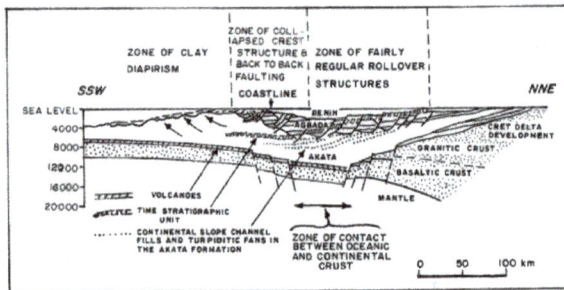

Figure 2. Schematic dip section of the Niger Delta

sediment supply. The interplay of subsidence and supply rates resulted in deposition of discrete depobelts. When further crustal subsidence of the basin could no longer be accommodated, the focus of sediment deposition shifted seaward forming a new depobelt. Each depobelt is separate unit that corresponds to a break in regional dip of the delta and is bounded landward by growth faults and seaward by large counter-regional faults or the growth fault of the next seaward belt [18]. Five major depobelts are generally recognized, each with its own sedimentation, deformation, and petroleum history. The northern delta province, which overlies relatively shallow basement, has the oldest growth faults that are generally rotational, evenly spaced with increased steepness seaward. The central delta province has depobelts with well defined structures such as successively deeper roll over crests that shifts seaward for any given growth fault. Lastly, the distal delta province is the most structurally complex due to internal gravity tectonics in the modern continental slope.

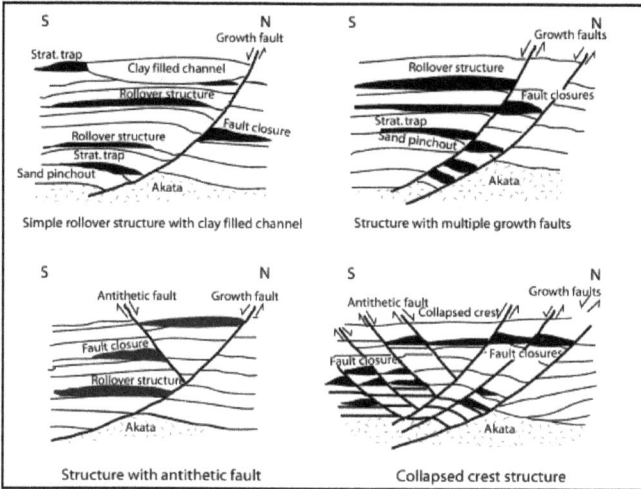

Figure 3. Principal Types of oilfield structures in the Niger Delta with schematic indications of common trapping configurations (After Tuttle etal., 1999)

3. Magnetic data

The total intensity magnetic data (fig. 4) was flown at an elevation of 2500ft (762m) above sea level with flight line spacing of 2km. This is therefore a low resolution data sourced from geological survey of Nigeria. The magnetic anomalies are sourced overwhelmly from the basement. The main advantage of this data for this study is that cultural features such as railroad tracks, power transmission cables, metals from buildings, drill cores, storage tanks, steel well casings, oil pipelines and other metallic objects are not sources of anomalies in the data and therefore, cultural editing are not required. Large concentrations of cultural sources with particularly strong and pervasive magnetic fields such as cathodically protected pipelines can seriously mask the geologic information contained in aeromagnetic survey data [19]. Gridding of the data were done at 1km interval along the flight lines which is orthogonal to the regional geologic strike. The grid spacing is tight enough to capture the anomaly details and meet the objective of this study. All the magnetic maps were plotted with potent software with the colour interval in all the figures being the convention in magnetic studies. The magnetic highs are depicted with yellows, oranges and reds while purples, blues and greens represent magnetic minima (lows). Using colours on the aeromagnetic map further accentuates the effects of visualization of the magnetic fields. The gradient zones in the total magnetic intensity field data are shear zones. The shear zones are relics of basement tectonics and are early Precambrian plate boundaries. They trend NE-SW and are principal zones of weakness in the basement and reflects edges of basement blocks.

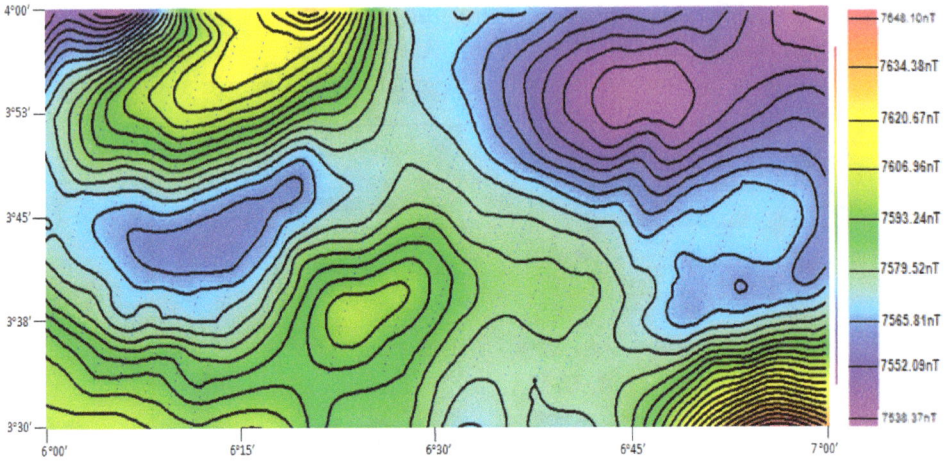

Figure 4. Total magnetic intensity data offshore Niger Delta. The gradient zones and elliptical contours reflects basement structures

The magnetic anomalies in fig.4 are as a result of total magnetization of rock and represent the vector sum of the induced and remanent magnetizations. The induced magnetizations are produced as a result of the interaction of magnetic minerals with the Earth's magnetic field. This is contrary to the remanent magnetization which acts independently of the Earth's present field. If remanent magnetization is significantly strong and acts in the direction opposite to the present field, it can generate isolated magnetic high at low latitude and produce a magnetic low at high latitude. If the induced magnetization acts in the direction of the Earth's field it produces a magnetic low in low latitude. Experimental work on rock magnetization has made it abundantly clear that contrary to the earlier belief, presence of permanent magnetization is often the rule than the exception, in the rocks of the Earth crust and permanent magnetization associates itself with induced magnetization to orient the polarization vector of the rock mass in some arbitrary direction [20]. The direction of this polarization vector influences appreciably the size and shape of the associated magnetic anomaly. The ratio of the strength of remanent magnetization to induced magnetization is known as Koenigsberger ratio. If the Koenigsberger ratio is greater than one, it suggests that the remanent magnetization played a dominant role.

The observed data was used to compute, by least squares, the mathematically describable surface giving the closest fit to the magnetic field that can be obtained within a specific degree of detail. We exploited the fact that the regional field is a first-order surface of the form:

$$T(x,y) = ax + by + c \qquad (1)$$

Where a, b and c are the coefficients and are computed so as to minimize the variation of the residual. This approach of computing the regional is suitable because higher order polynomials may be amenable to a large area over which the regional has many convolutions. The regional field was subtracted from the total intensity data to obtain the residual field data (fig. 5a).

In the total intensity data (lat. 4^0 $00'$N - lat. 3^0 $41'$N and long. 6^0 $00'$E - 6^0 $18'$E) an elliptical magnetic high and low trending E-W are separated by strong magnetic gradient. The low is closely flanked by a high trending NE-SW as shown in the total intensity and residual maps. The northeast sector is also characterized with elliptical anomaly trending E-W. The elliptical disposition is a pointer to dyke-like intrusives. The predominance of these lineaments striking NE-SW and E-W can be attributed to regional stresses in the basement. There is a high gradient in the southeast sector of the study area juxtaposed with elliptical anomalies. The elliptically shaped anomaly in the residual data has three small circularly shaped anomalies not revealed in the total intensity map. These are plug-like intrusives within the basement.

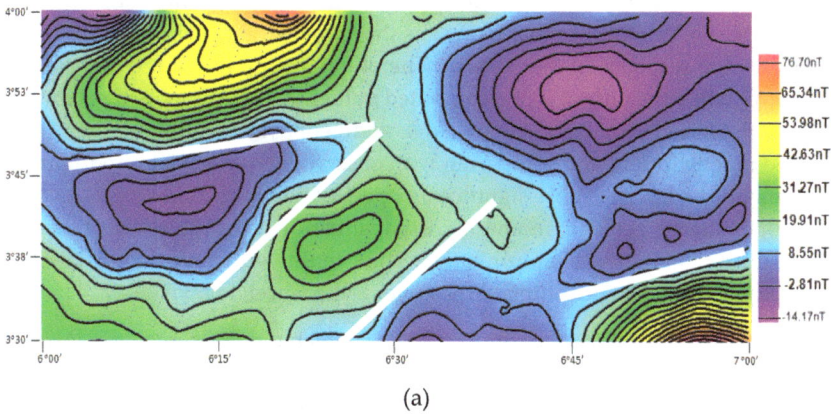

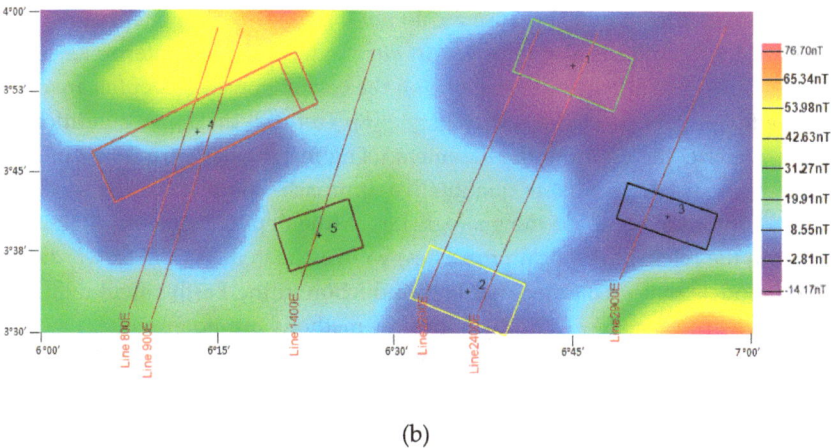

(a)

(b)

Figure 5. a. The Residual magnetic anomaly data showing some circular and elliptical contours not revealed in total magnetic intensity data. The white lines indicate shear zones. b. The location of profile lines 800E, 900E, 1400E, 2200E, 2400E and 2900E in the residual magnetic field data. The rectangular wire frames represents the magnetic sources.

4. 3D magnetic modelling and depth determination

For resource exploration purposes one of the most useful inferences that may be derived from analyses of potential field (magnetic and gravity) data is the depth to crystalline basement beneath sedimentary cover [21]. Most magnetic anomalies come from only a few rock types, such as volcanics, intrusives and basement rocks. Magnetic data therefore can be used to estimate depth to basement- a classic use for such data [22]. Generally, there are two approaches to potential field modeling: inverse and forward modeling. In magnetic modelling the inverse approach is whereby a 2D or 3D susceptibility or geometric model is computed to satisfy (invert) a given observed magnetic field. In this case, the input is the observed data while the output is the geologic model. That is, the observed data is used to draw conclusion about the physical properties of the system. Physics principle allows the means for computing the data values given a geological model. This constitutes forward modeling (problem).This implies that if one has the knowledge of the properties of a system one can predict the response of that system. Therefore, the input of a forward model is the geologic model while the output is the computed values. Forward modeling commences by erecting a model based on geologic knowledge and geophysical intuition, then calculating the predicted magnetic field and comparing with observations. The next important step is to iterate the model to fit. The most significant aspect of forward modeling is that it could show if the postulated geologic model is incompatible or compatible with potential field data. This reduces ambiguity in interpretation. Thus, in this study, we adopted 3-D forward modeling because the geologic setting of the Niger Delta is well known.

The 3D model constitutes a network or grid values which models a geologic surface represented as a surface of susceptibility contrast. The residual magnetic field data (fig. 5b) was used for modeling instead of the filtered/enhanced magnetic field data. It is not appropriate to model using filtered data, because we do not know if the component of the magnetic field removed by the filter is also removed in our model [23]. If an interpreter has two to three depth points, two at the edges and one on the basin floor, these depths are contoured with knowledge of the expected structural style [6]. To fulfill the above condition profiles were taken to model the depth to the basement using rectangular wire frame in fig. 5b. In our approach, we used a complete quantitative approach- complete in the sense that the three types of information about the geologic target (the depth, geometry/dimensions and the contrast in the relevant physical properties) were estimated. The 3D forward modeling is based on models that accommodated both induction in the Earth's field and remanent magnetization. Magnetics like other geophysical methods are non-unique. One way we adopted to reduce the ambiguity in interpretation is by using geometric simple body. In potential field modeling, popular geometric bodies usually exploited are ellipsoids, plates, rectangular prisms, polygonal prisms and thin sheets. In this study, we used rectangular prism model because of its simple shape and because it makes the process of modelling simple and stable. Thus, simple models were created using rectangular prism that conform regularly well with the data on the profiles and that are consistent with anomalies on the image of the observed field. Secondly, ambiguity is reduced because we know the geologic setting (rifting) of the Niger Delta. The most important element required for interpreting magnetic data is a geologic concept or structural model. We are never blind; even if the only data available in an area is magnetic data, we know the area is in rift setting or fore-land basin or along a passive margin. The data is no longer non-unique [23]. Another approach we used to account for non-uniqueness was to fix susceptibility and vary geometry until a reasonable fit was achieved. The modelled magnetic anomalies (figs. 6-9) resulted from lithologic and structural changes. Lithologic variation (igneous and metamorphic) usually produces the strongest magnetic signals. Amplitude of hundreds of nanoTesla is due to lithologic variations in the basement or igneous rocks within the sedimentary section while amplitude of tens of nanoTesla are related to basement structures [23]. The amplitudes of the anomalies modelled have been moderated by two factors. The main factor is that the basement rocks in the study area are buried by thick sedimentary sequence, thus their amplitude is moderated. The second factor is that high amplitude anomalies would be observed where basement structures are not present. In the study area there are sufficient basement structures (for example, faults, contacts and dykes). Thus, if a small anomaly caused by a large structure is superimposed upon a large anomaly caused by lithologic contrast, the two features may be inseparable.

Zones of lithologic contrast are often loci of structural disturbance [24]. Magnetic and gravity data have been traditionally thought of as regional screening tools capable of providing basin edges or basement mapping. In recent years, the application of these data has greatly expanded to include modelling of prospect-level targets. If detailed prospect-level quantification of the basement structure is required, a 3-D model would be more appropriate [25]. We exploited the algorithm of [20] based on magnetic anomalies due to rectangular prism-shaped bodies to determine depth to basement. This algorithm helped to meet our objectives because it considered both induction in the Earth's field and remanent magnetization. The parameters defining the prisms are shown in fig. 10. Six profiles (Line 800E, Line 900E, Line 1400E, Line 2200E, Line 2400E and Line 2900E) in fig.5b were modeled to obtain geometries (figs. 6-9) and physical properties of the basement sources. The attitude (orientation) of the body (sources) is affected by the manner in which the profiles cut the bodies. The shape of the magnetic anomalies in all the models were affected by the shape, depth of the sources, inducing and remanent field which varies in intensity and direction of magnetization [26]. Five discrete basement depth values were obtained from the modelled data and these values provided additional depth control offshore. A depth to basement at the adjacent onshore (fig. 11) gave a value of 12000m (fig. 12) by modeling body 6. These basement depth values which are equivalent to the thickness of sedimentary section in the study area contribute to basin modeling and put an upper limit on the thickness of source rocks, the base of which may not be well imaged from seismic information [25]

The range in values of magnetic susceptibility and remanent intensity reflects sources of basaltic and ultrabasic composition which may have utilized the tensional cracks in the fault system in the study area.

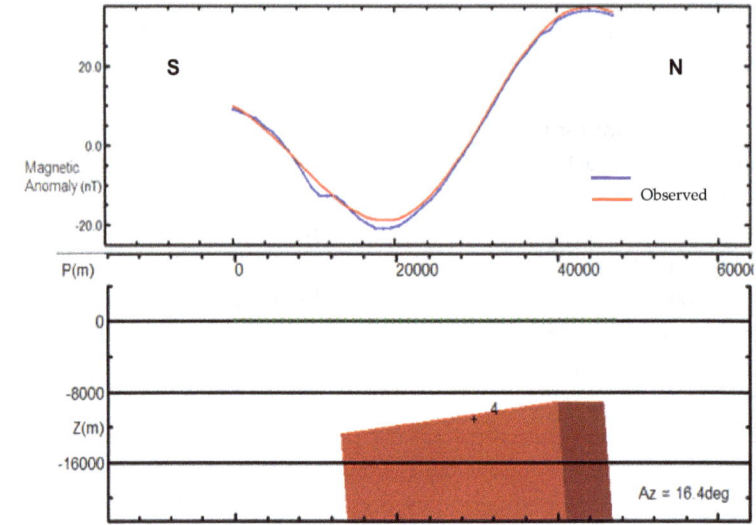

(a)

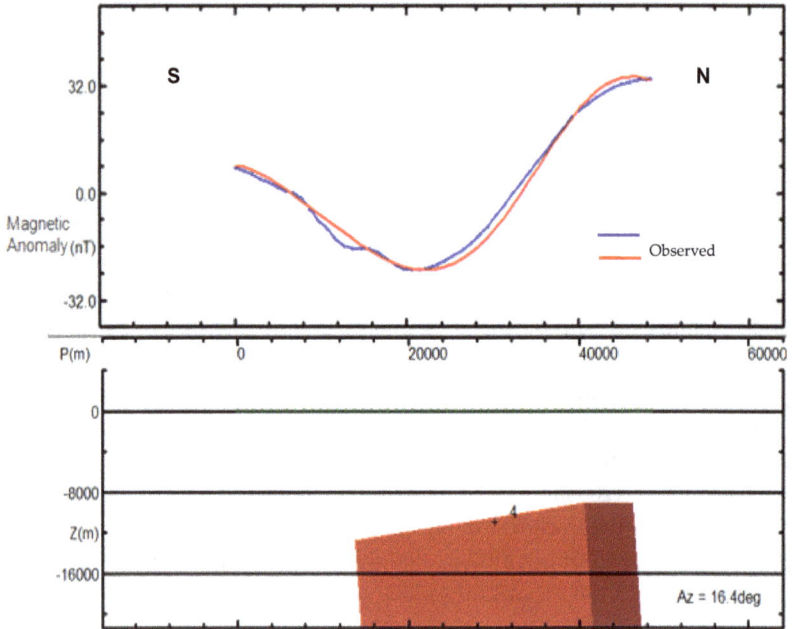

(b)

Figure 6. Modelling of (a) profile line 800E (b) profile line 900E showing dipping sources buried at depth 8,500m with a length of 28000m.

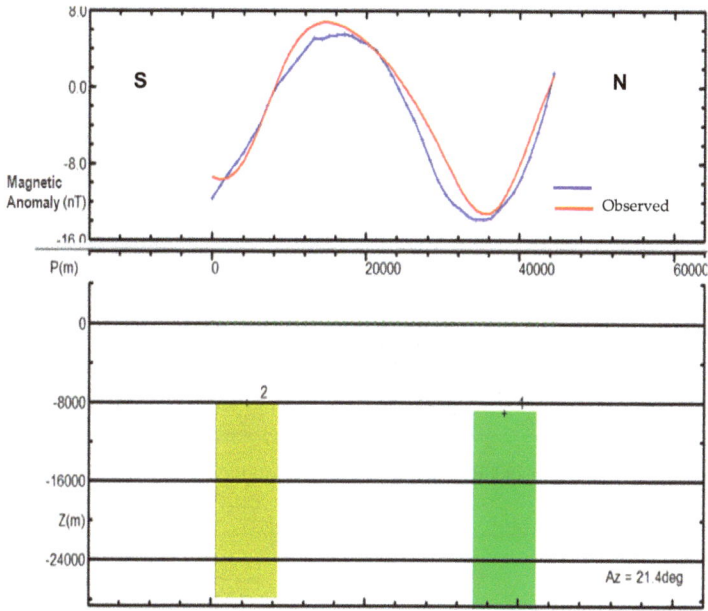

(a)

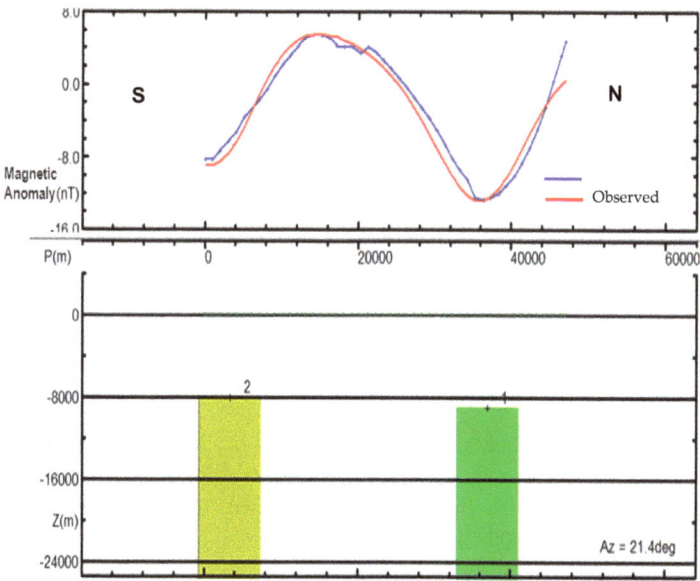

(b)

Figure 7. Modelling of (a) profile line 2200E (b) profile line 2400E. Magnetic signatures are due to remanence in the Earth's field

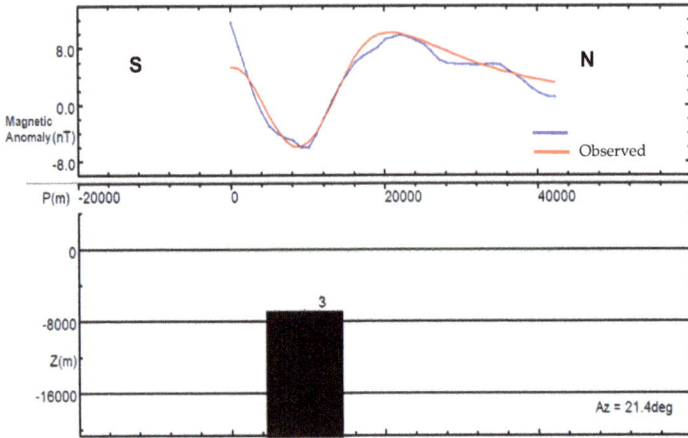

Figure 8. Modelling of Profile line 2900E which revealed a dyke-like source

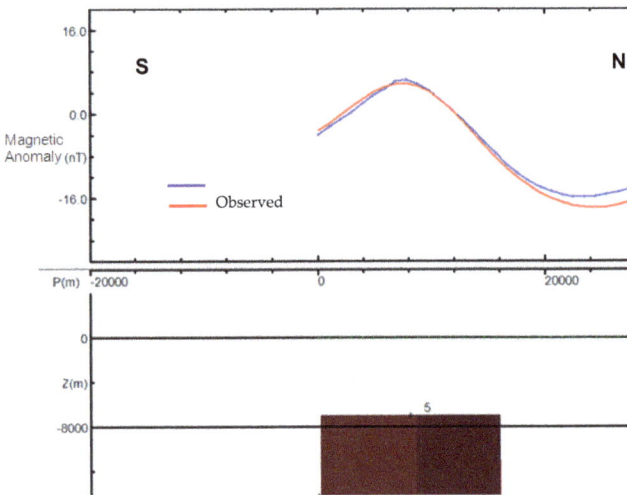

Figure 9. Modelling of profile line 1400E revealing a tabular body of length 16000m.

The magnetic profiles, Line 2200E and Line 2400E over bodies 1 and 2 show strong remanence (strong magnetic minima in the north flanked by moderate magnetic high in the south). This is manifested in the intensity of remanence (0.0600-0.1600Amp/m) and low susceptibility values of 0.007-0.008SI. These values point to a body of basaltic composition and the depth to the geologic body is 9000m. Bodies 3 and 4 modelled with profile Line 2900E and Line 800E/Line 900E respectively show signatures that are entirely due to induction in the Earth's field (strong magnetic lows) which is consistent with results from equatorial belt. The geophysical explanation of this magnetic low is that the susceptibility of the anomalous body is lower than that of the host rock. That is, a basaltic body intruded into the ultrabasic source of magnetic susceptibility, 0.017SI at a depth of 11,000m. Modelling of profile Line 1400E incorporated both induced and remanent magnetization. The remanent magnetization of body 5 is -0.3700Amp/m while the magnetic susceptibility is 0.008SI. Relatively strong high to the south and very moderate low to the north in the magnetic signature suggest remanence.

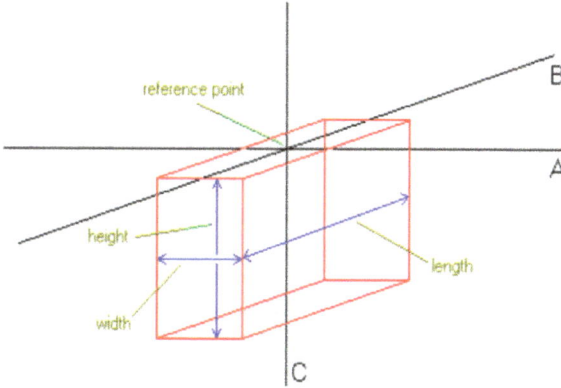

Figure 10. Rectangular prisms showing the parameters of the model

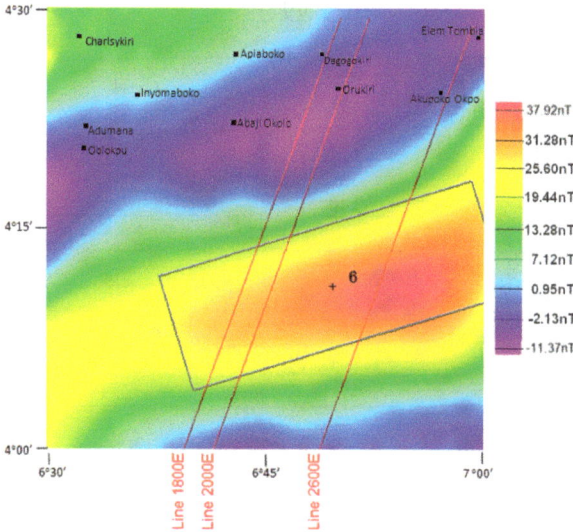

Figure 11. Location of profile lines 1800E, 2000E and 2600E on the magnetic data in the adjacent on-shore

The depth values obtained from the 3-D modelling were used to prepare magnetic basement depth map (fig. 13). A reasonable detailed basement structure map is an integral part of any regional geological or hydrocarbon evaluation process. Such a map identifies critical structural trends, the locations of the regions prominent structural prospects and location and geometry of the hydrocarbon deponcenters [27]. A magnetic basement low (thick sedimentary section) traverses the southwest and northwest sectors of the study area with a maximum sedimentary thickness of 11,736m. This is deep basement trough. At lat.3^0 $30^{'}$ -3^0 $41^{'}$N and long.6^0 $16^{'}$ - 6^0 $28^{'}$ E there is a basement high indicating structural high with a maximum thickness of 5,583m. This basement high is flanked either side by structural lows. In the northeast sector there is a basement high flanked by basement low. Thus, there is spatial relationship between paleotopographic highs on the Precambrian basement and structural and thickness anomalies in the overlying Tertiary sediments. Therefore, the depth to magnetic basement map (fig. 13) has located deep depocenter, high blocks and major sedimentary fairways in the study area.

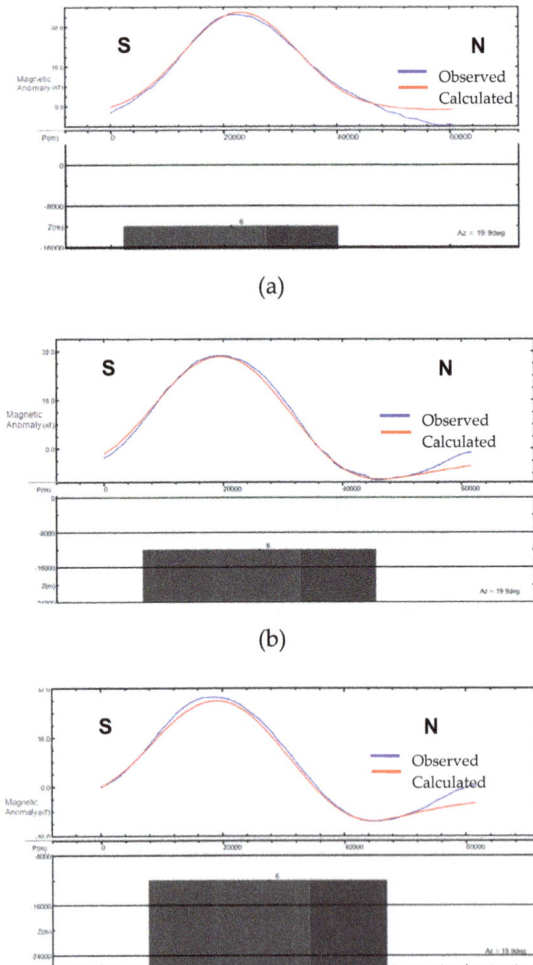

(a)

(b)

Figure 12. Modelling of profile line (a) 1800E, (b) 2000E and (c) 2600E revealed a sill-like body onshore

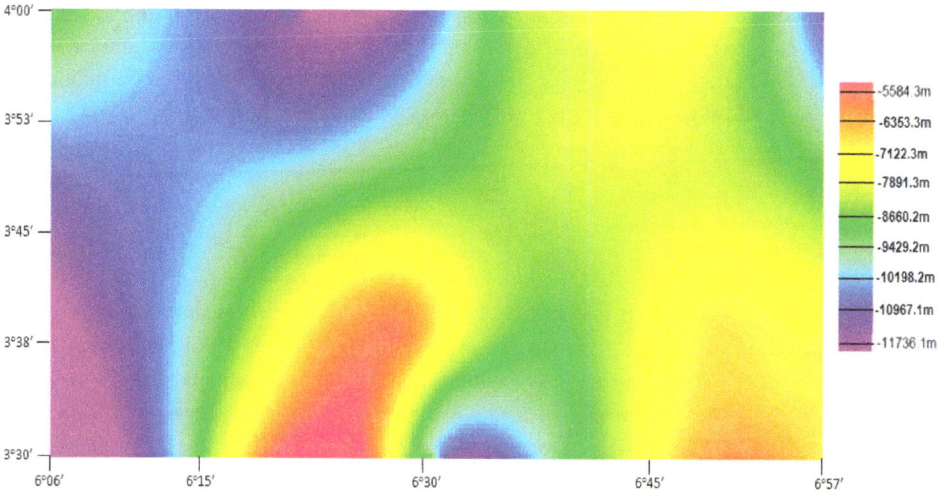

Figure 13. Depth to magnetic basement (thickness of sedimentary section), highlighting basement high, basement flanks and sedimentary fairways

5. Magnetic anomaly enhancement

The most important and accurate information provided by magnetic data is structural fabric of the basement. Major basement structures can be interpreted from consistent discontinuities and /or pattern breaks in magnetic fabric [1]. The basement structures manifest as shear zones, fault (brittle faults and domain fault boundaries) which are usually weak zones. These

basement structural features are lineaments and in most cases subtle. Subtle potential field lineaments could be gradient zones, alignment of separate local anomalies of various types and shapes, aligned breaks or discontinuities on the anomaly pattern. Subtlety of desirable lineament requires detail processing using a wide range of anomaly enhancement technique and display parameters [29]. Filtering and image processing of aeromagnetic data are essential tools in mineral exploration. Directional horizontal derivatives enhance edges (figs. 14 & 15a) while vertical derivative (fig. 15b) narrows the width of anomalies and so locate the source bodies more accurately [30].

The most commonly applied techniques include the horizontal gradient and analytic signal. Other methods for detecting edges of structures and linear features such as faults include tilt and diagonal derivatives. [31] and [32] gave expression for magnetic field horizontal gradient as

$$HG(x,y) = \sqrt{\left(\frac{\partial M}{\partial x}\right)^2 + \left(\frac{\partial M}{\partial y}\right)^2} \qquad (2)$$

Maxima in the horizontal gradient magnitude of the reduced-to-pole magnetic field are exploited to locate vertical contacts and estimate their strike directions; where M is the magnetic field. The analytic signal also reveals basement structure and uses its maxima to locate the outlines of magnetic sources and their edges. [33] defined the analytic signal from field derivatives as:

$$AS = \sqrt{\left(\frac{\partial M}{\partial x}\right)^2 + \left(\frac{\partial M}{\partial y}\right)^2 + \left(\frac{\partial M}{\partial z}\right)^2} \qquad (3)$$

While the horizontal gradient is less prone to noise because it calculates only the two horizontal derivates, it is not well suited to analyzing potential field data at low latitudes. This is because it requires reduction to the pole. Reduction to the pole is very unstable in magnetic equator (equatorial belt). The width of a maximum or ridge in analytic signal data is an indicator of depth of the contact as long as the signal arising from a single contact can be resolved [34]. While the analytic signal could be discontinuous, the enhancement is very handy at low magnetic latitude because it eliminates the problems inherent with reduction to pole (RTP) at low latitude.

One technique we find very useful is the directional horizontal gradient. It appears not to be popular but it is very effective in revealing basement features. This technique is simple and like the analytic signal (fig. 16), it can reveal N-S structures which are difficult to identify in equatorial belt. The directional horizontal derivative does not require reduction to the pole. [35] showed that the horizontal derivatives of a smoothly varying scalar quantity, $\varphi(x,y)$ measured on a horizontal surface can easily be determined using simple finite-difference methods. The horizontal derivatives of $\varphi(x,y)$ at point i,j are given approximately by

$$\frac{d\varphi(x,y)}{dx} \approx \frac{\varphi_i+1,j-\varphi_i-1,j}{2\Delta x} \qquad (4)$$

$$\frac{d\varphi(x,y)}{dy} \approx \frac{\varphi_i,j+1-\varphi_i,j-1}{2\Delta y} \qquad (5)$$

This can be performed in the Fourier domain. Thus,

$$F\left[\frac{d^n\varphi}{dx^n}\right] = (ik_x)^n F[\varphi], \qquad (6)$$

$$F\left[\frac{d^n\varphi}{dy^n}\right] = (ik_y)^n F[\varphi] \qquad (7)$$

Where, $(ik_x)^n$ and $(ik_y)^n$ are filters that transform a function measured on a horizontal surface into nth-order derivatives with respect to x and y respectively. We exploited the Fourier domain technique. This approach enhances anomalies with specific orientation and is very useful where subtle yet important trend need to be revealed but are obscured and complicated by trends in other directions [36]. This option enabled us to calculate gradient in the direction of greatest rate of change and the trend. Since the angle of the output grid tends to zero in equatorial belt then dx is the gradient to the east and dy is the gradient to the north. In fig. 14, N-S striking structures are clearly defined. E-W striking features in fig. 15 are not surprising at the magnetic equator. Generally, the interpretation of magnetic anomalies near the equatorial belt is difficult because the ambient (local) field is weak and horizontal. N-S striking structures are difficult to detect at the equatorial belt. Magnetic anomalies are generated when the flux density cuts the boundary of structures and if the structure strikes parallel with the field then in equatorial belt the flux stays within the structure and no anomaly is generated [34]. This effect can also be generated when magnetic field reduced to the equator (RTE) instead of reduced-to-the pole is carried out. In this case the N-S structures in RTE data are difficult to identify.

The enhancement maps show that the digitized aeromagnetic data is amenable to mathematical transformation, valuable tools for tectonic interpretation and resource exploration in the Niger Delta basin. In fig. 14 & fig. 16 the magnetic field defines a more N-S trending fabric. Some of the offsets and discontinuities in the gradient maps agree with changes in the total magnetic intensity and residual maps. This concurrence implies a major structural contact or faults and represents offsets in the basement which have controlled sedimentation patterns in the Niger Delta. The directional horizontal derivative maps and the analytic signal map show clear boundaries of major magnetized zones within the basement. The internal character and boundaries of the basement blocks and sub-domains are also revealed. Thus, the directional horizontal derivative data and the analytic signal map clearly demonstrate geophysical features and highlight trend directions of magnetic sources even though the aeromagnetic data is old and is of low resolution. Most of the important geologic features (faults and contacts) are reflected as lineaments in the magnetic data. A geologic lineament is a linear zone of weakness in the Earth's crust that may owe its origin to tectonic or glacial causes and often represents geologic features such as faults, dykes, lithologic contact and structural form lines [37]. Large-scale regional structures are revealed by low pass filtering. Comparing the low-passed magnetic data (fig. 17) with the total intensity data reveals the anomalies that survived the filtering. Principal orientations of magnetic field anomalies are revealed in the low-passed data and made the lineaments to be more pronounced indicating that the lineaments are associated with large scale features. The orientation of large scale features is E-W (the direction of the major domains) while the anomalies of short wavelength [short scale features] (fig. 18) are discordant with these major trends.

6. Discussion

The depth to magnetic basement map (fig. 13) has revealed a spatial relationship between the paleotopographic highs and lows in the Precambrian basement and structure and thickness anomalies in the overlying Tertiary sediments. The basement paleotopography suggests movement in the shear/wrench fault systems that were active before, during and after sedimentation. The residual and total intensity data revealed NE-SW trending boundaries crossing almost the entire study area. The NE-SW trending boundaries are shear zones and are related to primary NE-SW crustal block faulting that are related to the unique position of the Niger Delta during the opening of the South Atlantic at the boundary between the southern area of crustal divergence and the equatorial zones of crustal translation [18].This trending magnetic anomalies represent ductile healed basement structure of Early Proterozoic and earlier age. They predominate and obscure the desired subtle lineaments (brittle faults) trending N-S which were not revealed in the total intensity and residual magnetic maps. Appropriate processing using the directional horizontal derivative (fig. 14) and analytic signal map (fig. 16) clearly revealed the subtle anomalies. Specifically, the negative analytic signal in fig. 16 reflects zones of low magnetization which is a pointer to faults/fractures that are associated with possible depletion of magnetite. The northeast-southwest basement trends indicate possible extensions within the African continent of the Charcot and Chain oceanic fracture zones.

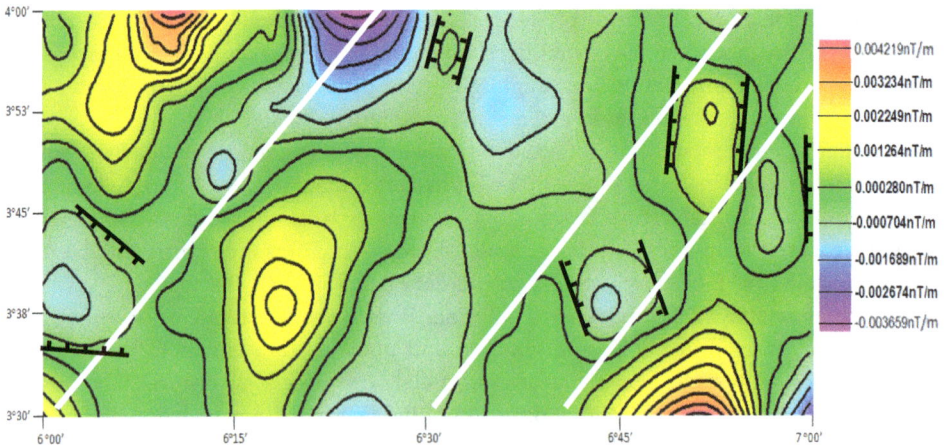

Figure 14. Directional horizontal derivative data highlighting subtle N-S structures and basement fault blocks. White lines are inferred accommodation zones

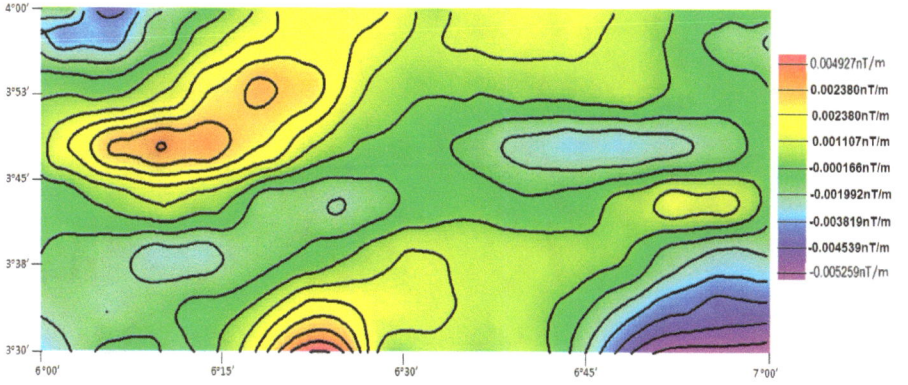

(a)

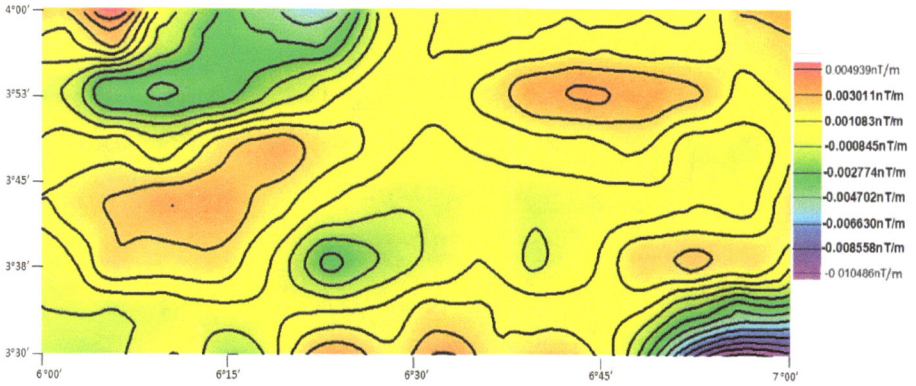

(b)

Figure 15. Directional derivative maps revealing E-W structures (a) data obtained by taking gradient (dy) in north direction (b) First vertical derivative

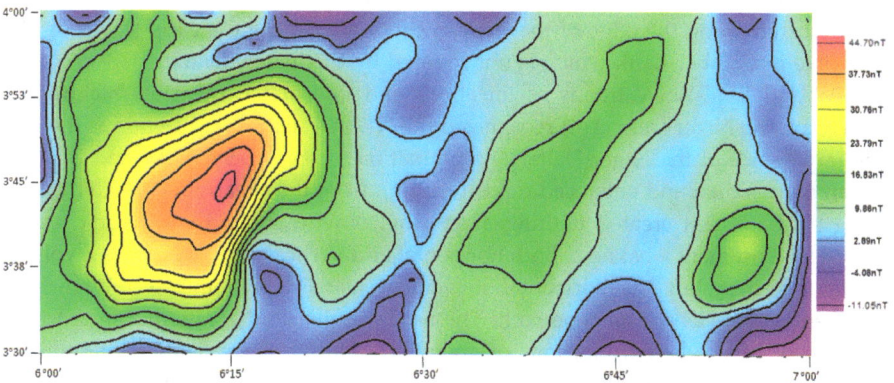

Figure 16. Analytic signal data highlighting basement block patterns and N-S structures

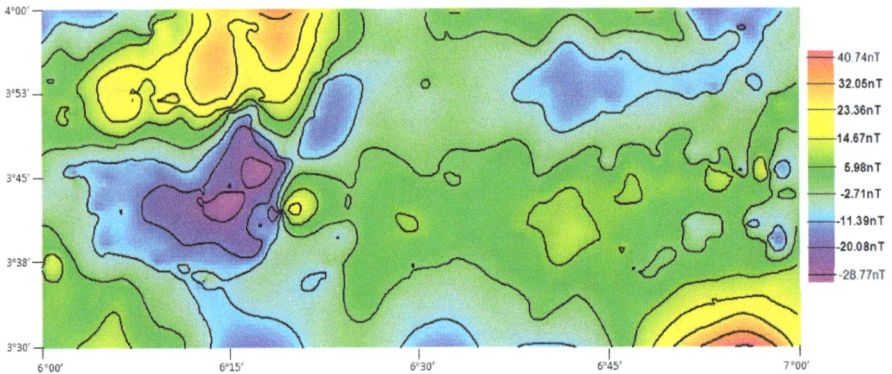

Figure 17. Low pass filtered data showing the major trends of the magnetic domains of the deep basement with discordant small scale structures

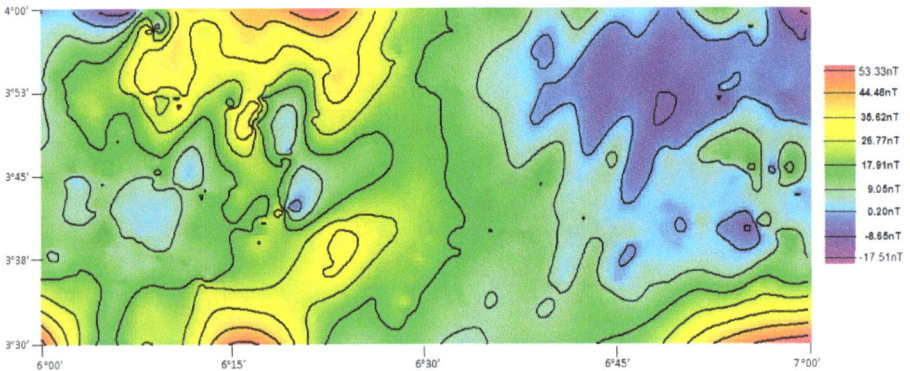

Figure 18. Filtered outputs of magnetic anomalies predominantly of short wavelengths which are discordant to the trend of the major magnetic provinces

The northwest-southeast trend (Romanche fault zone) equivalent is as a result of block faulting that occurred along the edge of the African continent during the early stages of divergence; visible in Calabar flank which is not covered by this magnetic data. [38] recognized the NE-SW and ENE-WSW trends as lineations and interpreted them as fracture zones trends beneath the Niger Delta. [14] recognized the NE-SW and NW-SE trends as the megatectonic framework of the Niger Delta. A combination of the NE-SW, E-W and N-S structures from the residual and enhanced maps resulting from the shear/wrench-fault tectonics involving the basement created faulting, fracturing, downwarp and epeirogenic warping along zones of basement weakness. Both horizontal and vertical movements are involved in wrench-fault system but the horizontal movement usually predominates. Wrench-fault system often appears as scissor-type fault. The Faults in this study were recognized from a combination of offsets and truncations of anomalies and steep gradients in the magnetic data. The strong shearing in the study area along a wrench fault system has vertical and horizontal displacements. The vertical displacement could be vividly seen as north-south striking structures in figs. 14 & 16 and horizontal displacement in the E-W striking structures (fig. 15). The N-S and E-W bounded fault blocks are secondary faults which must have influenced stratigraphy and major tectonic elements or as shears which controlled local features. [39] and [40] mapped family of faults with similar trends that control depositional history of the sedimentary basin in north-eastern Morocco and Potiguar rift basin in northeast Brazil respectively. The N-S trending structures were probably induced by a combination of differential subsidence across a fault zone and by local uplift due to wrench movements. These displacements created minibasin and arching of the basement (fig. 13) and block faults (fig. 14). The basement block boundaries are lineaments which affected deposition in the delta. Thus, sediment geometry in the study area is linked to subtle tectonic readjustment of basement blocks. These lineaments create conduits which aid the flow of fluids and may also act as barriers. The N-S and E-W structures in the enhancement maps are relatively weak structures and were created subsequent to the formation of dominant and stronger NE-NW and NW-SE trending anomalies which reflects the shape of the Niger Delta basin. Thus, the N-S and E-W anomalies represent the reactivated structures. These two trends in addition to the NE-SW trend form the three potential stress regimes responsible for the structural architecture of the study area. In individual mega-tectonic provinces these three trends are the dominant trends [41].

There are three evidences for reactivation. One of the evidences of reactivation is the arching up of the basement (fig. 13). During reactivation blocks within the basement may have moved along faults. The second evidence is that the N-S and E-W structures do not correlate with the basin shape. The third evidence is that when a thick sedimentary cover is forming pre-existing structures in the basement have potential to become reactivated. This have been demonstrated for areas that are evidently tectonically stretched such as shelves or basins on or adjacent to continental margins and in a slowly subsiding epicontinental basin, where pre-existing tectonic structures were reported to have been reactivated at times and subsidence is enhanced [42]. The N-S and E-W structures are bounded by faults. These faults are brittle in nature and may have developed by shear reactivation of a previously formed weak surface in a body of rock. In the upper crust of the Earth, roughly 10km in depth, rocks primarily undergo brittle deformation, creating a myriad of geologic structures [43].

In this study we opine that the basement structures are identified to play major role in sediment and hydrocarbon distribution in the Niger Delta in two ways: basement relief (basement highs and lows) and basement related faults. These two factors are episodic and appear to have controlled the trapping and migration of hydrocarbon in the Niger Delta. [8] identified two basic types of basement control on the overlying sedimentary section in Kansas: basement topographic control and reactivated basement faults or shear zones. Actual movement along the shear zones and lineament may be minimal but the minor change in topographic relief of the overlying sediments is an important control on deposition [44]. The embryonic faulted margins of the Atlantic are now the continental margins of West Africa and are prolific oil-producing regions. The faulted rift systems of Africa developed major sedimentary basins along its length and generated major oil provinces in Nigeria, Central

Africa and Sudan [45]. The residual map, the enhanced maps and the depth to basement map show structural characteristics and they are used in this study as evaluation tool in this hydrocarbon exploration setting. The shear/wrenching and the block faulting in the residual and enhanced maps represent offsets in the basement and controlled sedimentation patterns. The development of the delta has been dependent on the balance between the rate of sedimentation and the resulting sedimentary patterns appears to have been influenced by the structural configuration and tectonics of the basement [18].

The depth to basement map is characterized by structural highs flanked by structural lows. The structural low represents syncline/depocenter/subbasin. The structural high anomalies are interpreted in this study to be the focal points for the migration of oil and gas while the regional (lows) structural anomalies are the generating depocenters. Thus, structural high (positive) anomalies near structural low (negative) anomalies are the preferred targets in hydrocarbon exploration. Thus, the shear/wrench system is reflected as a series of geometrically arranged downwarp, epeirogenic uplift that may be subjected to continuous adjustment and compressional stress [44]. The uplifted blocks created the arches while downdropped ones produced the depocenters and we therefore opine that the flanks of the basement highs and basement lows are attractive sites for oil and gas accumulation. Oil and gas generated in such regional lows will migrate updip, where possible onto adjacent structural highs. Structural highs located between two adjacent basement lows offer special attractions for oil and gas migration from both sides [46]. We strongly opine that the basement structures from the residual map, the enhanced maps and the depth to basement map are as a result of multiple deep-seated tensional and shear/wrench faulting within the basement and that jostling of basement blocks have strongly influenced deposition in the Niger Delta basin. The aftermath of the basement motion in conjunction with the impact of differences in topographic relief in the sedimentary section during the Tertiary gave rise to the generation of the structural lows and structural highs. Subsequent migration of hydrocarbon was aided by fault induced by basement faulting. The basement blocks jostling beneath the Niger Delta may have created fracture pattern that may have enhanced or reduced porosity and permeamibility. Basement faults are known to have commonly influenced the distribution of hydrocarbon traps and mineralization zones in sedimentary cover [47]. [48] linked oil pools in lower productive beds of sedimentary cover to faulted zones in crystalline basement in known platform hydrocarbon fields.

7. Conclusion

The directional horizontal derivative data, the analytic signal data and filtered maps reveals the magnetic field lineaments and anomaly fabric that could be related to the basement faults beneath the Niger Delta basin. E-W striking structures are brittle faults/fractures which are usually subtle but are well highlighted even in the total intensity data probably because they are mineralized or associated with dykes. The N-S structures in the study area are due to extensional faulting in the Precambrian crystalline basement giving rise to alternating system of downwarp and epeirogenic uplift that may have pushed up the Tertiary sediments. Hence, the sediment geometry in the Niger Delta can be correlated to subtle tectonic readjustment of basement blocks beneath the sedimentary section. The downwarp in this study represents syncline/depocenter/subbasin while the structural high anomalies are interpreted to be the focal points for the migration oil and gas.

Author details

A.A. Okiwelu and I.A. Ude
Geophysics Research Unit, Department of Physics, University of Calabar, Nigeria

8. References

[1] J. Li and I. B. Morozov, Geophysical structural patterns of the crystalline basement of the Eastern WCSB, Canadian Society of Geophysicists Convention, 2007, p.672-675

[2] X. Li, On the use of different methods for estimating magnetic depth. The leading edge: The meter reader (coordinated by John W. Peirce), vol. 22 (11), pp.1090-1099, 2003

[3] M. Smith and P. Mosley, Crustal heterogeneity and basement influence on the development of the Kenya rift, East Africa, Tectonics, vol.12 no.2, pp. 591-606, 1993

[4] B. L. Crawford, P. G. Betts and L. Ailleres, An aeromagnetic approach to revealing buried basement structures and their role in the Proterozoic evolution of the Wernecke Inlie, Ykon Territory, Canada, Tectonophysics, 490, pp.28-46, 2010.

[5] J. F. Meyer Jr., "The compilation and application of aeromagnetic data for hydrocarbon exploration in interior Alaska", in Geologic applications of Gravity and Magnetic: case Histories" (Eds. Gibson. I. and Millegan. P. S.) (SEG Geophysical Reference No.8 and AAPG Studies in Geology, No. 43), Tulsa, 1998, pp. 37-39.

[6] P. S. Millegan and D. E. Bird, "How basement lithology changes affect magnetic interpretation" in Geologic applications of Gravity and Magnetic: case Histories" (Eds. Gibson. I. and Millegan. P. S.) (SEG Geophysical Reference No.8 and AAPG Studies in Geology, No. 43), Tulsa, 1998, pp. 40-48.

[7] H. V. Lyatsky, Magnetic and Gravity Methods in Mineral Exploration: the Value of Well-Rounded Geophysical Skills, Recorder (Canadian Society of Exploration Geophysics), pp.30-35, 2010.

[8] S. P. Gay Jr., "Basement control of selected oil and gas fields in Kansas as determined by detailed residual aeromagnetic data", in Geophysical Atlas of selected Oil and Gas fields in Kansas, Kansas Geological Survey Bulletin 237, 1995, pp. 10-16.

[9] G.E. Leblanc and W.A. Morris, "Aeromagnetics of southern Alberta within areas of hydrocarbon accumulation", Bulletin of Canadian Petroleum Geology, vol. 47(4), pp. 439-454, 1999.

[10] W. Jinqi, Relationship between tectonic evolution and hydrocarbon in the foreland   of the Longmen mountains, Journal of Southeast Asian Eart Sciences, vol.13, Nos.3- 5, pp. 327-336, 1996.

[11] A.S. Alsharhan and M. G. Salah, Tectonic implications of diapirism on hydrocarbon accumulation in the United Arab Emirates, Bulletin of Canadian petroleum Gelogy, vol. 45, No. 2, pp.279-296, 1997.

[12] H. Doust and E. Omatsola, , Niger Delta, in Edwards, J. D. and Santogrossi, P.A., eds., Divergent/passive Margin Basins, AAPG Memoir 48: Tulsa, American Association of Petroleum Geologists, pp. 239-248, 1990

[13] M. L. W. Tuttle, R. R. Charpentier and M. E. Brownfield, The Niger Delta petroleum system: Niger Delta province, Nigeria, Cameroun and Equatorial Guinea, Africa, U. S. Geological Survey Open-file Report-95-50H, 31P., 1995.

[14] J. Hospers, , Gravity field and structure of the Niger Delta, Nigeria, West Africa: Geological Society of American Bulletin, vol. 76, pp. 407-422, 1965

[15] C. M. Ekweozor and E.M. Daukoru, Northern delta depobelt portion of the Akata-Agbada petroleum system, Niger Delta, Nigeria, in, Magoon, L.B., and Dow, W.G. eds., The Petroleum System--From Source to Trap, AAPG Memoir 60: Tulsa, American Association of Petroleum Geologists, pp. 599-614, 1994

[16] A. A. Avbovbo, Tertiary lithostratigraphy of Niger Delta: American Association of Petroleum Geologists Bulletin, vol. 62, p. 295-300, 1978

[17] K. C. Short and A.J. Stäuble, Outline of geology of Niger Delta: American Association of Petroleum Geologists Bulletin, vol. 51, pp. 761-779, 1965

[18] B.D. Evamy, J. Haremboure, P. Kamerling, W.A. Knaap, F. A. Molloy and P.H. Rowlands, Hydrocarbon habitat of Tertiary Niger Delta: American Association of Petroleum Geologists Bulletin, vol. 62, p. 277-298, 1978.

[19] J. D. Philips, R. W. Saltus and R. L. Reynolds, Sources of magnetic anomalies over a sedimentary Basin: Preliminary results from the coastal plain of the Arctic National wildlife Refuge, Alaska, in Geologic applications of Gravity and Magnetic: case Histories" (Eds. Gibson. I. and Millengan. P. S.) (SEG Geophysical Reference No.8 and AAPG Studies in Geology, No.43), Tulsa, 1998,pp.130-134.

[20] B. K. Bhattacharyya, Magnetic anomalies due to prism-shaped bodies with arbitrary polarization, Geophysics, vol. 29 no.4, pp.517- 531, 1964.

[21] P. R. Milligan, G. Reed, T. Meixner and D. FitzGerald, Towards automated mapping of depth to magnetic/gravity basement-examples using new extensions to an old method, Australian Society of Exploration Geophysicists 17th Geophysical conference and Exhibition, Sydney, 2004, 4p.

[22] R. I. Gibson, Magnetic frequency-depth relationship, in Geologic applications of Gravity and Magnetic: case Histories" (Eds. Gibson. I. and Millegan. P. S.) (SEG Geophysical Reference No.8 and AAPG Studies in Geology, No. 43), Tulsa, 1998, pp. 59-63.

[23] D. E. Bird, Primer: Interpreting magnetic data, American Association of Petroleum Geologist Explorer, 18 (5), pp. 18-21, 1997.

[24] R. I. Gibson, Magnetic susceptibility contrast versus structure, in Geologic applications of Gravity and Magnetic: case Histories" (Eds. Gibson. I. and Millegan. P. S.) (SEG Geophysical Reference No.8 and AAPG Studies in Geology, No. 43), Tulsa, 1998, pp. 79-81.

[25] J. M. Jacques, M.E. Parsons, A.D. Price and D. M. Swhwartz, Draw upon recent work to provide evidence as to why gravity and magnetic survey data can still provide vital geological clues for oil and gas exploration, First break, vol. 21, 2003, pp. 57-62.

[26] S. Williams, T. D. Fairhead and G. Flanagan, Realistic models of basement topography for depth to magnetic basement testing, SEG International Exposition and 72 annual meeting, Utah, 2002, 4p.

[27] M. Alexander, J. C. Pratsch, C. Prieto, Under the northern Gulf basin: basement depths and trends, Abstract of paper presented at the 1998 SEG 68 annual meeting, New Oleans, LA, 1998, 3P.

[28] J. Li and I. Morozov, Aeromagnetic mapping of the Williston basin basement, 2006 CSPG-CSEG-CWLS Convention, 2006, pp. 65-68.

[29] H. V. Lyatsky, D.I. Pana and M. Grobe, Basement structure in Central and Southern Alberta: Insights from Gravity and Magnetic maps, Alberta Energy and Utility Board/Alberta Gelogical Survey (EUB/AGS) special report72, 83p., 2005.

[30] G.R. J. Cooper and D.R. Cowan, Filtering using variable order vertical derivatives, Computers and Geosciences, vol. 30, pp.455-459, 2004.

[31] L. Cordell and V. J. S. Grauch, Mapping basement magnetization zones from aeromagnetic data in the San Juan basin, New Mexico [Astract]. Program, SEG, 1982 Annual Meeting, 246-247

[32] L. Cordell and V. J. S. Grauch, Mapping basement magnetization zones from aeromagnetic data in the San Juan basin, New Mexico, in Hinze, W. J. ed., The utility of regional gravity and magnetic anomaly maps, Society of Exploration Geophysicists, pp.181-197, 1985.

[33] W. R. Roest, J. Verhoef and M. Pilkington, Magnetic interpretation using the 3-D analytic signal, Geophysics, vol. 57 pp. 116-125, 1992.

[34] GETECH GROUP. (2007). Advanced Processing and Interpretation of Gravity and Magnetic Data. 27. Available:
http://www.getech.com/services/advancedprocessingandinterpretation.pdf

[35] R. I. Blakely, Potential theory in gravity and magnetic applications. Cambridge University press, Cambidge, 1996, 441p.

[36] R. C. Brodie. (2002). Geophysical and remote sensing methods for regolith exploration, CECLEME open file report 144, pp.33-45. Available:
http://www.crcleme.org.au/Pubs/../OfR%20144/06Magnetics.Pdf

[37] M. Lee, W. Morris, J. Harris and G. Leblanc, An automatic network-extraction algorithm applied to magnetic survey data for the identification and extraction of geologic lineaments, The Leading Edge, vol.21, No. 1, pp. 26-31, 2012.

[38] O. O. Babalola and M. Gipson, Aeromagnetic anomalies and discordant lineations beneath the Niger Delta: Implications for new fracture zones and multiple sea-floor spreading directions in the "Meso- Atlantic" Gulf of Guinea Cul-De-SAC, vol.18 (6), pp.1107-1110, 1991.

[39] R. EI Gout, D. Khattach, M. R. Houari, O. Kaufmann and H. Aqil, Main structural lineaments of north-eastern Morocco derived from gravity and aeromagnetic data, Journal of African Earth Sciences, vol.58, pp. 255-271, 2010.

[40] D. L. De Castro, Gravity and magnetic joint modeling of the Potiguar rift Basin (NE Brazil): Basement control during Neocomian extension and deformation, Journal of South American Earth Sciences, vol.31 (2-3), pp. 186-198, 2011.

[41] J. Affleck, Magnetic anomaly trend and spacing patterns, Geophysics, vol. 28 (3), pp.379-395, 1963

[42] A. Wetzel, R. Alenbach, V. Allia, Reactivated basement structures affecting the sedimentary facies in a tectonically "quiescent" epicontinental basin: An example from NW Switzerland, sedimentary Gelogy, 157, pp. 153-172, 2003.

[43] B. A. V. Pluijm and S. Marshak, Earth structure: An introduction to structural Geology and tectonics (2n Ed.), W. W. Norton and company, New York, 2004, pp. 1- 476.

[44] D. L. Brown and D. L. Brown, Wrench-Style Deformation and Paleostructural influence on Sedimentation in and around a Cratonic Basin, Rocky Mountain Association of Geologists Symposium, 1987, p. 58-70.

[45] J. D. Fairhead, Regional tectonics and basin formation: The role of potential field studies in Phanerozoic Regional Geology of the world, Elsevier, 2012, pp. 330-341.

[46] C. Prieto and C. Pratsch, Gulf of Mexico study links deep basement structure to oil fields: Demonstrating deep structural control, Intergrated Geophysics Corporation Footnotes series on Interpretation, 2000, 4P.

[47] H.V. Lyatsky, G. M. Friedman and V. B. Lyatsky, Principles of practical tectonic analysis of cratonic regions in Lecture notes in Earth Sciences, Springer-Verlag, vol. 84, 1999, 369p.

[48] T. N. Plotnikova, Nonconventional hydrocarbon targets in the crystalline basement and the problem of the recent replenishment of hydrocarbon reserves, Journal of Geochemical Exploration, vol. 89, pp. 335-338, 2006.

Chapter 5

Geodynamic and Tectonostratigrafic Study of a Continental Rift: The Triassic Cuyana Basin, Argentina

Silvia Patricia Barredo

Additional information is available at the end of the chapter

1. Introduction

The stress and strain resultant from geodynamic processes control the uplifting and subsidence of different portions of the crust so sedimentary basins can be considered pieces of the earth which have suffered prolonged subsidence related to thermo-mechanical processes. According to these latter, a mechanical analysis of subsidence in any basin can be used for interpreting the regional distribution of depositional sequences; similarly the origin of the sedimentary sequences can be related back to the tectonic activity which controlled the insertion and evolution of the sedimentary basin [1]. Depositional sequences result from a complex interaction of the supply of sediments, the availability of the accommodation space (both with tectonic components), sea level variations (which also may have a significant tectonic influence) and climate variations. The first-order control on basin geometry is the deformation field resulted from the tectonic activity and is a fundamental control on sedimentation and the location of their resulting environments. For each basin, the geometry of the main faults or the lithospheric flexure should provide the final morphology of the trough and the resulting subsidence and thus, will control the sedimentation rate, grain size, channel migration, avulsion episodes and the development of flood plains and/or lakes.

The Cuyana Basin corresponds to a passive continental rift, *sensu* [2] developed during Triassic times as a consequence of the early Mesozoic breakup of Gondwana (Permian to Late Triassic-Early Jurassic) e.g. [3, 4]. It was located in its western southernmost portion over a wide belt of older accreted Eopaleozoic terranes, known as Gondwanides orogen. This latter was an extensive belt composed of contemporaneous orogens of Palaeozoic age and their related basins located along the southern Gondwana margin, it was firstly defined by Keidel [5] but du Toit [6] referred to as "Samfrau geosyncline". The Cuyana Basin is composed of several asymmetric half-grabens linked by accommodation zones that were partially or completely disconnected in the early rifting phase [7, 8]. This geometry and the coeval tectonic activity that was mostly characterized by recurrent extensional pulses are thought to be one of the major factors in the evolution of the sedimentary sequences and the complex environmental relationships that this basin exhibits. Its continental deposits can be considered as a whole a second-order depositional sequence divided into three third order-sequences related with regional to local processes [9].

Much of the research on this basin has been conducted in the Cacheuta depocenter where an intense hydrocarbon exploration and production has been held in the past decades [10 to 13]. Instead, the northernmost sections of the Cuyana Cuyana Basin, located in the San Juan province, have been studied in detail by [9, 14-21].

Interbasinal correlations along the whole basin were due to [7, 18, 22 to 24] who considered the separate half-grabens of San Juan and Mendoza provinces as a single trough. Age control in the basin traditionally has been based on biostratigraphy e.g. [20, 25 to 30]. The studied area is part of a fold and thrust belt where complex structures and diverse inversion tectonics phenomena gave place to inverted subbasins composed of normal faults with reactivated inverse displacement, new inverse faults, and fault propagation folds [8, 31]. Consequently, lithostratigraphy and biostratigraphy proved to be insufficient to understand sequences contrast along the basin, see [32] for discussion. By the 1990s modern stratigraphic tools have also been applied in an attempt to establish a more precise correlation among depocenters and even between the active and flexural margin of the Rincón Blanco trough, which are presently disconnected [8, 9, 20] but much uncertainty remained because of the lack of absolute age data in the basin as a whole. Presently, a more complete chronostratigraphic control has been achieved using the proposed cyclostratigraphy scheme of [13] with isotopic dating [29, 32 to 34] which permitted to arrive to an enhanced evolutionary model.

Classically, the Cuyana Basin has been considered as a passive rift developed as a consequence of extensional to transtensional forces resulted of the collapse of a Permian orogen and the beginning of the Gondwana breakup, the stratigraphic features represent the interplay between tectonics (subsidence-uplift) and sediment accumulation rates. The evolution of the Gondwanides along the southern portion of Gondwana was a key process through which the Triassic basins of Argentina developed. In this dominant geodynamic framework it is emphasized here that making correlations can highlight similarities and differences among basins and even within a given one, as it is the case of the Cuyana Basin. Subsidence analysis of the subsurface Bermejo and outcropping Ischigulasto-Villa Unión-Marayes and Cuyana basins (Figure 1) reveals the existence of notably various episodes of accelerated subsidence during Middle to Late Triassic, suggesting that all of them share a common tectonic history.

It is explored here these results in detail integrating geodynamic and tectonostratigraphic concepts to understand the rift evolution and the history of its infilling. The resulting data, permitted to understand the coeval deformation and the tectonic processes acting on the lithosphere in the western margin of Gondwana.

2. Geological setting of the Cuyana Basin

Related to the Gondwanan Orogeny, that affected the western margin of southern South America (Permian to the Late Triassic-Early Jurassic) and the beginning of the Mesozoic Gondwana breakup, a series of rifts were developed (Figure 1). They correspond to a complex system of rapidly subsiding, NNW-SSE trending narrow asymmetric half-grabens bounded by predominantly normal faults in one margin that parallels the grain of older Paleozoic structures e.g. [9, 12, 19, 36 to 38]. The border fault is sometimes segmented with segment boundaries marked by change in strike or fault overlap, relay ramps, transfer faults and rider blocks(Figure 2). The notable increase in thickness towards the center border fault suggests that the structures were syndepositionally active and thus it is proposed here the architecture and fill were influenced by the geometry and the displacement of the bounding normal faults.

Among them, the Cuyana Cuyana Basin is the largest Triassic rift basin of western Argentina and considered to extend over an area of more than 60.000 Km². It corresponds to a passive continental rift developed during differential intraplate stresses derived from a backarc extension acting on a normal crust and moderate thermal flux [9]. At present, it underlies a lowland segment of the Argentine foreland but several exposures are along both flanks of the Precordillera, in Mendoza and San Juan provinces (Figure 1). The rift is composed of several asymmetric half-grabens roughly triangular in cross section, and linked by accommodation zones [7, 8] which correspond to Cacheuta, Las Peñas-Santa Clara, Rincón Blanco, Puntudo General Alvear, Ñacuñan and Beazley (see Figure 1). Border faults (BF) of the Cuyana Basin strike obliquely to the maximum extension direction (NNE-SSW), have a stepping geometry and opposite dip directions (Cacheuta and Santa Clara-Rincón Blanco) causing the faulted and hinged margins sometimes shift from side to side of the rift basin as it is evident between Cacheuta and Santa Clara subbasins e.g. [7, 9, 11], or keep the same dip orientation as between Santa Clara, Rincón Blanco and Puntudo depocenters (Figure 2). Each segment of these master structures shows tips that are marked by a change in strike or fault overlap, relay ramps and rider blocks. Most of them were controlled by the normal to oblique reactivation of pre-existing zones of weaknesses in the crystalline basement, and thus main border fault exhibits N-S and NNW-SSE direction while accommodation and transfer zones between main Puntudo, Rincón Blanco and Cacheuta depocenters display a mostly NNE-SSW direction. Major segments do not become hard linked and the displacement was partitioned among several segments without being physically connected. In this way, intrabasinal highs could persist throughout synrift and postrift sedimentation along the whole basin. The lack of polarity reversals in the northern segment is attributed to the basement fabric control with fault reactivation of ancient N-S and NNW-SSW structures. In particular, within Rincón Blanco depocenter, the border fault is segmented and its closer spacing segments hard-linked with evident transfer of displacement from the southern to northern segments (Figure 2) [1, 8]. The restricted half-grabens were separated by left-oblique-slip (left lateral?), WNW (Az 112°) that are oblique to the border fault.

Figure 1. Triassic rift basins of central-western Argentina and the main subbasins. From [35].

The Cuyana Basin contains, approximately, up to 3700 meters of continental rocks of predominantly alluvial, fluvial, and lacustrine origin interbedded with tuffs of coeval volcanism with intraplate affinities e.g. [11, 39]. The notable increase in thickness towards the border faults suggests that these structures were syndepositionally active. Field studies and seismostratigraphic analysis show that changes in sequence geometry occurred in-phase with intra-continental elastic stress relaxation, as fault reactivation and probable basement reworking, during the quasistatic event on the western margin of Gondwana. These recurrent extensional pulses that controlled the extension along the Cuyana Basin impinged several different characteristic to the infilling of the Cerro Puntudo, Rincón Blanco and Cacheuta depocenters and thus they will be treated in separate items.

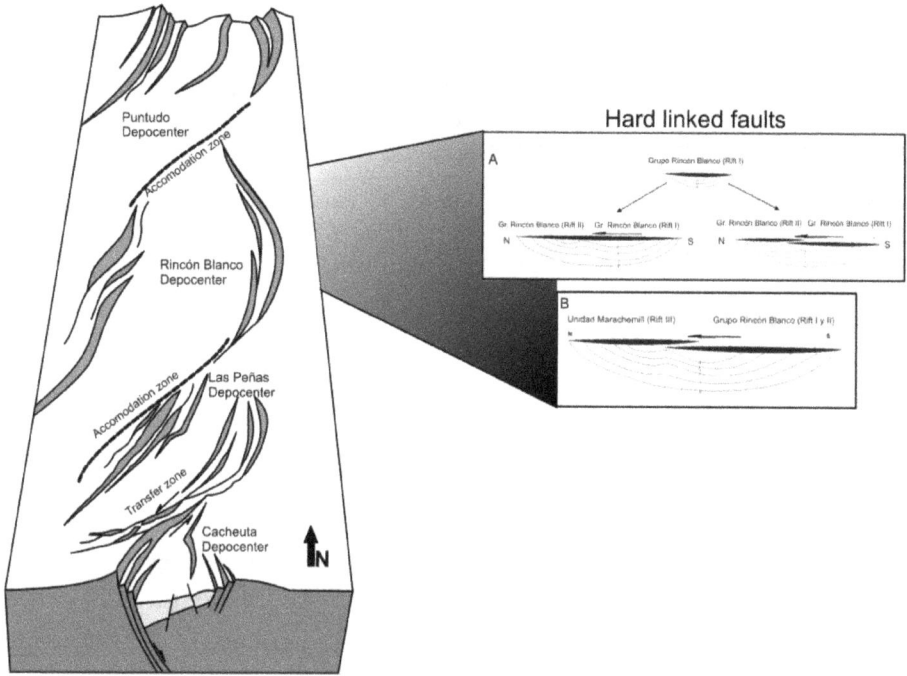

Figure 2. Cuyana Basin main depocenters. Border faults (BF) have a stepping geometry and opposite dip directions (Cacheuta and Santa Clara-Rincón Blanco) or keep the same dip orientation as between Santa Clara, Rincón Blanco and Puntudo depocenters. Major segments are related through accommodations zones which constitute intrabasinal highs. To the right, the scheme shows a plan view of the Rincón Blanco trough whose segments became hard linked. From Barredo & Ramos [1].

Extension took place along three phases, named Rift I, II and III and their corresponding synrift deposits by Barredo [9]. During Synrift I, the deposits were restricted to a series of partially isolated depressions. Thus, in proximal positions the alluvial coarse clastic facies are covered and interfingered with medium to fine-grained sandstones and tuffs interpreted as deposited in fluvial, and lacustrine/playa-lake settings e.g. [8, 9, 13, 20, 23, 26]. This first sequence is separated by a regional unconformity by the second depositional phase or Synrift II. It is a fining-upward sequence which consists, in general, of cross-bedded sandstones, black shales and tuffs, all related to braided to high sinuosity-river systems which grade into a widespread lacustrine setting. Finally, this second rifting stage passes upward to shallow lacustrine to fluvial sandstones, shales, and tuffs deposited during the early to late post-rift. This shallowing-upward succession is widespread in the basin and displays an onlap relationship with the underlying beds, directly overlying Paleozoic basement [8, 9, 13]. During this stage, there was a renewal of the subsidence in the basin. Recent investigations permitted to identify a the third extensional event during early Late Triassic (to Early Jurassic times?) probably associated with the opening of the Neuquén Basin, located farther south in northern Patagonia [1, 32]. Main evidences of this event are not regionally found, the best exposures outcrop in the Rincón Blanco Subbasin, however to the south, the subsurface Barrancas

Formation (Cacheuta depocenter) shares similar tectonostratigraphic characteristics with that of the Rincón Blanco but reveals certain uncertainties about its age. By the moment, this unit has been assigned to the Late Triassic [40] and even to Lower Jurassic [41]. The infill of Synrift III is represented by an alluvial-fluvial succession developed under renewed mechanical subsidence of the basin and marked semiarid conditions.

The final thermal relaxation of the basin seemed to have occurred during Jurassic and Cretaceous times and aborted during Tertiary times when the lithosphere undergone flexural subsidence induced by the Andean orogenic overloading and by sediment charge.

3. Puntudo Subbasin

The northernmost exposures of the continental Triassic Cuyana Basin crop out at the Cerro Puntudo locality on the western flank of the Precordillera in the San Juan province (Figure 3). These exposures record the sedimentation near the northern end of the basin related to a fault tip end. An angular unconformity separates de Cerro Puntudo succession from the underlying Choiyoi Group volcanics and it is tectonically truncated at top [34].

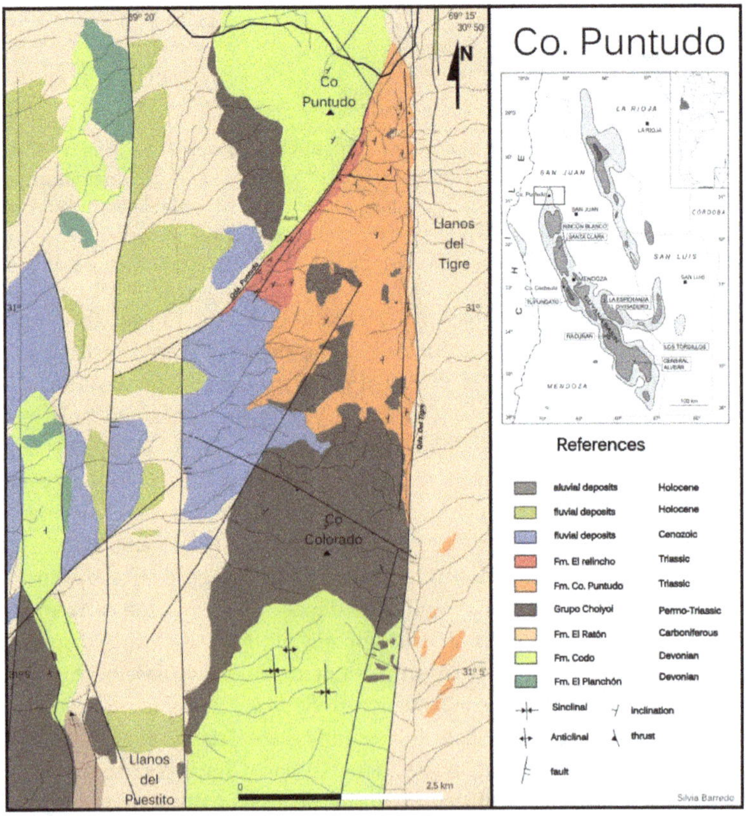

Figure 3. Geological map at the Co. Puntudo area. Modified from Mancuso [34]

The present structure consists of east verging (Andean) thrusts with wavelengths of ~1 km. These thrusts are responsible of the truncation and suppression of the west El Relincho facies (El Puntudo fault) (Figure 3). Triassic basin inversion was positive and topographic uplifting and gently folding of the ancient trough has been observed. The deposits are folded in northeast striking asymmetric syncline with its eastern flank well developed and the western almost truncated. The anticline of the eastern margin is interpreted as a result of reactivation of ancient rider blocks as a post depositional anticline during reverse motion of the structure. West-northwest faults seem to follow the Paleozoic fabric and are parallel to the Rincón Blanco Sub-basin transfers and thus, genetically related [9].

The extensional structure consists of a north-south striking fault (border fault?) located to the east. This assumption is based on the thick fan-conglomerates with main paleocurrents showing a marked southwesternward component of the flow parallel to the basin margin and towards the fault zone centre, where subsidence favoured the insertion of a shallow lake. This flow direction can be linked southernward with the axial sediment system of the Rincón Blanco Sub-basin. The border fault is presumed to be a west-dipping structure, probable lystric in cross section, with highest displacement to the center (displacement estimates are of 400 meters). On the basis of the maximum sequence thickness the "border fault" has been 12 km long. A notable reduction of the sedimentary record to the north and south can be extracted of the onlapping of the synextensional strata over basement rocks, particularly in the southern part where at the Cerro Colorado the Choiyoi volcaniclastics are exposed. Paleocurrents in this area points to a northward flow and thus, has been interpreted as an accommodation zone (transfer) that kept this depocenter probably disconnected from the Rincón Blanco Sub-basin.

The Triassic column reaches an exposed thickness of approximately 400 meters and has been divided into two units, the Cerro Puntudo and the El Relincho formations [42] arranged in two of the three synrift cycles recognized for the Cuyana Basin. The first tectosedimentary sequence or Synrift I corresponds to the Cerro Puntudo Formation. It is a fining-upward succession which starts with a thick package of 400 m alluvial fan conglomerates and cross-bedded coarse sandstones (Figure 4). This succession is mainly composed of an alternation of massive red clast-supported conglomerates and subordinated sandstones which passes upward to a braided fluvial system dominated by red and reddish-brown conglomerates and sandstones. Sheet-floods related to ephemeral fluvial deposits, characterized by red fine-grained sandstones, mudstones, and light-coloured tuffaceous limestones, are interfingered and cover the braided fluvial beds. The upper one-third of this first cycle comprises 75 m thick well-stratified and laterally persistent gray micritic and stromatolitic limestones, reddish brown mudstones, red fine-grained sandstones with rippled lamination, and thin levels of green tuffs interbedded. This part of the section was interpreted as deposited in a relatively shallow carbonate-rich lake with marked cyclicity recognized by a succession of retrograding and prograding events. Coeval volcanism is represented by the interbedded tuffs. To the southeast, the lacustrine deposits are dominated by siliciclastic facies made of reddish brown mudstones, red fine-grained sandstones, and green tuffs which were interpreted as deposited in a mudflat environment e.g. [23, 24, 34]. The climatic conditions during the deposition of the first synrift phase were markedly seasonal, as represented by the cyclical nature of the lacustrine deposits, not associated to tectonic controls. Palynomorphs suggest the presence of a relatively diverse flora composed by at least riparian vegetation (ferns and lycopsids) associated to the lake margins and a forest (araucariaceans) in the fluvial floodplains [34]. Due to the lack of evidence of evaporitic facies and extensive desic-cation features humid climate conditions developed at least seasonally thus a subtropical seasonally dry climate might be assume for the upper part of the first Synrift phase. A U-Pb SHRIMP zircon age of 243.8 ± 1.9 Ma (Anisian) obtained from juvenile magmatic zircons in a tuff interbedded with the lacustrine beds constrained the deposition of the Synrift I at Cerro Puntudo to the Early-lower Middle Triassic [34].This age is consistent with palynological data obtained from equivalent lacustrine levels.

The second tectonosedimentary sequence or Synrift II, is represented by the El Relincho Formation. It starts at the point where the lacustrine deposition is abruptly interrupted by the instauration of a coarse alluvial setting. The fining-upward El Relincho succession (approx. 140 m thick) is dominated by green clast-supported conglomerates exhibiting clasts of the underlying Cerro Puntudo sequence. The succession passes upward into reddish brown, cross-bedded, coarse to medium -grained sandstones of braided fluvial origin. The upper limit of the Triassic succession at Cerro Puntudo area is tectonically truncated

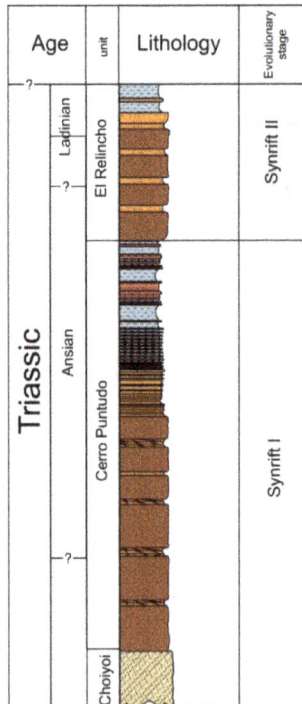

Figure 4. Generalized stratigraphic column of the Puntudo subbasin infilling showing the rifting epi-sodes recognized across the whole Cuyana Basin. Modified from Mancuso [34].

4. Rincón Blanco Subbasin

The Triassic Rincón Blanco subbasin is located in the Precordillera fold and thrust belt at 31° 24' – 31° 33' south (Figure 5). It is 40 km wide and stretches approximately ~60 km N-S. The subbasin was mainly developed over Ordovician-Silurian distal platform deposits and Carboniferous glacimarine diamictites and turbidites nevertheless, to the north the column overlies Permian marine platform clastics. The present geometry of the Rincón Blanco active margin is a north–south tight asymmetric syncline. The structural analysis indicates that the east margin has been truncated by a series of west verging back-thrusts since Cenozoic times. Basin marginal deposits have in places been removed by erosion or preserved in fault inliers. The western margin has undergone east verging Andean thrusts (Cenozoic) with wavelengths of 2 km, presumably older than the back-thrusts, that has separated de outcrops from the flexural margin of the depocenter. Only, the south-eastern portion of the active margin remains practically unaffected, although some inversion of earlier extensional structures has been noted (Figure 6). Flexural margin is represented by isolated outcrops and subsurface deposits and is presently separated by a basement high from the active margin. Inversion was positive and topographic uplifting and gently folding of the ancient troughs has been observed, in particular in the Cerro Bola-Cerro Amarillo region (southernmost outcrops). Classic basin-inversion structures consist of anticlinal folds and reverse faults, like the BF segment exposed in Cerro Amarillo. Post depositional anticlines and synclines formed during reverse motion along the reactivated faults. Minor inverse faults area associated with the backtrhusts, they are composed of fault bend, fault propagation, and detachment geometries. In the flexural margin they have also uplifted the ancient intrabasinal highs where sequences are drastically thinned or basement exposed (Figure 7).

The spatial arrangement of the architectural elements permitted to interpret this Triassic depocenter as an asymmetric trough with a hinged margin. Basin-scale morphology could not be determined precisely because Triassic outcrops are tectonically truncated except for the western margin that could be modelled in subsurface [31]. Faults have displacement magnitudes of 1000 m and sole into a subhorizontal detachment surface at about 10000 m deep, below which deposits are undeformed. Using the architecture and some features of the sedimentary record it was established that the border fault (BF) was composed of right-stepping, west-dipping, normal to oblique faults, separated by a transfer zones (see Figures 2 and 5) became hard-linked. In cross section they are lystric and in plan view sinuous with highest displacement toward the center (displacement estimates are of 2 km). The resulting half-grabens are supposed to have gradually grown in depth and length through time surrounded by uplifted footwalls as a consequence of the absolute upward motion during fault displacement and isostatic adjustment. They correspond to Cerro Amarillo/Rincón Blanco and Marachemill troughs, in the active margin; Barreal and Agua de Los Pajaritos, in the flexural margin.

Kinematic indicators and the orientation of diabase dikes suggest a NNE-SSW (Az 30°) extension direction for the depocenter and a N-S to NNW-SSE border faults strike [8,9]. Local transfers zones correspond to west-northwest (Az 112°), left lateral strike-slip faults interpreted to have been formed to accommodate differential extension within the depocenter (Figure 6). These faults can be associated with the Paleozoic fabric which consists largely of generally northwest-striking and west-northwest striking reverse and tranpresional faults [37, 43]. Intrabasinal highs produced by rotation blocks made up internal elevations relative to the rift flank and trough and, consequently, the flexural margin contained much thinner sequence of synrift strata (~1800 meters) [31] than the active margin (3000 meters).

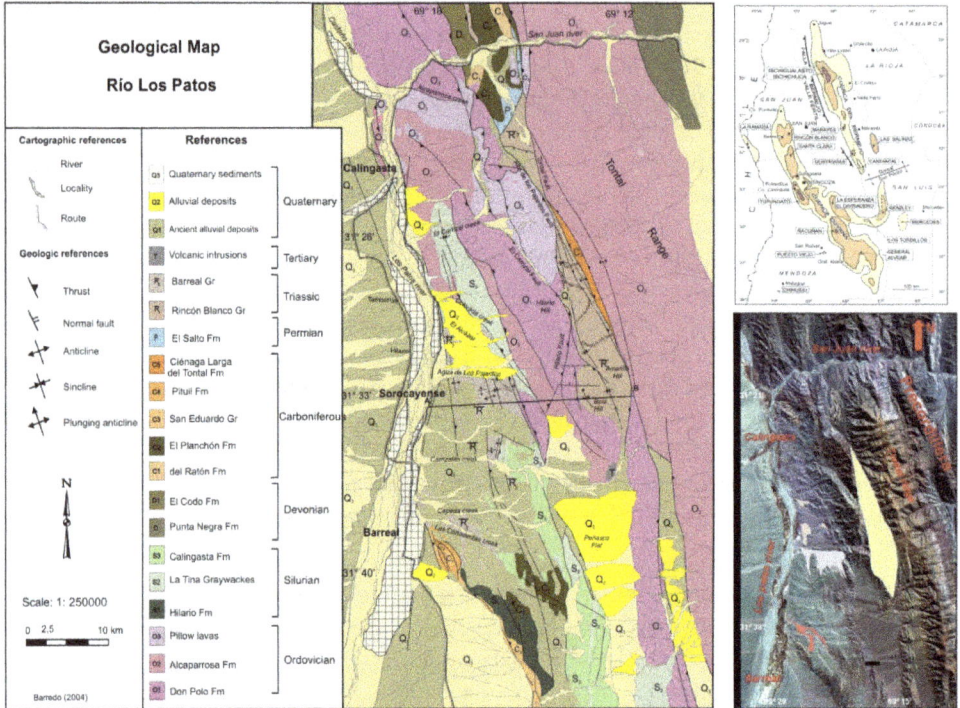

Figure 5. Left, geological map of the Rincón Blanco depocenter with detailed information from the flexural margin. Right, location map of the Rincón Blanco half-graben outcrops in the Cuyana Basin. The satellite image shows the distribution of the main depocenters, subsurface is not included. From Barredo [9].

The infilling consists of almost 3000 m coarse conglomerates interfingered with sandstones, shales, tuffs, tuffaceous mudstones and bimodal volcanic rocks composed of rhyolites and rhyolitic tuffs and ignimbrites associated with the Choiyoi uppermost effusives [39, 45]. Alkaline basalts are also present but only in the flexural margin [9, 44]. The upper limit of the Triassic succession is marked by a regional uncomformity with the overlying Cenozoic foreland deposits. Three packages of genetically linked units bounded by regional extended unconformities of third order are associated with the three rifting stages (Synrift I, II and III) of the Cuyana Basin. These rifting stages beeing related with the acceleration of faults subsidence during Middle to Late Triassic. They comprise the Rincón Blanco and

Marachemill groups (active margin) and the Sorocayense Group (flexural margin) (Figure 7). Their evolution was mainly controlled by tectonic pulses but sediment supply and climate was also important [1, 9].

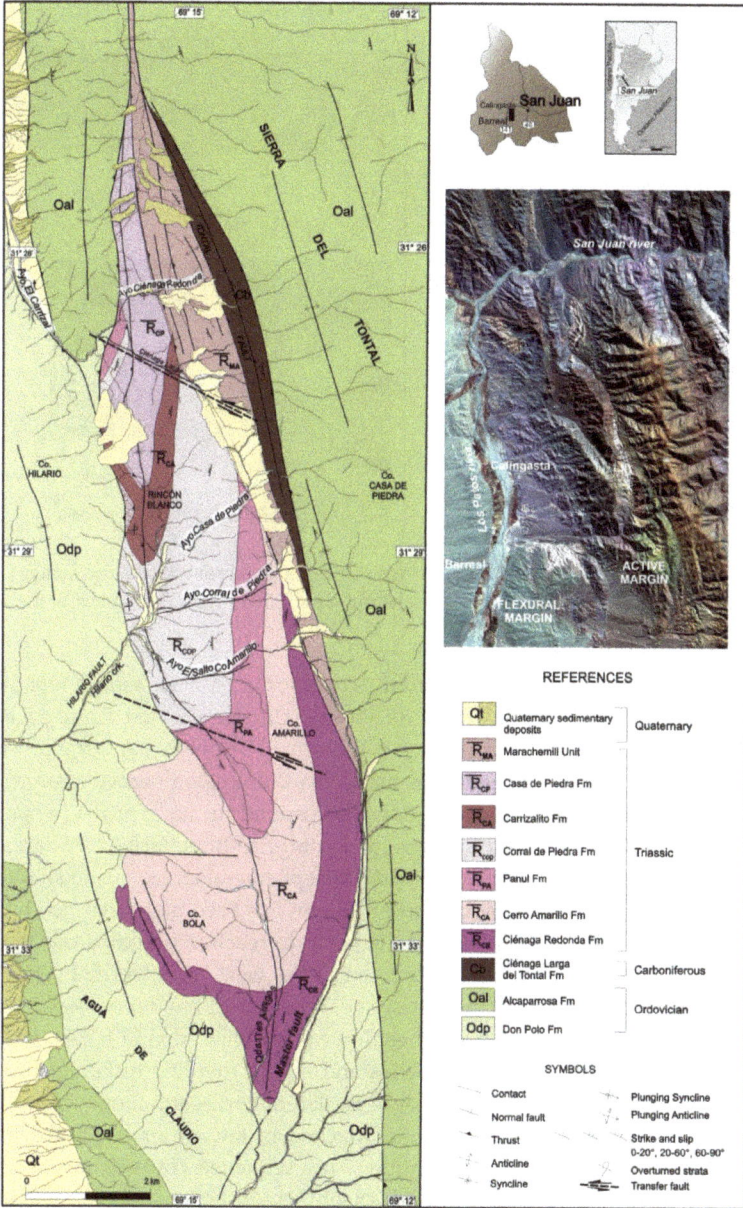

Figure 6. Detailed geological map of the active margin of the Rincón Blanco depocenter with inferred transfers faults. From Barredo [8,9].

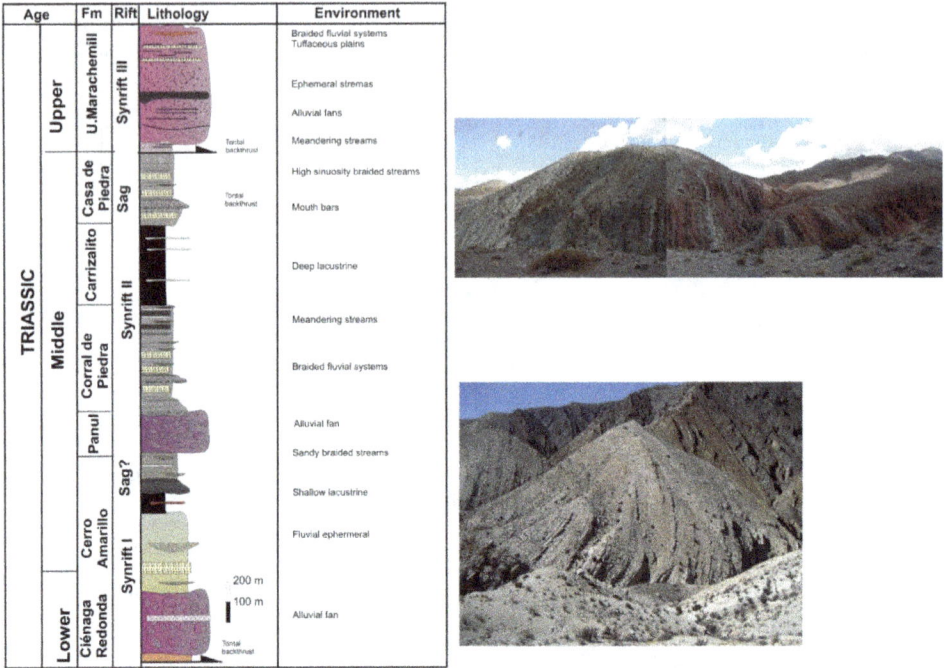

Figure 7. Stratigraphic column of the Rincón Blanco active margin. From Barredo & Ramos [1]. Upper right, north view of the coarse facies of the Marachemill Unit. Lower right, north view of the deep lacustrine facies of the Carrizalito Formation.

In the active margin, the Rincón Blanco Group is composed of, from base to top: the Ciénaga Redonda, Cerro Amarillo, Panul, Corral de Piedra, Carrizalito, and Casa de Piedra formations [15]. The trace of the contact between the synrift sediments and the prerift rocks is concave and well exposed to the south, in the Cerro Bola region (active margin). The Ciénaga Redonda and Cerro Amarillo formations correspond to the Synrift I, when the basin constituted a simple fault-bounded through, and comprise three facies association, alluvial fans, alluvial fans braidplain settings dominated by ephemeral streams, and shallow, mostly ephemeral lacustrine deposits. A maximum of 1200 m has been measured nearby the Amarillo and Bola hills (Figure 7) decreasing to the north to reach in outcrops less than 100 meters. This thickness diminishing is also observed to the west. The Synrift I starts with conglomerates and breccias interpreted as cohesive debris flow of fan deposits and un-cohesive debris flow developed in non-canalized sheetflows (Ciénaga Redonda Formation). Interfingered with these facies, there are ignimbrites and scarce rhyolitic tuffs and tuffs with an estimated age of 246 Ma [32]. Overlying, there are laterally discontinuous sheet-like beds of massive and horizontally bedded sandstones stacked in few meters with alternating conglomerates of the Cerro Amarillo Formation. This sequence corresponds to short-lived, wide, poorly canalized flows or low energy sheetflows in ephemeral streams of fans and bajadas. More distally, these systems are associated with silty-sandstones, massive to parallel mudstones with desiccation cracks and anhydrite lenses alternating with black (lignitic?)

shales deposited in a shallow lacustrine environment. This succession is covered by sandstones, shales and tuffs of braided fluvial origin with a predominantly axial paleoflow influx.

The Synrift II is separated from the underlying units by a regional unconformity and is composed of amalgamated conglomerates with angular basement derived lithologies, sandstones and ryolithes from the underlying Triassic units [9]. This sequence passes upwards to lenticular bodies of massive to trough-cross conglomerates and pebbly sandstones with erosive bases. They correspond to alluvial fan and braided fluvial facies association. The volcaniclastic content is notoriously high compared with the underlying units and consists of tuffs, and tuffaceous sandstones and conglomerates (Panul Formation). Fining–upward tabular sequences of fine conglomerates to sandstones with trough-cross bedding, lag deposits, and erosive bases follow. They grade to medium fine-grained, moderately to well sorted tabular sandstones/tuffaceous sandstones, reworked tuffs, and thick overbank mudstones at top of meandering fluvial origin (Corral de Piedra Formation). The whole succession passes vertically and laterally into the lacustrine facies of the Carrizalito Formation composed of thick massive mudstones/tuffaceous siltstones with volcaniclastics, mostly chonitic, levels. Upward these beds are covered by thinly stratified silty sandstones and massive mudstones related to the deeper facies of the lake. They are associated with massive or horizontally laminated marls (30 cm in thickness) and pale-grey organic levels which consist of laminated bituminous shales (oil-prone coals) with micritic calcite. The lake deposits are covered by clast supported planar to trough-cross bedded lenticular conglomerates and sandstones interbedded with laminated mudstones and limestones interpreted as deltaic mouth bars deposits (basal levels of the Casa de Piedra Formation). Upward these facies are covered by lenticular conglomerates and sandstones with trough-cross bedding and erosive bases of fluvial origin. They grade to fine-grained sandstones/tuffaceous sandstones and thick overbank mudstones which at top interfinger with ash fall tuffs.

A U-Pb Shrimp age on tuffaceous beds at the base of the lake levels of the Corral de Piedra Formation, in the Rincón Blanco depocenter, have yielded an age of 239.5±1.9 Ma [32]. This age constrains the initial deposits of Synrift II to the late Anisian (Middle Tiassic) which coincides with the recently date of 239.2 ± 4.5 Ma obtained by Spalletti [29] for the Potrerillos Formation (initial infilling of the Synrift II at the Cacheuta subbasin).

Finally, to the north-east margin of the basin a series of almost 900 meters of mainly volcaniclastic and clastic rocks were interpreted as the Marachemill Unit and included in the (Synrift III) [1, 32]. It is in tectonic contact with the Rincón Blanco Group through the Tontal backthrust (see Figure 6). Three facies associations characterized this unit: proximal to distal alluvial fans, ephemeral streams and fluvial systems. The first depositional environment comprises thick massive, mud-rich matrix-supported red conglomerates and breccias composed of volcaniclastic and sedimentary clasts mostly from the underlying Rincón Blanco Group. These facies interfinger with several ignimbrites and rhyolitic tuffs levels and tuffaceous sandstones. Poorly canalized flows and sheet floods developed in ephemeral environments follow upwards. They are characterized by coarse massive to horizontal laminated sandstones with erosive to sharp bases and shale intraclasts and are laterally associated with flood plain sandstones and shales with oxidized organic matter and desiccation cracks. Volcaniclastics are abundant and correspond to ash fall tuffs and pyroclastic flows. The third depositional environment corresponds to a braided fluvial system with conglomerates to granular sandstones in frequently amalgamated bodies. The conglomerates are poorly sorted with sub-angular to sub-rounded clasts from the Ordovician basement. Overlying beds correspond to tabular sandstones with planar cross stratification which grades to fine to medium sandstones with small scale trough-cross stratification and ripple-cross lamination capped by thin beds of massive or rippled greenish grey siltstones, tuffs, and mudstones with invertebrate burrows and mudcracks. Paleocurrents show an east-northeast predominant inflow. Ash-fall deposits are abundant constituting thick tabular massive, sometimes laminated, deposits and reworked levels that can be interpreted as tuffaceous plains with isolated poorly developed fluvial channels. U-Pb zircon age of 230.3 ± 1.5 Ma (SHRIMP) has been obtained for the Marachemill Unit suggesting an early Carnian age for its deposition [32].

The Sorocayense Group represents the infilling of the Agua de los Pajaritos and Barreal groups which correspond respectively to the north and south separated depocenters of the flexural margin [18,26]. The first is composed of: Agua de Los Pajaritos, Monina, Hilario, El Alcázar and Cepeda formations e.g. [46] (see Figures 5 and 8) and the other by: Barreal, Cortaderita and Cepeda formations [18]. The northern Agua de Los Pajaritos consists of a basal alluvial-fluvial sequence with abundant subaerial volcanic tuffs and volcaniclastic rocks overlain by deep lacustrine facies and fluvio-deltaic facies (Monina Formation) [46, 47]. Clast-supported conglomerates and breccias dominate at the base and are interpreted as non-cohesive debris with subordinate matrix-supported paraconglomerates (cohesive debris flow). Non-canalized and channelized levels follow, the lower ones correspond to sheet-flows and are characterized by massive sandstones with high energy directional structures; the second and most dominant upward, consist of lenticular thin bedded conglomerates associated with extended and shallow channels. These facies are followed by finning upward sequences of cross-bedded mostly tabular sandstones, primary tuffs, reworked tuffs, shales and bituminous shales, deposited in a high sinuosity river setting with well developed flood plains (frequently obliterated by fallout thepra). These facies laterally interfinger and are covered by sandstones, marls with algae lamination and massive and bituminous shales. Sandstone and siltstones with wave and climbing ripple stratification and tangential cross stratification correspond to deltaic mouth bars. At top and intercalated are white and green fallout tuffs and chonites, massive or thinly laminated and sometimes indurated with siliceous replacement. The clastic dikes cut these lacustrine facies. The fluvial-deltaic and fluvial strata of Hilario Formation follow. They consist of thickening and coarsening upward units characterized by laminated mudstone and thin sandstone beds which pass into amalgamated sandstones beds with small scours and cross-bedded sets of sandstone and clast-supported conglomerates, sandstones and thin coaly layers arranged in a series of fining upwards units. Finally, the facies are replaced by lenticular shaped fining upward coarse tabular sandstones and shales of a fluvial high sinuosity environment. This clastic and volcaniclastic sequence resembles that of the Synrift II of the active margin (Rincón Blanco Group) and has been preliminary correlated with it [46]. Sandstones with erosive bases associated with clayed mudstones, tuffs, marls, limestones, and massive and rippled laminated sandstones of high sinuosity fluvial and palustrine environments are considered the base of the El Alcázar Formation. This succession unconformably overlies the Synrit II and has been assigned to the early post-rift [32]. Tabular multicolored shales with tuffs, subordinates sandstones, and massive lenticular conglomerates are interpreted as lacustrine facies with alternating mouth bar and turbidites deposits. It is gradually replaced by lenticular sand and conglomerate bodies associated with mudstones, tuffs of floodplain origin sourced from outside the rifted basin. Lava flows mostly tholeitic interfingers these facies [44]. A renew of the extensional regime (Rift III) gave place to the deposition of the fan derived clastics of the Cepeda Formation lateral equivalent to the Marachemill Unit of Rincón Blanco depocenter [1].

The Synrift I in the Barreal depocenter encompasses distal fans and fluvial braidplain environments followed by shallow lacustrine shales, mudstones, and tuffaceous mudstones and deltaic sandstones [48] (Figure 8). Braidplain environments consist of clast-supported, massive, sometimes normally graded lenticular conglomerates and breccias with basal sharp and irregular contacts associated with sandstones that show parallel or low angle planar cross stratification. These facies are followed by massive and thin laminated shales, fine current-rippled sandstones laterally related to silty to fine-grained sandstones interbedded in a heterolithic arrangement with coal fragments, interpreted as lacustrine deposits. This first synrift sequence is included in the lower half of the Barreal Formation.

The second synrift phase (Synrift II) is marked by the deposition of braided fluvial systems with lenticular to tabular beds of poorly sorted conglomerates with subordinates sandier lenses and more tabular planar cross stratified sandstones associated with fining upward channel fill sequences.

This succession is covered by lacustrine deep facies with bituminous shales and thin subordinate sandstones of mouth bars and turbidites deposits, and finally lenticular conglomerates and coarse sandstones of fluvial systems, probably high sinuosity sandy braided system (upper half of the Barreal Formation). The post-rift facies are represented by the Cortaderita Formation beds. It starts with tabular shales, tuffs, and massive and rippled wave laminated sandstones of lacustrine origin. This facies are covered by lenticular pebbly mouth bar sandstones that pass upward to conglomerates and sandstones arranged in lenticular bodies of sandy braided fluvial system, sometimes obliterated by cinder fall out. The succession finishes with the instauration of a high sinuosity river system with well developed flood plains. The Synrift III (Cepeda Formation) consists of amalgamated clast and matrix supported conglomerates of alluvial proximal fans composed of volcaniclastics and siliclastics mostly from the underlying Cortaderita Formation; subordinate sandstones to the top of the conglomerates were interpreted as braided ephemeral plains. Sheet floods and poorly canalized flows developed in ephemeral environments with coarse massive to horizontal laminated sandstones with erosive to sharp bases and shale intraclasts. Volcaniclastics are abundant and correspond to ash fall tuffs. Upward, braided fluvial deposits of frequently amalgamated poorly sorted conglomerate bodies with sub-angular to sub-rounded clasts with subordinate sandstones cover the ephemeral facies. It passes upward to tabular sandstones with planar cross stratification which grade to fine to medium sandstones with small scale trough-cross stratification and ripple-cross lamination. To the top, massive or rippled green-ish grey siltstones, tuffs, and mudstones with invertebrate burrows and mudcracks domi-nate; this beds were interpreted as mudflat deposits.

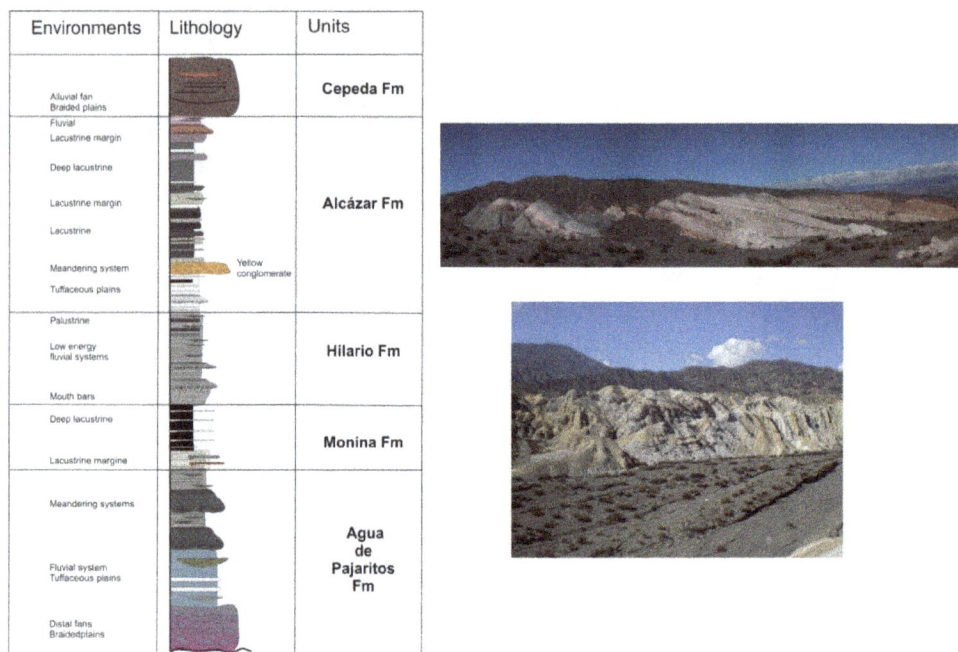

Figure 8. Stratigraphic column of the Agua de los Pajaritos depocenter (flexural margin). From Barredo [46]. Upper right, a south view from the El Alcázar postrift facies of clastic and pyroclastic materials deposited in shallow lake and fluvial environments. Lower right, east view the deep lacustrine Monina Formation.

Paleofloristic analysis based on the macrofloral remains preserved nearly in the whole col-umn on both the flexural margin and active depocenters indicate the presence of evergreen subtropical floras adapted to seasonally dry climatic conditions in the Synrift I and lower part of Synrift II sequences. Thus, they do not indicate a marked climatic shift in this interval although floristic information from the uppermost part of the successions in this part of the Cuyana Basin is very fragmentary e.g. [26, 27].

5. Cacheuta Subbasin

The Cacheuta Sub-basin outcrops constitute the southernmost exposures of the Cuyana Basin. The best outcrops are located in the southern flank of the Cerro Cacheuta and in the nearby Cerro Bayo in the Potrerillos locality, west of Mendoza city (Figure 9). The eastern margin of the sub-basin was developed over Precambrian crystalline basement conversely; on the western margin the Triassic succession overlies Silurian-Devonian turbidites, Cambrian-Ordovician limestones, and the volcanics of the Permian-Triassic Choiyoi Group complex. A regional uncomformity determines the upper limit with the overlying Jurassic?–Cenozoic deposits.

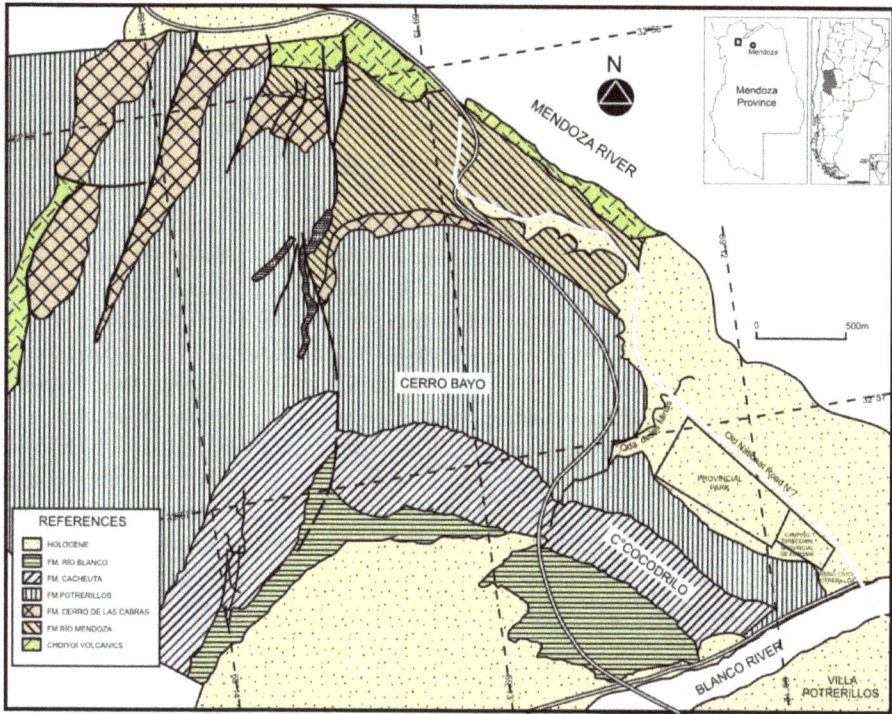

Figure 9. Geological map of the Cacheuta Subbasin outcrops at Potrerillos area. Modified from [49].

The Cacheuta Group represents the infilling of this through and is composed of, from base to top, by: the Río Mendoza, Cerro de las Cabras, Potrerillos, Cacheuta and Río Blanco formations. These facies characterizes two episodes of rifting followed by a period of post-rift subsidence. Unconformabling overlying these sequences, there is another unit, the Barrancas Formation of clear affinities with upper extensional events registered in the northern portion of the Cuyana basin. Up to the moment, it has been assigned a Late Triassic age by Rolleri & Fernández Garrasino [40] and a Lower Jurassic age by Reigaraz [41].

The structure consists of a series of N-NW to N-S trending asymmetric folds, broken locally by reverse high angle faults displaying either eastward or westward vergence [50] (Figure 9). These structures end against the Precordillera province to the north and extend far southeast to the San Rafael Block (see Figure 1). Tertiary structures encompass compresional Andean thrusts and positive inverted Triassic faults composed of folds and faults. Basement involved faults were mildly inverted and they still preserve extensional features. Minor antithetic and synthetic faults are release structures formed during bending. Compresional high angle inverse faults (60º) associated with the reactivation of Palaeozoic structures are also found.

The subbasin was narrow and asymmetric roughly triangular in cross section e.g. [13]. The more steeply inclined basin margin is located to the west and consists of a predominantly normal-slip fault/s associated with the border fault, which generally trends north–northeast. It is a steep dipping basement-involved structure with a hinged margin to the east. It coincides with an important Ordovician suture, the Valle Fértil Lineament [11, 12]. In cross section, it is lystric and in plan view it has been interpreted as being sinuous with highest displacement toward the center e.g. [7, 13]. Surface and subsurface information show that faults have displacement magnitudes of thousand meters and sole into a sub-horizontal detachment surface. Synrift sequences display reverse drag or rollover folds in subsurface. Strike-slip minor faults with a west-northwest strike and left-lateral slip can be observed in surface and subsurface. They were interpreted as inverted Permo-Triassic oblique-slip normal faults with dextral displacement. Intrabasinal highs produced by rotation blocks made up internal elevations relative to the rift flank and trough. In this way several sub-depocenters with local control were developed and limited by oblique structures to the border fault. They correspond to transtensional structures with north-northwest to northwest trending sinistral high-angle faults (60º and 90º to the northeast); additionally, northwest dextral lateral structures were also found. The Potrerillos area corresponds to a displacement transfer zone in which sinistral strike-slip meso-scale faults play an important role. It was a topographic high that separated this sub-basin from the Las Peñas through to the north. In this case, longer north–northeast-striking border fault (BF) of the Cacheuta depocenter is connected by shorter segments to the northern depocenter border by an opposite dipping fault. It has been interpreted as a soft linked connection as faults transferred displacement from one segment to another without being physically connected [9].

The Triassic column in this depocenter is of approximately 3000 m thickness (Figure 10). During the early depositional phase (Synrift I), the succession (approx. 800-1000 m thickness) is characterized by reddish conglomerates of alluvial fan facies (Río Mendoza Formation) related to the active margins of the rift. They laterally interfinger with multicolored mudstones, fine-grained sandstones, and tuffs of ephemeral-fluvial and playa-lake origin deposited basinward. In the depressed areas of the subbasin, shallow lake facies were accumulated characterized by the deposition of relatively thin oolitic grainstone beds and stromatolitic limestones interbedded with tuffs (Cerro de las Cabras Formation). This first sequence is separated by a regional unconformity by the second depositional phase (Sinryft II) [9, 13, 40]. The second synrift sequence (Synrift II) is a fining-upward succession (Potrerillos and Cacheuta formations) mainly represented by lower energy facies and fine-grained deposits than the underlying sequence. It was accumulated on a smoother relief due to the infilling of the depocenter and the deposits reach up to 1200 m in thickness. The Potrerillos Formation is characterized by fluvial conglomerates at the base intercalated with light greenish cross-bedded sandstones, and light tuffaceous sandstones of perennial braided river origin; this fluvial deposits grade basinward to greenish-grey laminated siltstone and sandstones interbeded with black bituminous shales and tuffs, related to high-sinuosity river systems. This facies laterally interfinger and are covered by the widespread lacustrine black shales of the Cacheuta Formation. In the maximum transgression of the lake, the lacustrine facies overlaps the fluvial deposits of the Potrerillos Formation to the basin borders. Finally, the post-rift phase in the depocenter is characterized by the red sandstones, mudstones, and tuffs of the Río Blanco Formation. This succession (up to 1000 m) has an onlap relationship with the underlying beds and represents the instauration of a fluvial-deltaic system over the lacustrine black shales. The Barrancas Formation (100 m) completely developed in subsurface, is composed of distal alluvial to fluvial facies stacked in amalgamated bodies developed under a semiarid climate [51]. It unconformable overlies the Río Blanco Formation and is overlain by Middle - Late Jurassic basalts of Punta de las Bardas Formation.

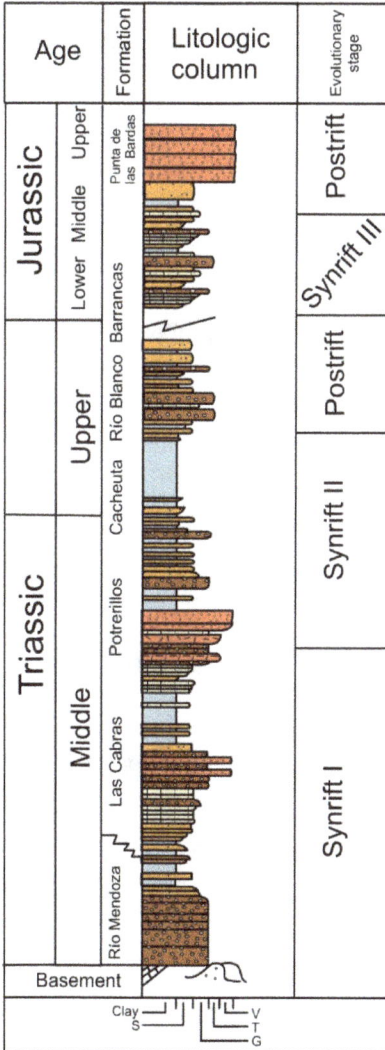

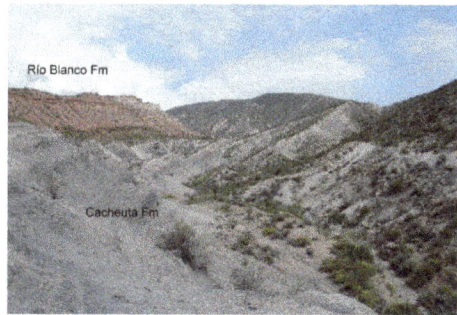

Figure 10. Proposed stratigraphic column of the Cacheuta sub-basin infilling showing the rifting episodes recognized in this work for the whole Cuyana Basin. Modified from Kokogian [13]. To the right, detail photograph of the Cacheuta lacustrine facies and the transition to the postrift facies of Rio Blanco Formation.

Paleofloristic analysis on the macrofloras preserved nearly in the whole column indicate the presence of subtropical floras adapted to seasonally dry climatic conditions. Moreover, the flora preserved in the Potrerillos and Cacheuta beds are evergreen forests in contrast with that represented in the ovelying Rio Blanco Formation that is characterized by a deciduous forest. This change indicated a climatic shift to dryer conditions during the deposition of the fluvial redbeds of the Rio Blanco unit e.g. [27].

Recently, U-Pb SHRIMP ages on tuffaceous beds of the top of Cerro de las Cabras and the base of the Potrerillos formations have constrained the initial infilling of the Cacheuta subbasin (Synrift I) and the beginning of the Synrift II to the early Anisian and the Anisian-Ladinian boundary, respectively [29, 33]. This age is coincident with that recently obtained from the lacustrine beds at the top of the first synrift stage in the northernmost exposures of the Cuyana Basin at Puntudo [34].

6. Geodynamic and tectonosedimentary evolution of the Cuyana Basin

The depositional sequences of the Cuyana Basin resulted from the complex interaction of the supply of sediments, the availability of the accommodation space (both with tectonic components), sea level variations and climate variations. The main control on basin geometry was the deformation field resulted from the tectonic activity along the margin of Gondwana. In this sense, fault geometry provided the final morphology of the trough and its resulting subsidence, which in term controlled the sedimentation rate, grain size, channel migration, avulsion episodes and the development of lacustrine environments.

The Gondwana continental margin has been created as a result of the accretion and amalgamation of different continental blocks and allochthonous terrains to the South American proto-margin since the Proterozoic times up to even the Early Paleozoic e.g. [52 to 55] (Figure 11). This tectonic cycle terminated as a consequence of the Sanrafaelic orogeny [56] with the formation of a magmatic arc in the Late Carboniferous-Early Permian times [58, 57]. Following this deformational event and during Middle Permian to Early Triassic times, an important silicic magmatism episode occurred associated with an extensional period. It was named Choiyoi Group [58] and was interpreted as the result of the collapsed of the Permian orogen e.g. [36, 4, 38, 59] probably because of the declining and/or cease of the long-lived subduction of the Panthalassan margin [59].

Additionally, and on the basis of the strain analysis, several authors suggested that these extensional forces could have been related to the beginning of the Pangea breakup e.g. [36, 60, 61]. Zerfass [62], proposed a regional uplift due to oblique compression for Lower to Middle Triassic times which is somehow in agreement with the geodynamic contest envisioned by Llambías [63] for the basement of the Neuquén Basin, except for the thermomechanical parameters that could have triggered this upwelling, for these latter authors suggest the probable influence of a thermal rise and/or slab brake off. According to Zerfass [62] scheme, the basal Talampaya and Tarjados formations, corresponding to the Ischigualasto Basin (see Figure 1), would constitute together a second-order sequence and could be separated of the tectonic history of the basin. By MiddleTriassic - Early Jurassic times however, the inception of a new juvenile magmatic arc associated with the renewal of the subduction processes, added extra intraplate extensional forces [63 to 65]. Consequently, rapidly subsiding, fault-bounded, narrow back arc-related troughs were formed and arranged in an *échelon* pattern like the Bermejo, Ischigualasto-Villa Unión-Marayes and Cuyana, basins (see Figure 1) e.g. [11, 32, 66]. Charrier [67], Criado Roque [68], Spalletti [26] among others, and more recently, Milani & De Wit [69] have proposed a transtensional origin for Triassic basins related to a sinistral shear zone along in the western plate margin. According to the latter authors, this transtensional regime could be related with the geodynamic processes occurring in a still active Gondwanides orogen. By Early to Middle Jurassic times subduction was completely restored and a huge extensional regime were developed thereafter in the backarc [3, 70, 71].

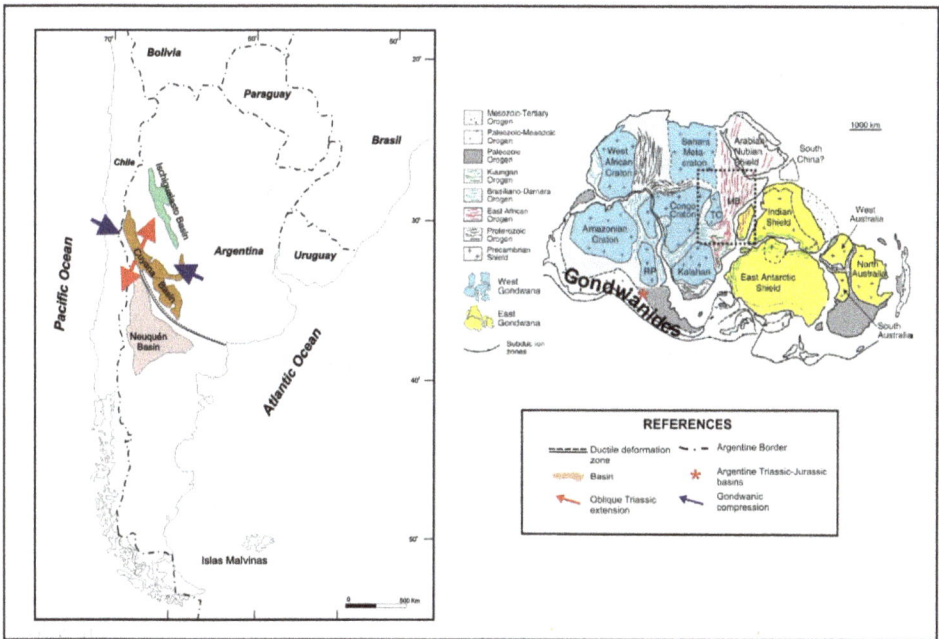

Figure 11. Regional Map of the distribution of the Triassic-Jurassic basins of Argentina. Arrows indicate stress direction, in blue main compressional regime during Upper Carboniferous-Lower Permian and in red oblique extensional regime during Middle to Upper Triassic times. Right, pre-break-up Gondwana configuration taken from Corti [82]. See red asterisk for basin location.

The Cuyana Basin is a NW trending rift whose Triassic record reaches 3700 meters [9]. Its origin has been associated with an oblique extensional regime (Az 35°) [9, 72] which reactivated the northwest-striking lithosferic weakness related to the ancient Gondwanan sutures e.g. [3, 53, 73] (Figure 11). In this scenario, work hardening of the lithosphere by repeated deformational events during the Palaeozoic along the Gondwanides led to an increase in the lithospheric strength and together with the obliquity of the extensional regime, gave place to a north to south trend o basin formation. Accordingly, the Bermejo Basin and

Ischigualasto-Villa Unión-Marayes basin opened first in Early-Late Triassic [74] followed by the Cuyana Basin, during the Middle Triassic [32] and south-westward in northern Patagonia, the Neuquén Basin during the Late Triassic–Early Jurassic times e.g. [63]. Similar extensional events have been recognized along the cratonic areas of southern Brazil by Hackspacher [75] and Zerfass [62].

An oblique subduction along the Panthalassan margin has been proposed by Martin [76] for the Permian–Triassic times which account for the obliquity of the extension as lateral upper crust shear. Considering that the Gondwanides basement is composed of a series of different crustal domains with distinct mechanical characteristics [77] and which together are different from that of the cratonic region, the stress transmission from the this belt to the backarc and foreland regions could have had shear components with different magnitude along strike. In this sense, several authors like Llambías & Sato [59, 38, 78], among others, proposed that during Permian times continental crust block rotation along the Gondwana margin took place leading to a reorganization of plate boundaries and plate kinematics, especially along the Triassic.

In the particular case of the Cuyana Basin, Middle Triassic extension took place under a relative high thermal flux of about 70 mv/m^2 [31, 39] which affected a normal to thinned crust of 30 km thick. Magmatism was due to decompression and melting of the asthenospheric mantle driven by intraplate stresses and lithospheric thinning and probable by the influence of a thermal rise during slab brake off [63, 79]. It constituted an alkaline bimodal suite initially acidic because of crustal melting, and then basaltic and probably sourced from the base of the lithosphere while subduction was in progress by Middle to Upper Triassic [39].

Faults reflect the historical sequence of changing stress regime during Triassic times in the Gondwana margin. Some of these structures reactivated ancient preexisting weakness zones like the Paleozoic sutures within the Gondwamides. In the beginning of extension the limited crustal extension led to the formation of a series of restricted half-grabens separated by transfer faults which apparently follow a the Paleozoic fabric e.g. [9, 12, 37] or intrabasinal highs. Extension persisted up to Jurassic (Cretaceous?) times with at least three remarkable pulses of fault reactivation which are well preserved in the Rincón Blanco subasin; while in the Cacheuta depocenter there are still doubts. The Barrancas Formation has been considered of Late Triassic age by Rolleri & Garrasino [40] using cyclostratigraphic concepts or Early to Middle Jurassic age by Regairaz [41] on the basis of field relationships. The existence of Late Jurassic to Early Cretaceous basalts (Punta de las Bardas Formation) over unconformable overlying this unit made several authors to considered Barrancas Formation as Jurassic. In any case, enhanced extension during westward migration of the magmatic arc [63] and the final insertion of the subduction complex with negative trench roll back and probable extra thermal flux governed the deposition of this unit and de evolution of Punta de las Bardas basalts. Further north, the Marachemill Unit could be considered the initial signs of this renewed extensional event of the Gondwana margin.

The Cuyana Basin was filled with a tectonically induced second-order thick pile of continental deposits arranged into three third-order sequences. These are composed of alluvial, fluvial, lacustrine, deltaic and pyroclastic deposits [1, 32]. The switch from major fluvial to lacustrine environments along the basin is interpreted to reflect a change in tectonic activity. These third-order depositional sequences can be correlated with those of the Santa María Supersequence (Paraná Basin-Southern Brazil) [62].

Depocenters in the Cuyana Basin are half-grabens roughly triangular in cross section. The border fault (BF) consists of a network of mainly normal to oblique-slip faults which are in plan view soft linked and separated by strike-slip minor faults (Cacheuta and Rincon Blanco/Santa Clara) or by breached relay ramps (Rincon Blanco-Puntudo). Only between the southern and northern segments are inversion of segment polarities. Fault-displacement folds were formed and locally influenced sedimentation, with synrift units thickening in the synclinal lows and thinning onto the highs in the footwalls. This situation produced the wedge-shaped sedimentary units that can be traced through the depocenters in the field and subsurface. The high dip angle of the border faults controlled the significant amount of throw, especially of the Cerro Amarillo area (Rincón Blanco depocenter) with an initial infilling of 1200 meters. The footwall of the border faults were uplifted in response to absolute upward motion coupled with the isostatic unloading. In the particular case of the Rincón Blanco depocenter this topographic high, with the slope to the east, gave place to the erosion of the basal Choiyoi deposits, so common in the Puntudo and Cacheuta depocenters. These shoulders prevented sediment inflow and streams entered the basin along the hinged margin but also axially mostly sourced by the transfers (Agua de Los Pajaritos and Barreal formations). The rift was not connected with the sea and so eustasy played no role in the basin infilling evolution, drainage systems were in fact controlled by local (mostly lake) base levels. In this scenario, the relationships among incremental accommodation space (mostly associated with tectonic subsidence), sediment+water supply and short-time climatic influences determined which depositional system predominated. When the sediment supply exceeded the incremental accommodation space (basin overfilled), during first stages of rifting, alluvial to braidplain deposition predominated with strata progressively onlapping the hanging-wall (Cerro Puntudo Formation, Ciénaga Redonda and Río Mendoza formations). These deposits occupied a narrow band close to the main faults and transfers, and were coeval with the latest tholeitic magmatism (Choiyoi Group) from a Cisuralian Magmatic arc [33,39].

The gradual faults growth in length and displacement drove depocenters to increase in depth, length, and width and thus, the incremental accommodation space through time. In this way, they were underfilled with small drainage systems entering from footwall region between border faults or from the fault tips where footwall uplift was minimal. The main fluvial systems were mostly axial or longitudinal (Cerro Amarillo, Cerro de Las Cabras and Cerro Puntudo formations). When the incremental accommodation space significantly exceeded the sediment supply and water input, shallow hydrologically closed lacustrine deposition (playa-type) predominated (Rincón Blanco and Cacheuta depocenters) and a carbonate-rich lake in Puntudo depocenter all of them with interbedded tuffs. These lakes were located close to the faults and subjected to climatic base-level fall-to-rise turnarounds and thus show a marked cyclicity. Climate was semiarid and seasonally humid. When the incremental sedimentation rate equalled the accommodation space rate climatic induced or, less probable, tectonic induced (early post-rift?) fluvial systems developed over these lakes. According to U-Pb zircon dating this tectono-sequence can be constrained to the base of the Anisian (early Middle Triassic) [32].

In Middle Triassic (Rift II) a significant reactivation of the faults gave place to the development of a regional unconformity that can be correlated across the three depocenters. It has been interpreted as being an intra-Triassic tectonic event, probably associated with the acceleration of the subduction rate on the west margin as a consequence of the reinsertion of the convergence [63]. It was responsible of the high sediment supply that gave place to the deposition of the alluvial-fluvial facies of El Relincho, Panul and the base of Potrerillos formations (Synrift II initiation). Fluvial to deep lacustrine facies developed when the accommodation space rate succeeds the sedimentation rate in a basin that widened significantly as it is suggested by numerous onlaps observed in all the troughs. Climate was sub-tropical and seasonally dry and lakes were balanced and deep (meromictic) with deltaic and fluvial deposition occurring around the margins. Climatic induced lake-level cycles are clearly seen in the stacked parasequences bound by lacustrine flooding surfaces which are superimposed on variable rates of subsidence of the rift border fault zone. Upward, the lacustrine–fluvial transition resulted from a decrease in incremental accommodation space and/or an increase in sediment supply once extension slowed in the post-rift stage (Casa de Piedra, Cortaderita and El Alcázar formations). The shoaling tendency of the lakes during this stage was due mainly to the basin growth which forced to spread the water over broader regions, drowning extended regions with the development of palustrine environments. Important ash fall tuffs and reworked volcaniclastics interfinger these postrift facies along the whole basin. This second tectono-sequence corresponds to the Anisian (upper Middle Triassic [32].

Another extensional event took place in Early Late Triassic (Rift III) [32]. At present, best exposures associated with this event are found in the Rincón Blanco Subbasin (Cepeda and Marachemill formations) [1, 32]. In the Cacheuta depocenter a tectosedimentary similar unit is represented by the Barrancas Formation, firstly assigned to Upper Triassic by [10]. However and on the basis of field relationships, Regairaz [41] proposed a Lower to Middle Jurassic age. Presently, there is no absolute dating for this unit but several authors like Rolleri & Fernandez Garrasino [40] using the sequence stratigraphy concepts could demonstrate its Triassic affinities. On a regional scale, the lack of Jurassic - Cretaceous strata in the northern portion of the Cuyana Basin (Puntudo and Rincón Blanco depocenters) likely suggests that it stopped subsiding while the southern portion of the basin were still active contemporaneously with the opening of the Neuquén basin with the deposition of the Barrancas and Punta de las Bardas formations [32]. Yet, it is worth mentioning that Jurassic levels have been described in nearby latitude in the Precordillera by Coughlin [80] and Milana et al. [81] which suggest that further studies are needed. In Southernmost Brazil and Uruguay, the Triassic Sanga do Cabral Supersequence was divided in three sequences [62] that can be also correlated with these stages. Proximal fans and braided ephemeral to perennial streams of semiarid, seasonally humid (?) climatic regime developed as a consequence of the lost of accommodation space and a high sediment supply. This tectono-sequence was assigned an early Carnian age (lower Late Triassic) in the Rincón Blanco trough by Barredo [32] but taking into account the evolution of the Barrancas Formation and Punta de las Bardas Basalts, of the southern Cacheuta depocenter, it could have spanned up to Jurassic (Cretaceous?) times. Subduction was definitely restored during upper Late Triassic-Lower Jurassic under extensional conditions associated with the westward migration of the arc e.g. [63]. The new extensional regime would have governed the evolution of the upper section of the Cuyana Basin whose final thermal relaxation should have occurred during Middle-Late Jurassic and Cretaceous times and aborted during Tertiary times when the lithosphere undergone flexural subsidence induced by the Andean orogenic overloading and by sediment charge.

7. Conclusion

The Gondwana continental margin has been created as a result of the accretion and amalgamation of different continental blocks and allochthonous terrains to the South American proto-margin during Paleozoic. In this scenario, Triassic continental rifts of Argentina settled sometimes controlled by the inherited fabric. Hence, a deep geodynamic analysis is necessary if we want to understand the evolutionary history of these basins. The uplifting and subsidence of different portions of the crust will control the insertion and development of sedimentary basins and the regional distribution of their infilling. In the Cuyana Basin, depositional sequences resulted from a complex interaction of the supply of sediments, the availability of the accommodation space (both with tectonic components), sea level variations (which also may have a significant tectonic influence) and climate variations, being the first-order control on basin geometry the deformation field resulted from the tectonic activity. Its origin has been associated with an oblique extensional regime which in some cases reactivated the northwest-striking lithosferic weakness related to the ancient Gondwanan sutures. The resulting sedimentary environments display significant differences across subbasins which made bio and lithostratigraphic correlations quite imprecise. Still, recent isotopic dating permitted to arrive to an enhanced evolutionary model when combined with thermo-mechanical analysis, cyclostratigraphy and tectonostrigraphic tools. A more precise model of the Cuyana Basin evolution now predicts that it was filled with a tectonically induced second-order thick pile of continental deposits arranged into three third-order sequences. These are composed of alluvial, fluvial, lacustrine, deltaic and pyroclastic deposits separated by key stratigraphic surfaces or sequence boundaries resulting from lacustrine flooding and/or forced regressive surfaces. The Synrift I contains volcaniclastic, alluvial-fluvial and saline lake deposits when accommodation exceeded sediment and water input. The Synrift II starts with alluvial–fluvial followed by deep lacustrine and deltaic environments under wetter climatic conditions. The Synrift III encompasses volcanic and volcaniclastic alluvial and fluvial dominated environments developed under semiarid conditions. These three stages are associated with the evolution of the Gondwana margin during Permo-Triassic times, when lithosphere undergone a wide spread extensional regime due to the almost cease of subduction, the collapse of the ancient Permian orogen and the reorganization of intraplate stresses during Pangea breakup (Rift I and II). The inception of a new juvenile magmatic arc associated with the renewal of the subduction processes, added extra intraplate extensional forces which gave place to a third regional scale event, the Rift III, which begun in Late Triassic and extended up to Middle Jurassic. The final thermal relaxation took place during Cretaceous when lithosphere undergone flexural subsidence induced by the Andean uplift.

Author details

Silvia Patricia Barredo

Department of Petroleum Engeneering, Instituto Tecnológico de Buenos Aires (ITBA), Buenos Aires, Argentina

Acknowledgement

The author wants to acknowledge discussions specially with Dr Victor Ramos and Don Dr Pedro Stipanicic over the development of her PhD Thesis. My views from the Triassic systems highly benefited from their critical observations.

My particular thanks to Laura Giambiagi and Maisa Tunik for their assistance in the field, to Claudia Marsicano and Guillermo Ottone for sharing their paleontological knowledge and for providing great inputs to this study, and Luis Stinco for his critical view about the manuscript. To all my colleagues at conferences and meetings about extensional tectonics and hydrocarbon exploration. A great part of this study has been supported by UBACYT X182: 2008-2010 (VAR.) and PIP CONICET 5120 (EGO.). Additional financial support was provided by the Consejo Nacional de Investigaciones Científicas y Técnicas (CONICET). Finally, I want to highlight the great support from the Instituto Tecnológico de Buenos Aires (ITBA) to my research and for providing the means to perform it through "Proyecto ITBA-70" *Geodinámica de cuencas productivas y de potencial generador"*.

8. References

[1] Barredo SP, Ramos VA (2010) Características tectónicas y tectosedimentarias del Hemigraben Rincón Blanco: Una Síntesis. Revista de la Asociación Geológica Argentina 65 (1): 133-145.

[2] Allen PA, Allen JR (2005) Basin Analysis: principles and applications. Second edition. Blackwell Scietific Publication, 549 p., Oxford.

[3] Uliana MA, Biddle KT (1988) Mesozoic-Cenozoic paleogeographic and paleodynamic evolution of southern South America. Revista Brasileira de Geociências, 18: 172-190.

[4] Zeil W (1981) Volcanism and geodynamics at the turn of the Paleozoic to Mesozoic in the Central and Southern andes. Zentralblatt für Geologie und Paläontologie 1(3/4): 298-318. Stutgart.

[5] Keidel J (1916) La geología de las Sierras de la Provincia de Buenos Aires y sus relaciones con las montañas de Sudáfrica y Los Andes. Ministerio de Agricultura de la Nación, Sección Geología, Mineralogía y Minería, anales, XI (3): 1-78.

[6]   duToit AL (1937) Our wandering continents. Oliver and Boyd: 336 p. Edinburgh.

[7]   Legarreta L, Kokogian DA, Dellape D (1993) Estructuración terciaria de la Cuenca Cu-
      yana: ¿Cuánto de inversión tectónica? Revista de la Asociación Geológica Argentina 47:
      83-86.

[8]   Barredo SP (2005) Implicancias estratigráficas de la evolución de las fallas normales del
      hemigraben Rincón Blanco, cierre norte de la cuenca Cuyana, provincia de San Juan. In:
      Cazau LB, editor. Electronic proceedings of the VI Congreso de Exploración y Desarro-
      llo de Hidrocarburos: 9 p.

[9]   Barredo SP (2004). Análisis estructural y tectosedimentario de la subcuenca de Rincón
      Blanco, Precordillera Occidental, provincia de San Juan. [PhD Thesis]. Buenos Aires,
      Argentina. Unpublished: 325 p.

[10] Rolleri EO, Criado Roque P (1968) La cuenca triásica del norte de Mendoza.
      Proceedings of the III Jornadas Gelógicas Argentinas (Comodoro Rivadavia, 1966) 1: 1-
      79.

[11] Ramos V.A., Kay S.M. (1991) Triassic rifting and associated basalts in the Cuyo Basin,
      central Argentina. In: Harmon, R.S. Rapela C.W., editors. Andean magmatism and its
      tectonic setting. Geological Society of America Special Paper 265, 79-91.

[12] Ramos VA (1992) Control geotectónico de las cuencas triásicas de Cuyo. Boletín de
      Informaciones Petroleras, 5: 2-9.

[13] Kokogian DA, Seveso FF, Mosquera A (1993) Las secuencias sedimentarias triásicas. In:
      Ramos VA, editor. Relatorio XII Congreso Geología Argentina and II Congreso de Ex-
      ploración de Hidrocarburos, Geología y Recursos Naturales de Mendoza (Buenos Ai-
      res): 65-78.

[14] Groeber P, Stipanicic PN (1953) Triásico. In: Groeber P, editor. Mesozoico. Geografía de
      la República Argentina: 1-141 (Sociedad Argentina de Estudios Geográficos, GAEA, 2.
      Buenos Aires.

[15] Borrello A.V., Cuerda A.J. (1965) Grupo Rincón Blanco (Triásico San Juan). Comisión de
      Investigaciones Científicas. Provincia Buenos Aires, Notas: 2 (10):3-20. La Plata.

[16] Quartino BJ, Zardini RA, Amos AJ (1971) Estudio y exploración geológica de la Región
      de Barreal-Calingasta. Provincia de San Juan, República Argentina. Asociación Geológi-
      ca Argentina, Monografía 1: 184 p.

[17] Stipanicic PN. (1972) Cuenca triásica de Barreal. In: Leanza AF, editor. Geología
      Regional Argentina, Academia Nacional de Ciencias de Córdoba: 537-566.

[18] Stipanicic PN (1979) El Triásico del Valle del río Los Patos (Provincia de San Juan).
      Segundo Simposio de Geología Regional Argentina, Academia Nacional de Ciencias de
      Córdoba, II: 695-744.

[19] López-Gamundí OR (1994) Facies distribution in an asymmetric half-graben: the north-
      ern Cuyo basin (Triassic), western Argentina. 14th International Sedimentological Con-
      gress, Abstracts: 6-7. Recife

[20] Spalletti LA (1999) Cuencas triásicas del Oeste Argentino: origen y evolución. Acta
      Geeológica Hispánica, 32 (1-2)(1997): 29-50.

[21] Barredo SP, Stipanicic PN (2002) El Grupo Rincón Blanco. In: Stipanicic PN, Marsicano
      CA, editors. Léxico Estratigráfico de la Argentina III: 870 p.

[22] Yrigoyen MR, Stover LW (1969). La palinología como elemento de correlación del Triásico en la cuenca Cuyana. proceedings of the IV Jornadas Geológicas argentina 2: 427-447. Buenos Aires.

[23] Strelkov EE, Alvarez LA (1984). Análisis estratigráfico y evolutivo de la cuenca triásica mendocina - sanjuanina. Proceedings of the IX Congreso Geología Argentina III: 115-130. Bariloche.

[24] Sessarego HL (1988) Estratigrafía de las secuencias epiclásticas devónicas a triásicas, aflorantes al norte del río San Juan y al oeste de las sierras del Tigre, Provincia de San Juan. [PhD Thesis]. Buenos Aires, Argentina. unpublished: 330 p.

[25] Stipanicic PN, Bonetti MIR (1969) Consideraciones sobre la cronología de los terrenos triásicos argentinos. Proceedings of the I Simposio Internacional Estratigrafía y Paleontología del Gondwana, Mar del Plata. UNESCO, Ciencias de la Tierra 2: 1081-1120. París.

[26] Spalletti LA (2001) Modelo de sedimentación fluvial y lacustre en el margen pasivo de un hemigraben: el Triásico de la Precordillera occidental de San Juan, República Argentina. Revista de la Asociación Geológica Argentina 56 (2): 189-210.

[27] Artabe AE, Morel EM, Spalletti LA (2001). Paleoecología de las de las floras triásicas argentinas. In: Artabe AE, Morel EM, Zamuner AE, editors. El Sistema Triásico en la Argentina. Fundación Museo de La Plata "Fracisco P. Moreno": 199–225. Argentina.

[28] Morel EM, Artabe AE (1993) Floras mesozoicas. In: Ramos VA, editor. Geología y Recursos Naturales de Mendoza. Proceedings of the XII Congreso Geológico Argentino y II Congreso de Exploración de Hidrocarburos (Mendoza), Relatorio, 2(10): 317-324. Argentina.

[29] Spalletti LA, Fanning CM, Rapela CW (2008) Dating the Triassic continental rift in the southern Andes: the Potrerillos Formation, Cuyo Basin, Argentina. Geologica Acta 6, 267-283.

[30] Marsicano CA, Barredo SP (2004). A Triassic tetrapod footprint assemblage from southern South America: palaeobiogeographical and evolutionary implications. Palaeogeography, Palaeoclimatology, Palaeoecology 203, 313-335.

[31] Zamora G, Cervera M, Barredo SP (2008) Geología y potencial Petrolero de un bolsón intermontano: Bloque Tamberías, Provincia de San Juan. In: Schiuma M, editor. Trabajos técnicos, VII Congreso de Exploración y Desarrollo de Hidrocarburos: 397-407. Argentina.

[32] Barredo SP, Chemale F, Ávila JN, Marsicano C, Ottone G, Ramos VA (2012) U-Pb SHRIMP ages of the Rincón Blanco northern Cuyo rift, Argentina. Gondwana Research (21): 624-636. DOI: 10.1016/J.Gr.2011.05.016.

[33] Ávila JN, Chemale F, Mallmann G, Kawashita K (2006) Combined stratigraphic an isotopic studies of Triassic strata, Cuyo Basin, Argentine Precordillera. Geologial Society of America Bulletin v. 118: 1088-1098. Doi: 101130/B25893.1

[34] Mancuso AC, Chemale F, Barredo SP, Ávila JN, Ottone G, Marsicano C (2010) Age constraints for the northernmost outcrops of the Triassic Cuyana Basin, Argentina. Journal of South American Earth Sciences, 30 (2010): 97-103. doi: 10.1016/J.Jsames.2010.03.001

[35] Stipanicic PN, Marsicano CA (2002). Léxico Estratigráfico de la Argentina, Volumen III: 870 p. Asociación Geológica Argentina. Buenos Aires.

[36] Uliana MA, Biddle KT, Cerdan J (1989) Mesozoic-Cenozoic paleogeographic and geodynamic evolution of southern South America. Revista Brasileira de Geociencias 18: 172-190.

[37] Japas MS, Cortés JM, Pasini M (2008) Tectónica extensional triásica en el sector norte de la Cuenca Cuyana: primeros datos cinemáticos. Revista de la Asociación Geológica Argentina 63 (2): 213-222.

[38] Kleiman LE, Japas MS (2009) The Choiyoi volcanic province at 34°S - 36° S (San Rafael, Mendoza, Argentina): Implications for the Late Paleozoic evolution of the Southwestern margin of Gondwana. Tectonophysics 473: 283-299.

[39] Barredo SP, Martínez A (2008) Secuencias piroclásticas triásicas intercaladas en la Formación Ciénaga Redonda, Rincón Blanco, Provincia de San Juan y su vinculación con el ciclo magmático gondwánico del Grupo Choiyoi. Proceedings of the 12th Reunión de Sedimentología: 91-92. Buenos Aires

[40] Rolleri EO, Fernández Garrasino C (1979) Comarca septentrional de Mendoza. Proceedings of the Simposio de Geología Regional Argentina, Academia Nacional de Ciencias de Córdoba I: 771-809. Argentina. Córdoba.

[41] Regairaz AC (1970) Contribución al conocimiento de las discordancias en el área de las Huayquerías, Mendoza, Argentina. Proceedings of the IV Jornadas Geológicas Argentinas, Mendoza, 1969, 2: 243-254.

[42] Mombrú CA (1973) Observaciones geológicas en el Valle de Calingasta-Tocota. Provincia de San Juan. YPF Unpublished: 50 p. Buenos Aires.

[43] Japas MS, Salvarredi, J, Kleiman LE (2005) Self-similar behaviour of Triassic rifting in San Rafael, Mendoza, Argentina. Proceedings of the Gondwana 12: 210. Argentina.

[44] Rossa N, Mendoza N (1999) Manifestaciones volcánicas en la cuenca triásica de Barreal Calingasta, San Juan. Proceedings of the XIV Congreso Geologico Argentino II: 171 - 174. Salta, Argentina.

[45] Martínez A, Barredo SP, Giambiagi L (2006) Modelo Geodinámico para la evolución magmática Permo-Triásica entre Los 32º Y 34º LS, Cordillera Frontal de Mendoza, Argentina. Proceedings of the XIII Reunión de Tectónica - San Luis 4 p. ISBN 978-9871031-49-8.

[46] Barredo SP, Tunik M, Pettinari G, Giambiagi L, Zamora G (2010) A new stratigraphic synthesis of the Agua de Los Pajaritos depocenter, flexural margin of the Rincón Blanco Subbasin. Proceedings of the 18Th International Sedimentological Congress.

[47] Baraldo JA, Guerstein PG (1984) Nuevo ordenamiento estratigráfico para el Triásico de Hilario (Calingasta, San Juan). Proceedings of the 9th Congreso Geológico Argentino 1: 79-94. San Juan. Argentina.

[48] Bonati S, Barredo SP, Zamora Balcarce G, Cervera M (2008) Análisis tectosedimentario preliminar del Grupo Barreal, cierre norte de la Cuenca Cuyana, provincia de San Juan. In: Schiuma M, editor. Trabajos Técnicos, VII Congreso de Exploración y Desarrollo de Hidrocarburos: 409-420. Mar del Plata. Argentina.

[49] Massabie AC (1986) Filón Capa Paramillos de Uspallata, su caracterización geológica y edad, Paramillos de Uspallata, Mendoza. Primeras Jornadas sobre Geología de Precordillera. Asociación Geológica Argentina Serie A (2): 71-76. Buenos Aires.

[50] Dellape DA, Hegedus AG (1993) Inversión estructural de la cuenca Cuyana y su relación con las acumulaciones de hidrocarburos. Proceedings of the XII Congreso Geología Argentina y II Congreso de Exploración de Hidrocarburos III: 211-218, Mendoza, Argentina.

[51] Zencich S, Villar HJ, Boggetti D (2008) Sistema petrolero Cacheuta-Barrancas de la Cuenca Cuyana, provincial de Mendoza, Argentina. In: Cruz CE, Rodríguez JF, Hetchem JJ, Villlar J, editors. Sistemas petroleros de las cuencas andinas. VII Congreso de Exploración y Desarrollo de Hidrocarburos: 109-134. Argentina.

[52] Ramos VA (1988) The tectonics of the Central Andes, 30 to 33 S latitude. In: Clark S, Burchfiel D. editors. Process in Continental Lithospheric Deformation, Geological Society America, Special Paper 218: 31-54. Boulder.

[53] Ramos VA (2000) The Southern central Andes. In: Cordani UG, Milani EJ, Thomaz Filho A, Campos DA, editors: Tectonic Evolution of South America. Proceedings of the 31 [st] International geological Congress: 561-604. Brasil.

[54] Dalla Salda, de Barrio R, Echeveste H, Fernández, R (2005) El basamento de las Sierras de Tandilia. In: de barrio R, Etcheverry R, Caballé M, LLambías, E, editors. Geología y Recursos Minerales de la Provincia de Buenos Aires. XVI Congreso geológico Argentino. Relatorio: 31-50. La Plata, Argentina.

[55] Pankhurst RJ, Rapela CW, Saavedra J, Baldo E, Dahlquist J, Pascua I., Fanning CM (1998). The Famatinian magmatic arc in the central Sierras Pampeanas: an Early to Mid-Ordovician continental arc on the Gondwana margin. In: Pankhurst RJ, Rapela CW, editors. The Proto-Andean Margin of Gondwana. Geological Society, London, Special Publications, 142:343–367.

[56] Caminos R, Azcuy CL (1996) Tectonismo y diastrofismo. 3. Fases diastróficas neopaleozoicas. In: Archangelsky S, editor. El sistema Pérmico en la República Argentina y en la República Oriental del Uruguay. Academia Nacional de Ciencias: 265-274. Córdoba, Argentina.

[57] Relledo S, Charrier S (1994) Evolución del basamento paleozoico en el área Punta Claditas, región de Coquimbo, Chile, (31°-32° S). Revista Geológica de Chile 21 (1): 55-69.

[58] Stipanicic PN, Rodrigo F, Baulies OL, Martinez C G (1968) Las formaciones presenonianas en el denominado Macizo Nordpatagónico y regiones adyacentes. Revista de la Asociación Geológica Argentina 23: 76-98..

[59] Llambías E, Sato A (1995).El batolito de Colangüil transición entre orogénesis y anorogénesis. Revista de la Asociación Geológica Argentina 50 (1-4): 111-131.

[60] Charrier R, Pinto L, Rodríguez MP (2007). Tectonostratigraphic evolution of the Andean Origen in Chile. In Moreno T, Gibbons W, editors. The Geology of Chile. The Geological Society: 21-114, London.

[61] Giambiagi L, Bechis F, García V, Clark A (2009). Temporal and spatial relationship between thick- and thin-skinned deformation in the thrust front of the Malargüe fold and thrust belt, Southern Central Antes. Tectonophysics 459: 123-139.

[62] Zefrass H, Chemale FJr, Schultz C L, Lavina E (2004). Tectonics and sedimentation in Southern South America during Triassic. Sedimentary Geology, 166: 265 – 292.

[63] Llambías EJ, Leanza HA, Carbone O (2007) Evolución Tectono-magmática durante el Pérmico al Jurásico temprano en la cordillera del Viento (37°05′S – 37°15′S): Nuevas evidencias geológicas y geoquímicas del inicio de la Cuenca Neuquina. Revista de la Asociación Geológica Argentina, 62 (2): 217-235.

[64] Hervé F, Fanning CF (2001) late Triassic detrital zircons in meta-turbidites of the Chonos Metamorphic Complex, Southern Chile. Revista Geológica de Chile 28(1): 98-104. Chile.

[65] Rapela CW, Pankhurst RJ, Fanning CM, Greco LE (2003). Basement evolution of the Sierra de la Ventana Fold Belt: new evidence for Cambrian continental rifting along the southern margin of Gondwana. Journal of the Geological Society, London, 160: 613-628.

[66] Rincón MF, Barredo SP, Zunino J, Salinas A, Reinante SME, Manoni R (2011). Síntesis general de los bolsones intermontanos de San Juan y La Rioja. In: Kowlowsky E, Legarreta L, Boll A, editors. Cuencas Sedimentarias Argentinas. XIII Congreso de Exploración y Desarrollo de Hidrocarburos: 321-406. Mar del Plata.

[67] Charrier R (1979) El Triásico de Chile y regiones adyacentes de Argentina. Una reconstrucción paleogeográfica y paleoclimática. Comunicaciones 26: 1-37, Santiago de Chile.

[68] Criado Roque P, Mombrú CA, Ramos VA (1981). Estructura e interpretación tectónica. In Yrogoyen M, editor. VIII Congreso Geología Argentina, Geología y Recursos Naturales de la Provincia de San Luis. Relatorio: 155-192, San Luis.

[69] Milani EJ, De Wit MJ (2008) Correlations between the classic Paraná and Cape Karoo sequences of South America and southern Africa and their basin infills flanking the Gondwanides: du Toit revisited. Geological Society, London, Special Publications 294: 319-342.

[70] Somoza R (1996) Geocinemática de América del Sur durante el Cretácico: Su relación con la evolución del margen pacífico y la apertura del Atlántico Sur. Proceedings of the XIII Congreso Geológico Argentino 2: 401–402.Buenos Aires, Argentina.

[71] Mpodozis C, Ramos VA (1990) The Andes of Chile and Argentina. In: Ericksen GE, Cañas Pinochet MT, Reinemud JA, editors. geology of the Andes and its relation to hydrocarbon and mineral resources. Circumpacic Council for Energy and Mineral resources, Earth Sciences Series11:59-90. Houston.

[72] Barredo SP, Stinco LP (2010) Geodinámica de las cuencas sedimentarias: Su Importancia en la localización de sistemas petroleros en Argentina. Revista Petrotecnia. Instituto Argentino del Petróleo y del Gas (IAPG) (2): 48-68.Argentina.

[73] Ramos VA (2009) Anatomy and global context of the Andes: Main geologic features and the andean orogenic cycle. In: Ka, SM, Ramos VA, Dickinson W, editors. Backbone of the Americas: Shallow Subduction, Plateau Uplift, and Ridge and Terrane Collision, Geological Society of América, Memoir 204, p. 31-65.

[74] Milana JP, Alcober O (1994) Modelo tectosedimentario de la cuenca triásica de Ischigualasto (San Juan, Argentina). Revista de la Asociación Geológica Argentina 49: 217-235.

[75] Hackspacher PC, Ribeiro LFB, Ribeiro MCS, Fetter AH, Hadler Neto JC, Tello CES, Dantas EL (2004) Consolidation and Break-up of the South American Platform in

Southeastern Brazil: Tectonothermal and Denudation Histories. Gondwana Research 7 (1), 91–101.

[76] Martin MW, Kato TT, Rodríguez C, Godoy E, Duhart P, Mc Donough M, Campos A (1999) Evolution of the late Paleozoic accretionary complex and overlying forearc-magmatic arc, south central Chile (38°-41°S): constraints for the tectonic setting along the southwestern margin of Gondwana. Tectonics, 18 (4): 582-605.

[77] Vaughan APM, Leat PT, Pankhurst RJ, editors (2005) Terrane Processes at the margins of Gondwana. Geological Society, London, Special Publications, 246.

[78] Visser JNJ, Praekelt HE (1998). Late Palaeozoic crustal block rotations within the Gondwana sector of Pangea. Tectonophysics 287, 201–212.

[79] Pankhurst RJ, Rapela C, Fanning CM, Márquez M (2006) Gondwanide continental collision and the origin of Patagonia. Earth-Science Reviews 76, 235–257.

[80] Coughlin TJ (2000). Linked orogen-oblique fault zones in the Central Argentine Andes: Implications for Andean orogenesis and metallogenesis. [PhD Thesis]. University of Queensland, Queensland. Unpublished: 268 p.

[81] Milana JP, Bercowski F, Jordan TE (2003) Paleoambientes y magnetoestratigrafía del Neógeno de la Sierra de Mogna, y su relación con la Cuenca de Antepaís Andina. Revista de la Asociación Geológica Argentina, 58 (3): 447-473.

[82] Corti G, Wijk J, Cloetingh, S, Morley C (2007) Tectonic inheritance and continental rift architecture: Numerical and analogue models of the East African Rift system. Tectonics 26, TC6006, doi: 10.1029(2006TC002086

Role of the NE-SW Hercynian Master Fault Systems and Associated Lineaments on the Structuring and Evolution of the Mesozoic and Cenozoic Basins of the Alpine Margin, Northern Tunisia

Fetheddine Melki, Taher Zouaghi, Mohamed Ben Chelbi, Mourad Bédir and Fouad Zargouni

Additional information is available at the end of the chapter

1. Introduction

The Mesozoic and Cenozoic evolution of the northern edge of the African margin (Fig. 1), and particularly the northern Tunisia, fossilized successive paleogeographic and tectonic episodes. In fact, after rifting and extensional periods, which started at the end of the Paleozoic and continued during the Mesozoic [1-6], was settled the Alpine orogeny that results from the convergence movements between the African and Eurasian plates; it is induced by compressive tectonic stresses, beginning at least since the Tertiary intervals and probably the Late Cretaceous [7-24]. This orogeny has induced, on the Mediterranean edges, many mountains chains extend from the Apennines at the East to the Betic Cordilleras at the West.

The various geological works established in northern Tunisia [25-42,18,43-47], north-eastern Algeria [48-50,23] and in the Siculo-Tunisian strait [51-57], demonstrated that the NE-SW inherited fault networks have controlled sedimentation during the Tethyan rifting and have also controlled the structuring of the central and northern Atlas during the successive tectonic events.

This margin of northern Tunisia, including the Tell and the Tunisian furrow domains (Fig. 2), is limited to the East by the Zaghouan master fault, which appears to have effect on the sedimentation since the Jurassic [58,59,33]. It is inherited from NE-SW trending Hercynian master fault networks (Fig. 2) and their conjugate faults [60,61]. These lineaments correspond, from SE to NE, to the Zaghouan fault (ZF), the Tunis-Elles fault (TEF), the

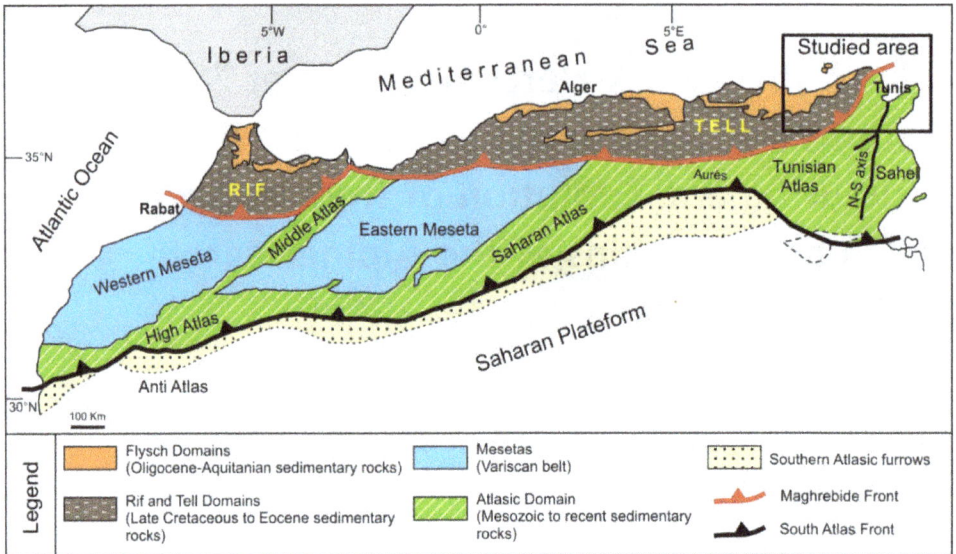

Figure 1. Geological sketch of the Maghreb (modified from Piqué et al. [4] and Frizon de Lamotte et al. [23]).

El Alia-Teboursouk fault (ETF), the Ras El Korane-Thibar fault (RKTF) and the Cap Serrat-Ghardimaou fault (CSGF). Movements of these master faults have effects on the sedimentary deposition and distribution since the Triassic rifting phase up to now.

Our study is mainly focused on northern Atlas (Fig. 2) where the structures tend to be well exposed and on the Tunisian Tell marked by sealed structures. These interpretations will be supported by seismic sections calibrated by petroleum well.

Indeed, based on the structural and paleogeographic zonations, developed during various geological phases in the Tunisian margin, we propose in this paper (i) to demonstrate the implication of the Hercynian deep structures, in the deformation of the Atlas. They have a role in the distribution of the sedimentary basins along the southern Tethyan margin during the Mesozoic and the Cenozoic times; (ii) attempt to clarify the kinematics and chronological relationship between the Tell and the northern Atlas, by proposing a coherent geodynamic model since the Tethyan rifting until the Cenozoic contractional periods, pointing out the role of the NE-SW tectonic lineaments; (iii) to trace a tectonic pattern of the northern Tunisian that contributes to a better comprehension of deformations affecting the Tellian and Atlasic domains in relation with the global geotectonic framework.

2. Stratigraphy

2.1. Triassic

The Triassic outcrops have often abnormal contact with Jurassic, Cretaceous, Paleogene and Neogene series in several localities of northern Tunisia (e.g. Ras El Korane, Jebel Ichkeul,

Jebel Lansarine-Sakkak, Jebel Chehid and Thibar). It is presented either as a diapiric structures [62, 31, 63, 64], or as sole of overthrust folds [32], or also as salt glacier structures [65-69]. The Triassic deposits are characterized by a chaotic aspect; it is consisted of Gypsiferous, argillaceous, carbonated and locally sandy facies, of varied color (Fig. 3).

In the Bizerte area (e.g. Bechateur), the Triassic appears rather stratiform. It is represented by black dolomitic limestones in metric layers becoming centimetric at the top of the succession [18]. The abundant fauna of lamellibranches indicates Carnian-Norian [70, 71]. The subsurface data show carbonated Triassic as well in Utique (W8 well), the Cap Bon (W9 well) and in the Gulf of Tunis (W5 well). The Late Triassic is predominantly evaporitic; its thickness is about 2500m.

2.2. Jurassic

The Jurassic outcrops largely in northern Tunisia [72, 33, 73-78]. Its thickness changes from 700m [76] to 900m [79]. It shows a Tethyan deep sea facies different from the Nara platform Formation defined by Burollet [26] in the Nara outcrop in central Tunisia. It is dominantly calcareous series with some marly intercalations (Fig. 3). In the Ichkeul and Ammar outcrops, the Jurassic shows a thickness series varied from 260m [73] to 480m [80].

At the level of the "Tunisian Dorsale", the lower limit of the Jurassic is not well defined due to the Late Triassic-Lias transition which still not paleontologically characterized [76].

2.3. Early Cretaceous

The Early Cretaceous largely outcrops at the Tunisian furrow [81]. The Jebel Oust lithostrat-igraphic section is considered as the most complete and richest of microfauna and macrofauna of the Tethyan area [82]. This section consists of fossiliferous marls and clays with micritic limestone intercalations and sandy and quartzitic recurrences (Fig. 3). This succession is approximately 2500m thick and its sedimentation has occurred in a subsiding marine environment [83, 82].

In subsurface, in the Gulf of Tunis, the north-eastern extension of the Tunisian through show Early Cretaceous deposits composed of clays, marls and limestones with some sandy layers [24].

2.4. Late Cretaceous

The Late Cretaceous outcropping in the Tunis area [103] is composed from the base to the top by two lithostratigraphic Formations (Fig. 3); the Aleg Formation (Turonian-Coniacian) composed of alternating limestones and marls rich in faunas (100m), marls and clays (50 to 180m). The Abiod Formation (Campanian-Early Maastrichtian), composed by marls and limestones; an argillaceous limestone bar of the zone with Globotruncana arca rugosa; limestone and marl alternations; a limestone bar (80m), marls locally gypsiferous and alternations of marl and limestone.

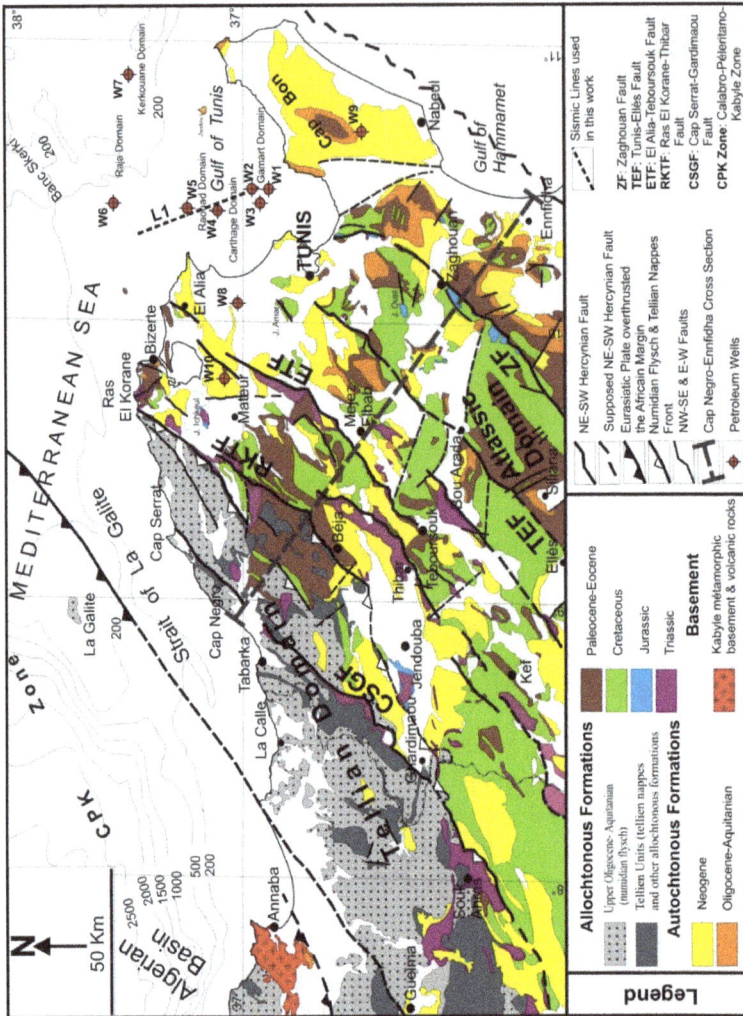

Figure 2. Study area location map of Geologic outcrops, cross section, seismic lines and petroleum wells (geologic outcrops are modified from 1/500.000 geological map of Tunisia; Castany, 1951).

In the Bizerte area, the Abiod Formation outcrop is composed by the first limestones bar attributed to Campanian, the intermediate marl and limestone alternations, the second limestones bar and the upper marl and limestone of Maastrichtian. This Formation makes more than 200m of thickness [18,81].

2.5. Late Maastrichtian-Paleocene

This serie is represented by a monotonous brown argillaceous package, which outcrops at several localities of northern Tunisia. This argillaceous deposits, corresponding to the El

Haria Formation, presents a lithological succession which starts with marl and limestone alternations, then brown to grey clays and finally of marl and grey limestone alternations (Fig. 3).

Compared to the Tunisian furrow, in the Bizerte area, the Tellian Paleocene is characterized by the presence of the "yellow balls". Its thickness is about 250m [18,84]. More to the South, in the Messeftine basin, the Paleocene was recognized at the W9 well, with a thickness of 800m [84]. Towards the West and the North-West, these deposits exhibit more significant thickness.

2.6. Eocene

It rests on the black marls and argillaceous limestones of the El Haria Formation; the Eocene is represented by two different series in northern Tunisia (Fig. 3): (i) a carbonated series of Early Eocene (Ypresian) known by two facies: the Bou Dabbous Formation with Globigerines and the El Garia Formation with Nummulites. In the transition zone, we have a mixed facies; (ii) an argillaceous to marly series of the Souar Formation (Middle Eocene).

In the Bizerte area, the Bou Dabbous Formation includes limestones containing Globigerines and schistous marls with glauconies and phosphates at the base. These marls are followed by relatively massive and bituminous limestones with Globigerines. This Formation is 100m thick [84].

To the West, in the Teboursouk area, the Djebba cross-section, to the West of the Gorraâ outcrop, shows a carbonated series of 70m thickness. It is composed, at the base by a carbonated term (10m) with Globigerines, then an upper carbonated term containing Nummulites.

To the South and the South-west of the Globigerines province, the Nummulites province appears represented, in its typical locality in Kef El Garia, by yellow bio-micritic limestones with Nummulites and other great Foraminifera. It makes 50m of thickness.

In the Bizerte area, the Souar Formation is composed by yellow marls and clays containing some limestones layers and yellow balls, typical of the Tellian series. Its thickness is about 300m [18].

Thickness of the Souar Formation, at Jebel Jebbas is about 500m [43] and becomes 800m thick at the Cap Bon peninsula [85,86].

2.7. Oligocene - Early Miocene

During this period, the Northern Tunisia is marked by two large basins including: a depocenter to the SE filled by the Fortuna Formation (600-800m) and a depocenter to the NW corresponding to the deposition of the Béjoua series [87] and the Oligo-Miocene Numidian Flysch succession (2500m; [88]). These two basins are separated by a "bald" zone [89, 24, 87] lengthened according a NE-SW direction, which seem coincides to the domes and diapirs zone of the Tunisian northern Atlas. The non depositional or erosional zone is inherited from the Eocene period where we have low deep depositional area separating the two basins (Figs. 2 and 3).

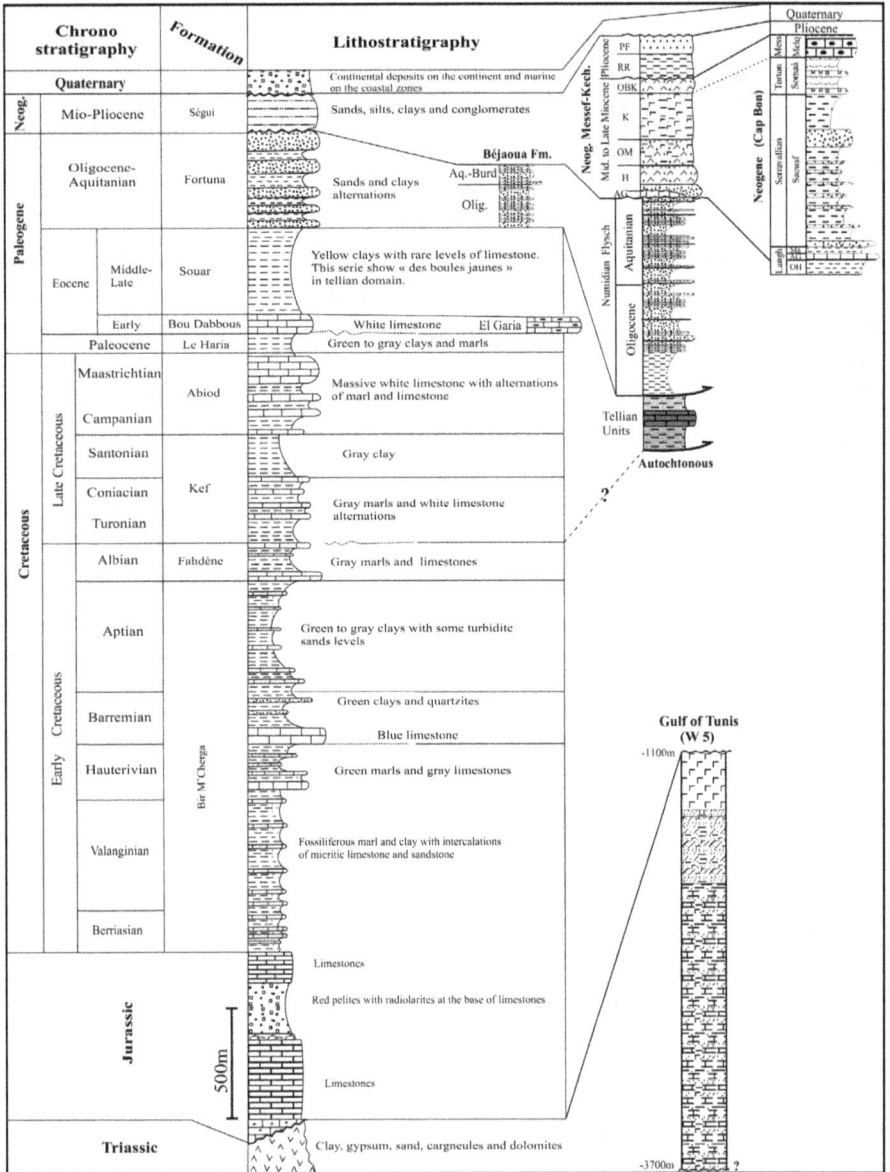

Figure 3. Synthetic column of geological series in northern Tunisia (AG- Ain Ghrab fm; H- Hakima fm; OM- Oued Melah fm; K- Kechabta fm; OBK- Oued Bel Khedim fm; RR- Raf-Raf fm; PF- Porto-Farina fm; OH- Oued El Hammam fm; Ma: Mahmoud fm)

At the West of the domes and diapirs zone (Bejaoua group), the Oligocene and Miocene successions show a stratigraphic continuity with the Eocene series. It contains a clayey and sandy lower part, a sandy intermediate part and a carbonated sandy upper part. Its thickness is approximately 100m [87].

The Oligo-Miocene Numidian Flysch succession, occupies the northwestern part of northern Tunisia. It is a thick allochthonous unit of turbidites including clays at the base and sandstones and shales at the top [32,90,89,46]. This succession consists of five turbiditic units filling the channel complexes with silexites at the top [89,88].

The Fortuna Formation occupies all southeastern basin of northern Tunisia. This Formation includes sandy, clayey and carbonated facies indicating a shallow marine environment [89]. This Formation is 600m thick in the Gulf of Tunis [24] and 800m in the Cap Bon [85,89].

2.8. Middle to Late Miocene

The Middle to Late Miocene series are well developed in the Gulf of Tunis [91,24]. The Middle Miocene starts with the Langhian Aïn Grab lumachellic limestone bars of 19m thick. This unit is marked by conglomerates at the base indicating the beginning of a major transgression, known on the scale of the country [26,85,92-94].

The following strata are marked by an important change facies and geographical distribution. In the Gulf of Tunis offshore area, the Saouaf Formation (Serravallian-Tortonian) and the Oued Bel Khedim Formation (Messinian) present facies and thickness variations [24]. They are composed of clays, silts and occasionally of sandy limestones at the base and marls and salts at the top. The thickness of these two Formations is about 1300m (Figs. 2 and 3).

In eastern Tunisia, the series are represented by the deep marine green clays of the Mahmoud Formation, rich with microfauna. Above, are deposited regressive continental series, resulting from the erosional strata [95] corresponding to the Segui Formation [58]. Whereas, to the coastline and offshore areas, the Beglia and Saouaf Formations (Serravallian) constitute the lateral equivalent of the segui Formation. It is a thick series that reach 1700m and composed of clays, sandstones and lignite alternations characterizing an internal shelf depositional environment [95].

In the Bizerte and Mateur areas, the upper Miocene deposits occupy the foreland basins of the Tellian domain. They fossilized marine, lagunal and detrital environment [26,71,96,18,94] and they are concentrating in five different depocenters corresponding to the Douimis, Jalta, Messeftine, Kechabta and El Alia basins [47], which are insulated either by morpho-structural ridges and emerged high zones. Due to the lack of stratigraphic markers, these deposits were subdivided in lithostratigraphic Formations [26]: Hakima, Oued El Melah, Kechabta and Oued Bel Khedim (Figs. 2 and 3). In Jalta, the Miocene deposits are completely continental.

2.9. Pliocene

The Pliocene is dominantly marine deposits outcropping at the East of the Mateur-Bizerte basins and on the southern edge of the Cap Bon Peninsula (Fig. 2). These series settle with unconformity on the various Miocene and ante-Miocene substratum. They are subdivided by Burollet [26] in two Formations: the Raf-Raf at the base and the Porto Farina at the top (Fig. 3). These two Formations outcrop at the south-east of the Douimis basin, with a first primarily argillaceous series (50m) and second predominantly sandy deposits rich in fossils (50m). In the El Alia-Ghar El Melah basin, Pliocene shows more significant thicknesses.

Towards the East, in the Gulf of Tunis, the Lower Pliocene deposits are represented by clays, sands and sandstones. They exhibit 300m of thickness with notable variations related to the structuring of the inhirited substratum and Triassic salt movements. Late Pliocene is essentially made up of sandy series of the Porto-Farina Formation. Its thickness is about 670m and it is sealed by the villafranchian series. From the current coastline, all along the master faults and at piedmont of the reliefs, Pliocene deposits becomes completely continental. These latter are integrated in a detrital sequence of the Ségui Formation, which is attributed to Late Miocene-Pliocene.

2.10. Quaternary

At the level of the depocenters and slopes of the reliefs, the Quaternary deposits are represented by continental facies; the marine ones are spread out over the entire eastern and northern coasts of northern Tunisia. The Early Pleistocene of the north of the Kechabta is composed of silts, continental sands and clays. This series locally made 200m (Fig. 2). The upper Quaternary (Late Pleistocene / Tyrrhenian) marine and eolian deposits are well developed all along the northern and north-eastern littoral of Tunisia [97].

3. Structural framework

The current tectonic framework of central and northern Tunisia [28,54,24] was guided at least by five NE-SW trending master faults (Figs. 4 and 5) that are associated with Triassic saliferous outcrops forming quite exposed ridges [71,98,63,33,18,99,43]. From the SE to the NW, we distinguish:

3.1. Zaghouan fault

It is marked by a relatively irregular layout (Fig. 4) and shows a constant tendency with overlapping towards the SE [100,28,101]. To the North, this master fault is associated to the Triassic and Jurassic outcrops [33]. Towards the south-east, this fault disappears before the Rouhia-Kalaâ Jerda graben. During the Mesozoic, this N40 master fault has bordered the "Tunisian trough" to the SE side, then it evolved to SE overlapping fault during the Cenozoic compressive phases [59,102]. It corresponds to the T2 transversal identified by Jauzein [28].

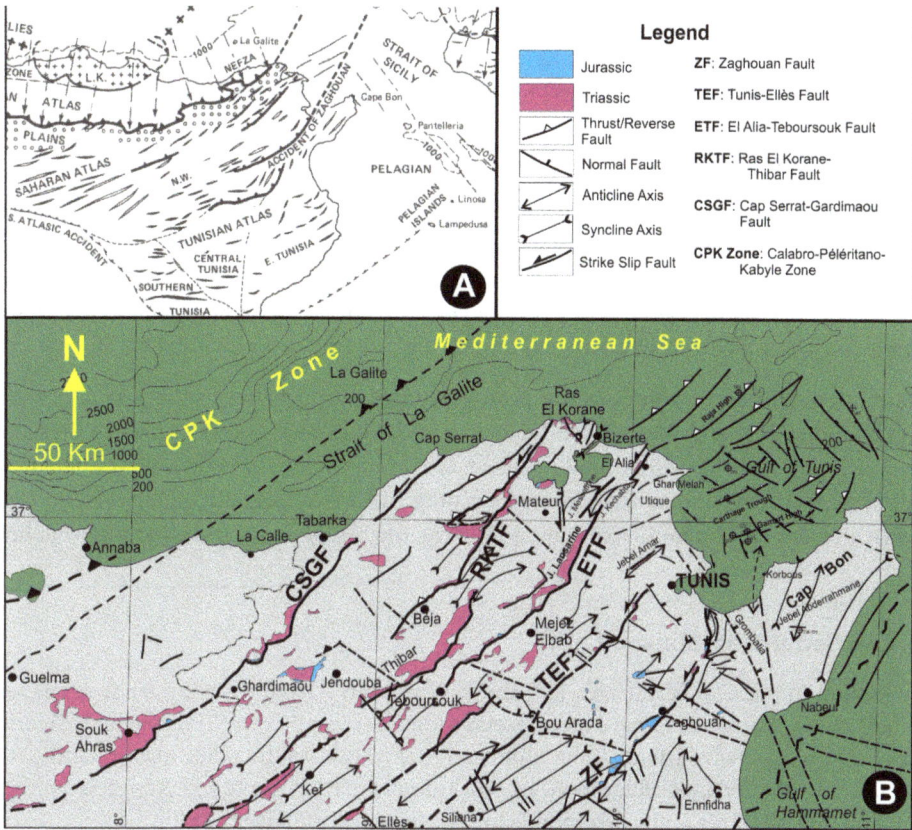

Figure 4. A- Atlasic and Alpin domains of Tunisia [101]. B- Tectonic map of northern Tunisia.

3.2. Tunis-Elles fault

This lineament of rectilinear layout (Figs. 4 and 5) shows a reverse movement with local overlapping and imbrications of the series [28]. This master fault has controlled distribution and evolution of the associated structures since the Triassic rifting phase [103, 43]. During the extensional periods, this fault has contributed in the individualization of the large subsiding basins and the delimitation of the Tunisian trough since the Aptian. During the end of Cretaceous and Cenozoic contractional periods, this fault has modeled the Tunisian Atlas following its sinistral overlapping movement. It induced the individualization of NE-SW poured south-eastern Atlasic folds and other transverse NW-SE folds strongly involved with sigmoid forms [104]. It corresponds to the T3 transversal [28].

3.3. El Alia-Téboursouk fault

It begins with the El Alia and Kechabta faults to the NE and continues with the fault delimiting the Sakkak-Lansarine diapir then the Teboursouk overlapping (Fig. 4) [31, 63,105-

107]. It corresponds to a sinistral left relay fault system. Towards the SW, it separates the Oulad Bou Rhanem graben from the Kalaâ Jerda graben [28] and bounds the Tebessa graben in Algeria [108]. It corresponds to the T4 transversal cited by Jauzein [28].

3.4. Ras El Korane-Thibar fault

It extends from the Kef Triassic o alignments to the SW to the Ras El Korane in the NE (Fig. 4), crossing the Thibar, Beja and Bazina structures [18]. The Ras El Korane-Thibar segment constitutes the NNE extension of the T5 transversal [28]. This fault, which borders the Bazina Triassic outcrops on the Eastern side [109] extends to the North and delimits the Ras El Korane Numidian deposits [71, 110, 111, 18, 112, 47].

This master fault corresponds to the paleogeographic limit between the Kroumirie and Mogods mountains and that of Hedil and Bizerte [71]. It bounds the Numidian Flysch in Ras El Korane and the Tellian units in Beja. It is a discontinuous and sinistral left relay fault system which is sometimes shifted by other later NW-SE faults. Dubourdieu [113] evokes a recent horizontal displacement towards the SW of about fifteen km on this lineament.

3.5. Cap Serrat-Ghardimaou fault

It is located in the Cap Serrat area [32] and continues in Algeria crossing Souk Ahras and Batna [113,28]. It appears to have an important role in separating the Tunisian and Algerian blocks that have evolved with some independence. [114]. As the other master faults, it is associated to the Triassic outcrops (Fig. 4). Moreover, it shows some Neogene volcanic extrusions [32, 45]. This fault corresponds to T6 transversal of Jauzein [28].

We can follow the extension of the majority of these faults in offshore. They affect the northern Tunisian plate such is the case of the Ras El Korane-Thibar and Cap Serrat-Ghardimaou faults (Fig. 2). Another lineament has the same direction being extended in offshore and could be attached to that, which delimits the Calabro-Peloritano-Kabyle zone (CPK) [29,35] to the SE (Fig. 2). The other faults affecting the North of Tunisia have NW-SE, E-W and N-S directions and they have played a significant role, beside the NE-SW master faults, on the distribution and evolution of the Mesozoic and Cenozoic basins.

These lineaments have subdivided the northern Tunisian margin into six compartments (Figs. 5 and 6) corresponding to the Enfidha-Cap Bon, Jebel Oust, Mejez El Bab, Mateur, Nefza and Tabarka. Within each compartment, the sedimentary floor is organized into several domains corresponding to grabens, half-grabens and horsts delimited by NW-SE, E-W and N-S faults related to the regional deformations.

Tertiary contractions on the northern Tunisian margin have also induced folds that have NE-SW Atlasic direction within the compartments. Some folds affecting the Neogene strata rather show near E-W directions.

Role of the NE-SW Hercynian Master Fault Systems and Associated Lineaments on the Structuring and
Evolution of the Mesozoic and Cenozoic Basins of the Alpine Margin, Northern Tunisia

141

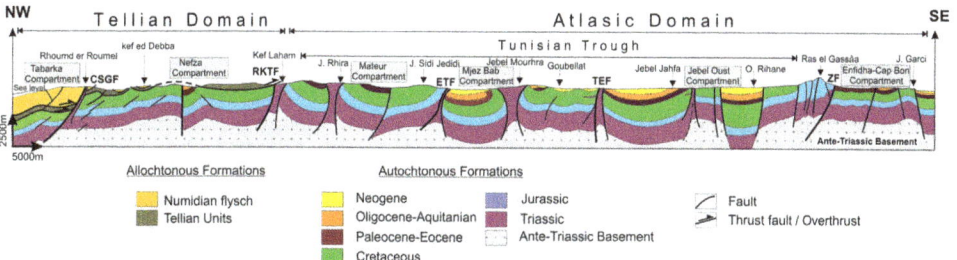

Figure 5. NW-SE geological cross-section crossing orthogonally structures of the Atlasic and Tellian northern Tunisian. The master faults subdivide this region into six compartments. (ZF: Zaghouan Fault; TEF: Tunis Ellès Fault; ETF: El Alia-Teboursouk Fault; RKTF: Ras El Korane-Thibar Fault; SGF: Cap Serrat-Gardimaou Fault. Location in Fig. 2.

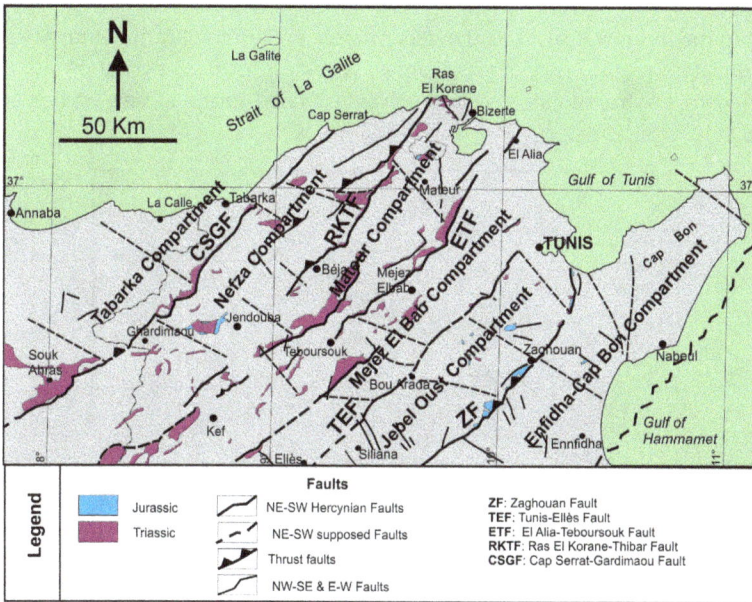

Figure 6. Structuring of the northern Tunisian margin into six compartments lengthened along NE-SW direction.

4. Geodynamic evolution

4.1. Introduction

Structuring and deformation of the substratum is one of the most significant parameters which induced the distribution and extension of Mesozoic and Cenozoic latter deposits.

Substratum of the Atlasic and Tello-Rifaine chain is well identified in Morocco, where it consists of Paleozoic strata deformed by the Hercynian orogenesis [4]. These strata are affected by N45°E to N70°E strike slip faults with high dip. In northern Algeria, the substratum seldom outcrops; it extends from eastern Morocco according to W-E direction parallel to the current South-Atlasic lineament (Fig. 7). In Tunisia, the Paleozoic substratum is little known, due to the outcrop missing, except the Permian of Jebel Tebaga in southern Tunisia and those encountered in petroleum wells on the Saharan platform in southern Tunisian.

In northern Tunisia, the ancient fracturing is distributed according the NE-SW master zone of faults, limited to the South by the Zaghouan master fault and to the North by the Cap Serrat-Ghardimaou master fault [100,32,33]. These old fractures are marked by Triassic and Jurassic intrusions and magmatic extrusions (Figs. 2 and 4). The NW-SE direction appeared especially in offshore Pelagian block of eastern Tunisia [4,93]. The current sedimentary and tectonic distribution results from the superposition of tectonic phenomena affecting Tunisia during Mesozoic and Cenozoic periods.

We present in the following sections the geodynamic evolution of sedimentary basins in the northern Tunisian margin, since Triassic times. We insist for each period on the role of the inherited Hercynian structures on basin evolution related to the nature and change of the regional tectonic constraints.

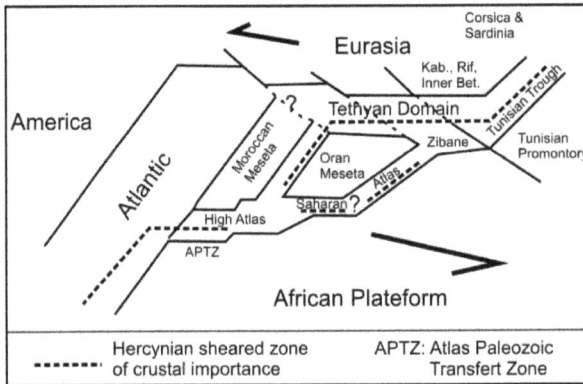

Figure 7. Atlasic domain of the Maghreb during Early Mesozoic [4].

We note the NE-SW orientation of the Hercynian deep faults in the Tunisian trough.

4.2. Triassic

At the Triassic period the paleogeography of Tunisia was dominated by an extended platform between the Saharan continent to the South and the Tethys to the North (Fig. 7). The Triassic series start with a continental detrital sedimentation, then evolves to marine carbonates and capped by evaporates characterizing a littoral environment. The paleogeographic changes were related to an unequally subsidence, which is particularly active in the central and the northern of Tunisia [45]. The Triassic rifting conducts to the paleo-Téthys [63,2]. Magmatic green rocks often accompanied the Triassic deposits of northern Tunisia.

Detrital deposits at the base of the Triassic indicate the beginning of a major transgression on the Hercynian unconformity. In the North of Tunisia, Triassic thickness exceeds 2500m. Due to the absence of well data reaching the base of the Triassic facies, we can't identify an appropriate underlying structure to this interval and thus there is no information on tectono-sedimentary control. The diapirs and domes zone, currently located between the Tellian domain and the "Tunisian Dorsale", have a NE-SW direction (Fig. 8). During the Triassic, this zone occupied an inherited horst of the Hercynian substratum and was delimited by two normal faults, which could correspond to the two known master faults of the northern Tunisian margin: the El Alia-Teboursouk and the Ras El Korane-Thibar faults. This high zone separates two subsiding domains; the north-western domain corresponding to the future Tellian-Numidian basin and the south-eastern domain corresponding to future "Tunisian trough". The bordering faults facilitated the migration of saliferous facies upwards. This can be explained by the particular frequency of the evaporate sediments in the zone located between these two major faults. Rises of the Triassic evaporates begin with Late Jurassic [6,13,116]; some authors believed Late Jurassic and Early Cretaceous [70,58].

4.3. Jurassic - Early Cretaceous

During the extensional Jurassic period, the southern Tethyan margin was structured into horsts, grabens, half-grabens and tilted blocks [117,118]. Since the Late Liassic, this active kinematics had amplified differential subsidence fossilized by thick series in depocenters and thin ones even condensed and/or with gaps on the highs [33,119,76].

Through the Atlasic domain, could exist a deep feature that controlled by the substratum structuring. The most obvious feature is that of the N-S Axis, which limits the deformed Atlasic platform to the West and the stable Sahel platform to the East. This master Axis extends to the North towards the Zaghouan master fault (Fig. 8) and has a continuous paleogeographic role from the Jurassic to Quaternary series [62,120,54].

In northern Tunisia, these old faults affecting the ante-Triassic substratum are not well expressed. However, from Jurassic and especially during Early Cretaceous, the N-S to NNW-SSE extension of the Tunisian margin induced genesis of subsiding basin (Tunisian furrow) delimited by the Zaghouan fault to the SE and that of Tunis-Elles to the NW [33,121,4,103,75,122,123,43]. This basin will receive an enormous accumulation of deposits, which exceeds locally 2000m for Barremian ([82]; Fig. 8). Nevertheless, near the Tunis-Elles fault, this stage is represented only by a few tens meters of limestone, marl and massive limestone.

These ancient listric faults will be reactivated and caused the collapse of the NW compartments. They generate a structuring into half grabens slightly tilted to the SE. At the same time, these NE-SW oriented structures are associated at the level of the sedimentary cover by N-S, NE-SW and NW-SE trending other fractures, which are guided by ascension of Triassic salt in extensional regime.

The Bou Kornine outcrop of Hammam-Lif is placed in a paleogeographic and intermediate structural position between two distinct paleostructural domains, belonging both to the

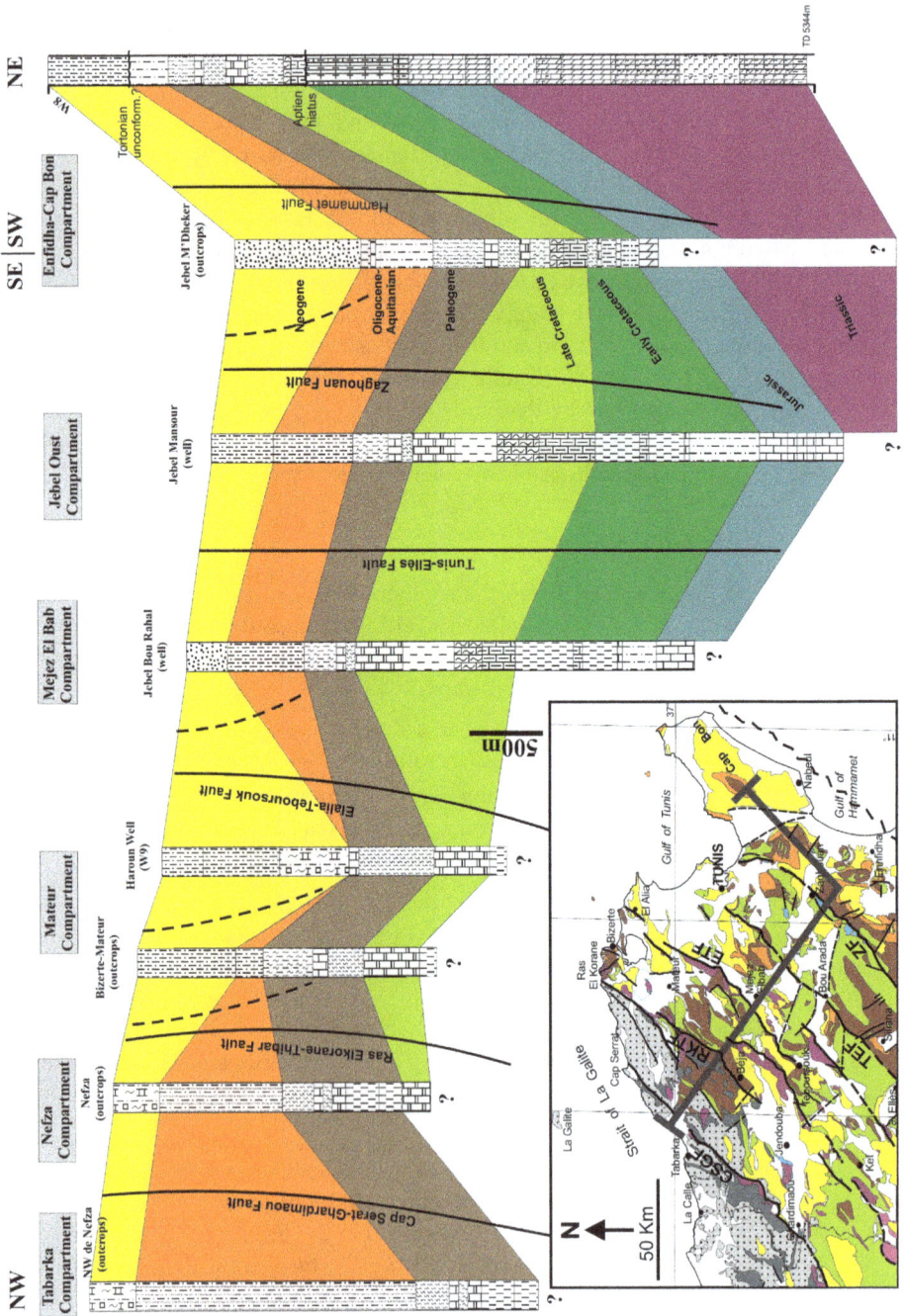

Figure 8. NW-SE and NE-SW Lithostratigraphic correlations of geological series outcropping in the different identified compartments.

Role of the NE-SW Hercynian Master Fault Systems and Associated Lineaments on the Structuring and
Evolution of the Mesozoic and Cenozoic Basins of the Alpine Margin, Northern Tunisia

145

Maghrebin margin of western Tethys. These domains are the "Tunisian Dorsale", corresponding to a carbonated platform and the "Tunisian trough", corresponding to a NE-SW deep graben with sedimentary pelagic filling [124].

Located on an active flexure zone between these two domains, the Jurassic deposits of Jebel Bou Kornine will record different stages of the geodynamic evolution of the margin [125]. Since Late Toarcian-Early Aalenian, synsedimentary tectonics of tilted blocks have a dominating role on the progressive erosion of the near carbonated platform, on the dynamics of the gravitation flow and on the installation of the four conglomeratic levels [125].

However, at the Tunisian trough, where differentiated grabens, sedimentation is thick and turbiditic [36]. The Tunisian furrow, characterized by upper Jurassic radiolarite deposits, has the main structural features of the Tethyan domain [117]. The Zaghouan lineament would have separated the Jebel Oust compartment from that of Enfidha-Cap Bon (Figs. 4, 5 and 6).

The Jurassic deposits remain unknown in the Gulf of Tunis because there are no drilled wells that reach them. The interpretation of seismic lines crossing this area shows structuring and geometry of the Jurassic limestones and marls above the Triassic carbonated strata (Fig. 9). These series are marked by downlap progradational structures on the sides of flanks. The Jurassic is characterized by condensed surfaces of unconformities, marked by high amplitude and good continuity reflections (Fig. 9).

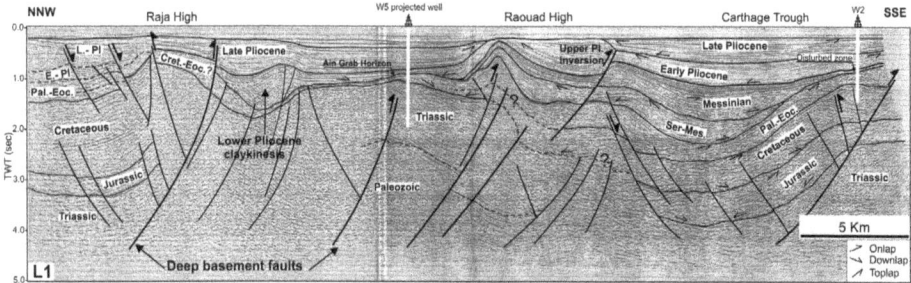

Figure 9. Interpreted seismic line L1 of the Gulf of Tunis, showing distribution of Mesozoic and Cenozoic deposits and its evolution towards the NE-SW and the associated Triassic ascensions [24]. E-Pl: Early Pliocene; L-Pl: Late Pliocene. Location in Fig. 2.

At the level of the northern Tunisian margin, depocenters, half-grabens and high zones lengthen preferentially according NE-SW direction since the Jurassic and especially the Early Cretaceous [63,126]. Triassic salt rising has been clearly emphasized since the Early Cretaceous extensional phase corresponding to an intracontinental rifting [4,3]. This deformation was very active, with formation of tilted blocks related to activity of synsedimentary normal faults. These structures have been induced by regional N-S transtensional event [112,123,24].

In the Gulf of Tunis, the Early Cretaceous is particularly thick in W6 well (2341m). However, it only presents 912m and 477m in W2 and W3 wells indicating structuring into low and raised zones under effects of bordering faults (Fig. 10). Differently to the underlying Jurassic

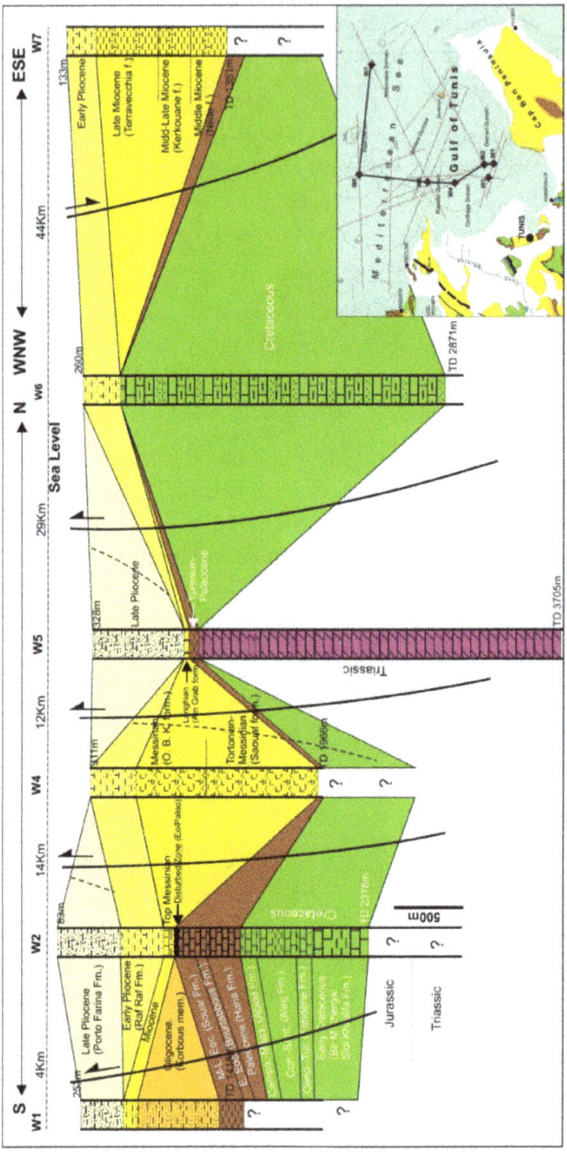

Figure 10. N-S and WNW-ESE Lithostratigraphic correlations of petroleum wells in the Gulf of Tunis, showing inversion of the subsidence at Top Cretaceous, Top Miocene and Top Early Pliocene [47].

reflectors, the Cretaceous deposits are thicker on the "Gamart" tectonic corridor and are considerably reduced towards the depocenter. This distribution seems to be related to the movements of the Triassic diapir, which caused the structural inversion and the tilting of high edges. Towards the "Raouad" raised structure, the sedimentary sequences are associated with retrogradational on laps and top laps (Fig. 10).

During the Berriasian, the Zaghouan fault delimits a subsiding basin, which occupies all the northwestern part of an uplifted zone of limestones with organic-detrital deposition [33]. This high zone constitutes the northwestern edge of the Enfidha compartment.

Locally, in Jebel Bou Rahal of the Mejez El Bab area, the El Alia-Teboursouk fault delimits the Triassic outcrops to the NW. It separates a NW basin with thick and continuous sedimentation, of a SE basin with reduction and gap of sedimentation (Barremian-Coniacian gap; [127]) in extensional and transtansional context.

The N-S extension induced appearance of other fractures orthogonal to the Hercynian strike slip faults already active since Jurassic [103,43]. Thus, the upper Aptian-Albian period is characterized by the occurrence of grabens directed close to NW-SE following the normal movement of faults that have the same direction [128].

In the Gulf of Tunis, this N-S to NNW-SSE extensional and transtensional direction recorded during this period has mobilized the NE-SW Hercynian faults, generating tilted blocks and opening grabens and horsts along the NW-SE trending faults (Fig. 10).

4.4. Late Cretaceous

A master change of the African plate movement compared to European plate has emphasized in Albian [102], related to beginning of the northern and southern Atlantic Ocean expansion. This displacement has caused movement of Africa to the North and stopped accretion of the oceanic lithosphere at the level of African-European Rift Zone.

During the Albian and Turonian, the Zaghouan master fault has separated a stable eastern platform from a deformed western platform represented by the "Tunisian furrow" [129]. This last, having a geosynclinal form, was structured into several compartments. This paleotectonic zonation had a great influence on the development and distribution of the varied facies during the Senonian.

In the "Grand Tunis" area [103], the tectono-sedimentary analysis of the upper Cretaceous series shows a significant instability of the sedimentary floor inducing a structuring into tilted blocks, which are bounded by NW-SE faults [103].

Differential movements of various faults exhibit the "keys of piano" architecture well express neighboring the Zaghouan fault [33] and that of Tunis-Ellès [43]. This extensional to transtensional period is accompanied by an intense halocinetic and magmatic activity [130-132]. Some E-W contractional pulsations were highlighted in the Tunisian furrow during the Late Albian-Cenomanian [43]. This transtensional event is also evident by the presence of slumped sandstones and synsedimentary N30- 40 trending normal faults, which affect the Cretaceous deposits of the Mejez el Bab area, at the level of El Alia-Teboursouk lineament [105] and in the "Grand Tunis" area (Jeriffet outcrop; [103]).

4.5. Paleocene

In northern Tunisia, several authors [58,129] distinguished two domains; the first where the Danian is present, whereas the other where it lacking; the limit between these two domains is roughly located on the Thala-Elles-Tunis line (Fig. 2). Paleogeographically, they correspond to two different basins where the El Haria Formation, of Late Maastrichtian-Paleocene age doesn't have the same stratigraphic significance [58]. The eastern Tunisia basin with less thick and locally condensed sedimentation and northwestern Tunisian furrow basin with very thick sedimentation.

Moreover, the limit between the two basins coincides perfectly with the Tunis-Elles fault (Figs. 2 and 8). This confirms the role of this fault on the control of deposition, which is well justified during the Paleocene (El Haria Formation) [42,43].

Furthermore, the Cap Serrat-Ghardimaou fault seems continued its impact on sedimentation by delimiting the Tellian facies to the NW, which is characterized by marl and limestone alternations from marls of northern facies of the Tunisian furrow to the SE [129].

The Tunis-Elles fault has controlled the Paleocene deposition; it clearly separates a low subsiding basin to the South-East from a subsiding basin to the North-West (Fig. 8).

The Ras El Korane-Thibar master fault (RKTF) induces an accumulation of more significant Paleocene series on the Western northern side than on the Eastern southern side. Thus, the thickness of this series reached 1200m in Bazina to the West of Mateur [109] and 800m in the Henchir Haroun (W10) petroleum well to the East of Mateur [47].

The distribution of Paleocene deposits on both sides of the NE-SW master faults indicates that has contributed to the installation of a tilted blocks and half-grabens associated with condensed series and hiatuses near the location of faults and a thick and argillaceous facies in the distal subsiding depocenters.

4.6. Early Eocene

During the Early Eocene (Ypresian), a high zone was individualized, which extends from the Kef area to the SW, towards the Mateur area to the NE. It corresponds to the extension of the Nummulitic limestone facies, which characterizes low subsiding platform. This high zone separates to the NW and SE two relatively deep provinces (Fig. 11A) with pelagic sedimentation corresponding to a limestone facies with Globigerines [133,34,134,102,135,136]. This high zone coincides with the domes zone. It is located, therefore, between two master faults: the El Alia-Teboursouk fault (ETF) to the SE and Ras El Korane-Thibar fault (RKTF) to the NW. The paleogeographic position of this high zone is related to the instability of the Triassic domes whose have rise of and pierced their covers [31,63,43].

This high zone, lengthened according to an NE-SW average direction, has an asymmetrical form with a southeastern margin with weak slope and a northwestern margin with steep slope. This asymmetry, inherited from previous period, was accentuated and reactivated again by Lutetian contraction [31,112,84,137].

In addition, the morphostructural ridges or underwater peaks [138] seem to have controlled the distribution of facies, following the movements of the NE-SW substratum faults, associated with the Triassic salt rising, which began since the extensional Cretaceous periods. These Triassic bodies length along the Hercynian lineaments, indicate their successive rejuvenations [40]. Moreover, in the Beja area, the Nummulitic limestone outcrops of the Early Eocene show a NE-SW privileged orientation [134].

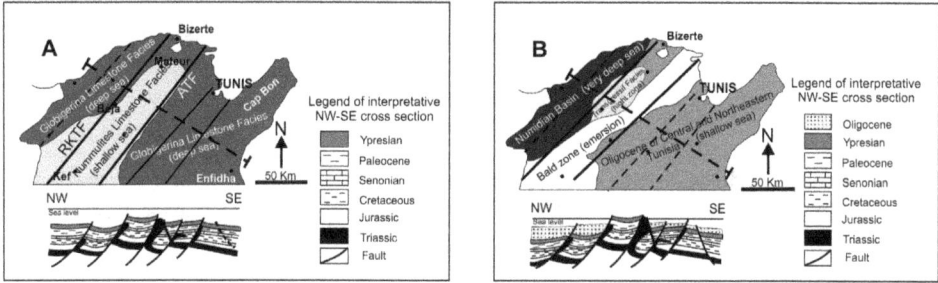

Figure 11. Paleogeographic and paleotectonic maps of northern Tunisia at the Ypresian (**A**) [129,133,34,134,135,84] and at the Aquitanian (**B**) [129,84,89,86]. Interpreted NW-SE cross-sections are based on the present work.

Under effect of the tectonic deformation, the Paleocene-Eocene deposits are unequally distributed in the Gulf of Tunis. Thickness of these Formations changes from 434m in W2 well, to 307m in W3 well. We note the absence of the Bou Dabbous and Souar Formations in W5 and W6 wells (Fig. 10). The seismic reflectors are bordered by an angular unconformity and are associated on the side of the "Gamart" structure by aggradational/retrogradational onlaps above marly and carbonated Maastrichtian seismic horizons. Reflectors are moderately continuous and associated with pinch outs on the "Raouad" uplift (Fig. 9). Deposits are marked, in W2 well by the Eocene breccias. These structures should indicate a slope of the sedimentary floor along the faulted zone (Fig. 9). The Paleocene-Eocene deposits, which are marked by a development in the center of the depression and the "Raouad" uplift, are reduced towards the "Gamart" high zones, where clays of El Haria and limestones of Boudabous are directly deposited on the Triassic evaporites (Fig. 9).

From the end of Cretaceous (Late Maastrichitian) and until the Middle Eocene, the NE-SW preexistent faults continue their effects on sedimentation in a contractional and transpressional regime.

The El Alia-Teboursouk and Ras El Korane-Thibar faults have controlled sedimentation during the Ypresian. Moreover, the NW-SE faults appeared above the Triassic bodies during the previous period, will express and we thus attend notable variations of the facies and thicknesses on both sides of these faults. The extensional movements testified by the NW-SE normal faults are integrated in a NW-SE regional contractional event [139,112,120,21,47,136]. This compressive constraint generated principally reverse displacements along the NE-SW ancient faults.

However, some authors [17,38,42,103,122,140,141] consider that this period constitute the continuation in time of the Mesozoic extension.

4.7. Middle to Late Eocene

During this period, there was the genesis of the Proto-Mediterranean following the movement of microplates towards the North. In the high zone, where underwater peak, identified during the Early Eocene, appear many gaps in the Middle Eocene and especially in the Late Eocene (Souar Formation) with the presence of several glauconitic levels (Fig. 8). Salaj [129] announce the absence of the Late Eocene, in most of this zone. This is related to the high structural position of this zone following rejuvenation of the Ras El Korane-Thibar and El Alia-Téboursouk faults [63,98] (Fig. 11A). Elsewhere, in the Tunisian furrow and at the level of the eastern Tunisian platform, the Middle and Late Eocene is very developed [86].

In addition, the Middle and Late Eocene, represented by marl and limestone alternations in the Mejez El Bab area, is much reduced [127]. These variations of facies accompanied sometimes by gaps and unconformities characterize the Lutetian contractional period [98,48,84,142-144,69,137,24].

At the outcrop scale, the witnesses of contractional tectonics are showed by the presence of (i) unconformity of the Oligocene on the Middle Eocene in the Jebel Sebâa outcrop and at the level of the Bizerte town [84]; (ii) unconformity of the Late Eocene at the level of the Bir Afou structure, which was formed during Late Maastrichtian-Early Eocene and the presence of synsedimentary reverse faults affecting marl and limestone alternations of the El Haria Formation of Late Maastrichtian-Paleocene age. These faults have N20, N40-50 and N70-80 directions [43], (iii) unconformity of the Oligocene on a folded substratum in the Enfidha area [144].

4.8. Oligocene - Early Miocene

In many localities of the Mejez El Bab, the marine Oligocene deposits, represented by clays and bioclastic sandy limestones with Nummulites, unconformably rests on the Triassic. It is surmounted by the Late Oligocene-Aquitanian continental deposits [127]. This tendency to emergence since the Kef area towards the Mateur and Bizerte areas is guided by the El Alia-Teboursouk and Ras El Korane-Thibar two master faults (Figs. 8 and 11B). During the Oligocene, we also attend to the appearance of a bald zone, which lengthens from the Kef area to the SW towards the El Alia offshore to the NE, passing by the Lansarine chain [98,89,24,87]. This bald zone separates two different basins, characterized by a clay-sandy deposition; (i) basins of Beja-Ghardimaou and subsiding and deeper Numidian basin to the NW and (ii) the less deep but subsiding Fortuna basin to the SE.

We think that this distribution is the result of the installation, in northern Tunisia, of an extensional event controlling the Oligocene-Early Miocene deposits as was announced by Piqué et al. [4].

At the regional scale, the Oligocene-Aquitanian basin is contemporary with reactivation of old lineaments, which appear as normal faults controlling the genesis of the half-grabens. At the local scale post Lutetian extensions are numerous and are fossilized by synsedimentary

Role of the NE-SW Hercynian Master Fault Systems and Associated Lineaments on the Structuring and
Evolution of the Mesozoic and Cenozoic Basins of the Alpine Margin, Northern Tunisia

151

normal faults [63,37,127,43]. The end of the Oligocene-Aquitanian is marked by a continental deposit characterized by coarse sands with cross-bedding stratifications. These deposits seem to be related to a total emergence of the majority of the study area.

We think that, in spite of the contractional constraints on the northern Tunisian and northern Algerian margin [15,145], induced by rotation of the Corso-Sarde block, we have always an extensional context. This event has induced the siliciclastic deposition in the Numidian basins [146,45] and of the middle Mejerda on the one hand, and the sandy deposits of the Fortuna Formation in central and north-eastern Tunisia on the other hand [147,89].

Moreover, several indices of NW-SE to N-S extensional deformation were announced (i) sedimentation is controlled by the activity of NW-SE and NE-SW faults, which delimited the different blocks, all along the Tunis-Elles zone [43]. The passage from the Eocene to Oligocene is marked by an inversion of subsidence following the reactivation of the NW-SE faults; (ii) several indices of N140 synsedimentary normal faults affecting the Oligocene sandstones of Korbous [147]; (3) the Ras El Korane-Thibar fault has moved in extensional mode and was at the origin of the clayey and sandy deposition on the NW side of the mega half graben; (4) furthermore, in the Téboursouk area, Perthuisot [63] highlighted an extensional event accompanied by N45 diapirism.

4.9. Middle Miocene - Quaternary

The structures recognized in the northern Tunisian margin (Figs. 2 and 4) result from the whole of the Eocene (Lutetian), Miocene (Tortonian) and Quaternary (Villafranchian) contractional phases, which followed the multiple extensional and transtensional episodes. These contractional phases induced tectonic inversions [24,47].

The evidence of major contractions are fossilized in the sedimentary sequences by angular unconformities as it is the case of the Neogene (post-Tortonian) deposits, which settled on the Oligocene-Aquitanian folded series at the Rmil outcrop in eastern Gaâfour [43] and on the Campanian-Maastrichtian (Abiod Formation) at the Mejez El Bab [127]. Moreover, an unconformity of the Pliocene marine on the underlying strata has showed in the Bizerte [18], in Messefftine and Kechabta outcrops (Menzel Bourguiba) [26,18,47] in the Cap Bon [37,93], overlappings to the SSE in the Lansarine chain [98], Quaternary deformations on the El Alia-Téboursouk fault near the Sloughia [148] and unconformity of marine Pliocene in the Gulf of Tunis [24].

During the Middle Miocene-Quaternary period, three types of Neogene basins (Fig. 12) in the Bizerte-Mateur area have been developed following their position compared to the raised zones delimited by NE-SW, N-S and NW-SE master faults [47]; they correspond to the (i) narrow, strongly subsiding synclines (Douimis, Kechabta and El Alia basins), (ii) lozenge-shaped basins (Messeftine basin) and (iii) trapezoidal basins (Jalta basin).

The Alpine and Atlasic contractional phases have caused the formation of the Tellian and Atlasic folds trending NE-SW as well as the installation of the overthrust folds at the northwestern end of Tunisia [32,38].

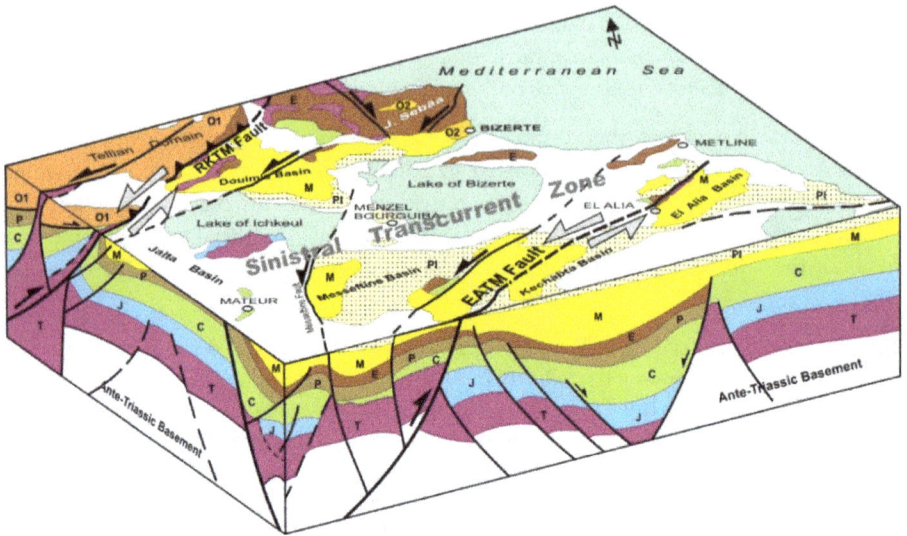

Figure 12. 3D Block diagram of the Neogene basins in the Tellian foreland domain [47]. EATM Fault: El Alia-Téboursouk master fault; RKTM Fault: Ras El Korane-Thibar master fault; Pl, Pliocene; M, Miocene; O1, Oligocene-Aquitanian (Numidian unit); O2, Oligocene-Aquitanian; E, Eocene; P, Paleocene; C, Cretaceous; J, Jurassic; T, Triassic.

These phases have also induced the reactivation of the old lineaments into reverse faults on the NE-SW direction and strike slip faults on the other directions. They also caused unconformity of the Miocene-Pliocene series (Ségui Formation) on the Oligocene sandy banks (Goubellat graben, [43]) and unconformity of the Middle Langhian Aïn Grab Formation on the folded limestones of the Ypresian during the Alpine phase [149].

From the end of the Messinian and during Early Pliocene, the transtensional events induced increase of the subsidence in the center of the depocenters, which received clayey sediments of the Raf-Raf Formation (Fig. 2). At the scale of the outcrop, this succession shows synsedimentary normal faults.

The contractional event begins again during Late Pliocene and Quaternary [31,150,15] and continues until the Actual [38,18,44,24]. Thus, several structures having participated in the Neogene evolution are reactivated under current tectonics [38]. In the Tellian continental domain, the Neogene basins are also deformed at the level of its levels.

In northern Tunisia, the distribution of epicenters of the earthquakes is oriented NE-SW according to the direction of the master faults [151]. The current movements of the Cap Serrat-Ghardimaou fault, for example, are characterized by a seismicity which expressed principally on the level of its active segment of Ghardimaou [152,153]. It is at the origin of several earthquakes; the last one dates from 17/09/1986. Its focal mechanism is related to sinistral strike-slip movement and the axis of sub horizontal shortening is oriented NW-SE to N-S [151].

The calculation of the composite mechanisms based on the seismic events recorded in the neighboring areas of the Ras El Korane-Thibar fault, to the NW of Garaât Ichkeul, shows sinistral strike slip movements with nodal plans of N45 direction and dip of about 70° towards NW. This plan can correspond to the Ras El Korane-Thibar fault and the axis of shortening is oriented N170 [151].

Other earthquakes are located along the NE-SW Zaghouan fault indicating its recent activity. This locally overlapping master fault seems continued in the gulf of Tunis. The composite mechanisms of the majority of the seismic events allow us to deduce an axis of pressure oriented NW-SE to NNE-SSW [152,154].

5. Synthesis and conclusions

The geodynamic evolution of the northern African margin during the Late Triassic and Jurassic was mainly guided by the reactivation of the first order NE-SW Hercynian faults, associated to a second order conjugate NW-SE, E-W and N-S faults. These lineaments have differentiated either lozengy basins in the Saharan Atlas [155] or high and subsiding zones in northern Algeria with basins lengthened according to a N50 direction [156,48,13] or rather the « pull apart » basins in the high and Middle Moroccan Atlas and in northern Tunisia [157,117,158]. Moreover, in the Apennines, the Alps and the Betic [159], the Triassic basins would have evolved in a strike slip mode, which is controlled by the Hercynian directions (Fig. 13).

The rejuvenation of the southern Tethyan margin faults is related to the movement of the African plate against the Eurasian plate along the E-W sinistral transforming fault [9,160,119]. The explanation of the subsiding NE-SW oblique Tunisian furrow, compared to the global direction of the N70° to W-E Maghrebin furrow is explained by the presence of N-S to NNW-SSE regional transtensional stresses during the Triassic and the Jurassic-Early Cretaceous.

This tectonic framework is explained, at the Mediterranean scale, by dextral to reverse dextral movement of the N70 master faults, separating the African and Eurasian plates [9,160].

The total closing of the Tethys related to the opening of the western Mediterranean in Early Miocene [16] has induced genesis and rising of the Atlasic chains, which constituted thereafter the structural units of Tunisia. Generally, the complexity of the geological structures increases towards the North of Tunisia, at the level of the Tellian domain, where are developed the overthrust folds [32] and where the Tethyan is closing.

The varied and oriented faults and folds affecting the northern Tunisian margin are the result of complex changes in geometry and style of movements of the African and Eurasian plates, which started since Triassic and continued until now.

During the Mesozoic, the northern Tunisian margin is characterized by tectonic instability highlighted by several variations of facies and thickness of series. The NE-SW and N70°E Hercynian lineaments have controlled deposition [4] and caused structuring into tilted blocks and compartments generally lengthened NE-SW. Each compartment is affected by other NW-SE and E-W conjugated faults.

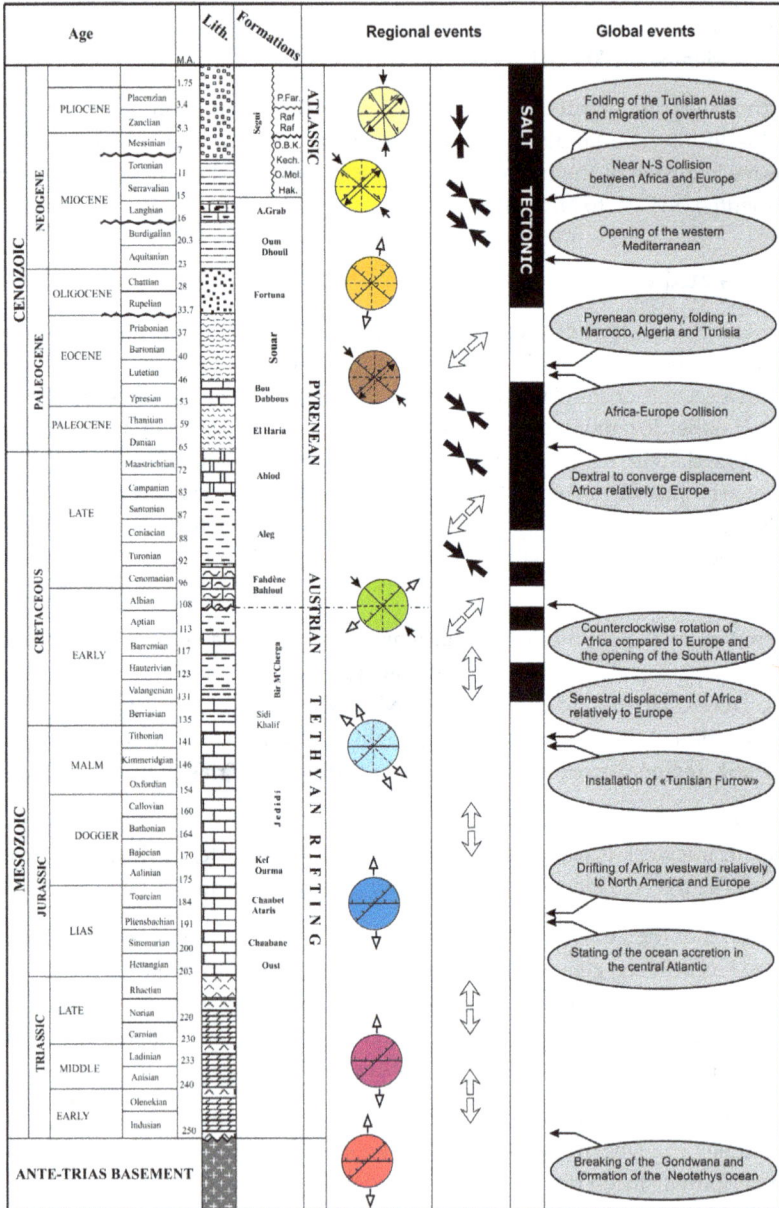

Figure 13. Table Summarizing the regional and global tectonic event that affected the northern Tunisian margin and western Mediterranean during the Mesozoic and Cenozoic [58,32,63,139,1,16, 164,168,33,160,2,169,119,37,165,127,40,18,4,166,84,120,75,103,123,54,21,43,45,125,154,24,47].

These faults have locally controlled the sedimentation and induced horst and graben structures lengthened orthogonally compared to the old structures [127,103,43,24]. This interpretation implies that the Tunisian furrow is an oblique « pull apart » graben, compared to the northern Maghrebian transform margin; that the bordering faults are very deep; that the Eocene inversion of the old bordering faults increase in Middle and Late Miocene, as in old Quaternary, generated reverse movements that controlled the rising of the central and northern Atlas and the Tellian Atlas.

Evolution of the Atlasic and Tellian domains was controlled by the Atlantic opening. Thus, at Late Liassic, the beginning of oceanic accretion in the central Atlantic induced drifting of Africa towards the East compared to North America and Europe [2,4,3] (Fig. 13).

During the Aptian and Albian, the opening of the Tunisian rift is related to the anti-clockwise rotation of Africa compared to Europe and the opening of the South Atlantic [2,4].

During the Cenozoic, the first effect of the Africa-Europe collision is marked by a clear folding phase in Algeria and Morocco at the Middle to Late Eocene [161,157]. In Tunisia, synsedimentary reverse faults are highlighted in the central and southern Atlas [37,120,142] as well as folds sealed by the Oligocene deposits [84,69,144], or by the end Eocene [137] in the northern Atlas (Fig. 13).

An Oligocene-Aquitanian extensional phase was highlighted in Tunisia [63,37,127,147,45,120, 162,24]. This phase was followed by upper Miocene major collision between Europe and Africa, resulting in the Tunisian Atlas and the installation of the overthrust folds poured to the South-East following a NW-SE contractional event [31,32,139,47]. The folding was largely amplified at Late Pliocene-Quaternary following the persistence of the NNW-SSE to N-S contractional regime [163,150,164,37,165-167,154].

These transcurrent movements have been evolved during various tectonic phases according to the geodynamic context. They controlled sedimentation in extensional context and they moved either in reverse or in strike slip of contractional event. Thus, these faults, which have at the beginning, the rather strong dips, tend to lean towards the West and the North-West following the NW-SE contractions. This is seen clearly for the most northern faults due to their proximity to the zone of contact between the African and Eurasian plates.

The NE-SW, NW-SE, E-W and N-S trending faults that have affected the North-African margin have evolved during the tectonic periods and controlled deposition in relation with (i) sinistral displacement of Africa compared to Europe during the Late Jurassic-Early Creta-ceous, following the opening of the southern and central Atlantic [160,2,4]; (ii) dextral to convergent displacement of Africa compared to Europe during Campanian-Lutetian [16,168,160]; (iii) collision of Africa against Europe since the Middle Miocene [32,137,168,169,54,47].

All the authors agree on the fact that faulting recognized in outcrop has related the effect of master deep lineaments. Movement of master faults are fossilized in sedimentary series by thickness and facies changes associated with complex structures and accentuated by salt tectonics along various orientations.

Author details

Fetheddine Melki and Fouad Zargouni
Department of Earth Sciences, FST, Tunis El Manar University, Tunis, Tunisia

Taher Zouaghi and Mourad Bédir
Georessources Laboratory, CERTE, Borj Cédria Technopole, University of Carthage, Soliman, Tunisia

Mohamed Ben Chelbi
Water Institute of Gabès, University of Gabès, Tunisia

6. References

[1] Zargouni F. Tectonique de l'Atlas méridional de Tunisie, évolution géométrique et cinématique des structures en zone de cisaillement. Thèse ès-Sciences, Université Louis Pasteur Strasbourg; 1985.

[2] Guiraud R., Maurin J.C. Le rifting en Afrique au Crétacé inférieur : synthèse structurale, mise en évidence de deux phases dans la genèse des bassins, relations avec les ouvertures océaniques péri-africaines. Bulletin de la Société Géologique de France, 1991; 5, 811-823.

[3] Guiraud R. Mesozoic rifting and basin inversion along the northern African Tethyan margin : an overview. In : Macgregor, D. S., Moody, R. T. J. & Clark-Lowes, D. D. (eds) 1998. Petroleum Geology of North Africa. Geological Society, London, Special Publication, 1998; (132), 217-229.

[4] Piqué A., Aît Brahim L., Ait Ouali R., Amrhar M., Charroud M., Gourmelen C., Laville E., Rekhiss F., Tricart P. Evolution structurale des domaines atlasiques du Maghreb au Méso-Cénosoïque ; le rôle des structures héritées dans la déformation du domaine atlasique de l'Afrique du Nord. Bulletin de la Société Géologique de France 1998 ; 169, 797-810.

[5] Jallouli C., Mickus K. Regional gravity analysis of the crustal structure of Tunisia. Journal of African Earth Sciences, 2000; 30 (1), 63-78.

[6] Zouaghi T., Bédir M., Inoubli M.H. 2D Seismic interpretation of strike-slip faulting, salt tectonics, and Cretaceous unconformities, Atlas Mountains, central Tunisia. Journal of African Earth Sciences 2005; 43, 464-486.

[7] Auzende J.M. La marge continentale tunisienne: Résultats d'une étude par sismique réflexion: Sa place dans le cadre tectonique de la Méditerranée occidentale. Marine Geology Research 1971; 1, 162-177.

[8] Auzende J.M., Bonnin J., Olivet J.L. La marge Nord-africaine considérée comme une marge active. Bulletin de la Société Géologique de France 1975; (7), 486-495.

[9] Biju-Duval B., Dercourt J., Le Pichon X. From Tethys ocean to the Mediterranean Sea. In: Biju-Duval, B. et Montadert, L. (Ed.), Structural history of the Mediter. basin. Split 1976, 1977; 143-164.

[10] Durand-Delga M., Fontboté J.M. Le cadre structural de la Méditerranée occidentale. XXVIè Cong. géol. Int., Colloque C5: Géologie des chaînes alpines issues de la Téthys. Mémoire du Bureau de Recherche Géologique et Minière, Paris 1980; 115, 65-85.

[11] Cohen, C.R., Schamel, S., Boyd-Kaygi, P., 1980. Neogene deformation in Western Tunisia: origine of the eastern Atlas by microplate-continental margin collision. Geological Society of American Bulletin, Part I 91, 225-237.

[12] Obert, D., 1981. Etude géologique des babors occidentaux (domaine tellien, Algérie). Thèse es Science, Paris VI, 1-635.

[13] Wildi, W., 1983. La chaîne tello-rifaine (Algérie, Maroc, Tunisie): Structure, stratigraphie et évolution du Trias au Miocène. Revue de Géologie dynamique et de Géographie physique, Paris 24, (3), 201-297.

[14] Bouillin, J.P., 1986. Le " bassin maghrébin ": une ancienne limite entre l'Europe et l'Afrique à l'Ouest des Alpes. Bulletin de la Société Géologique de France (8), 547-558.

[15] Letouzey, J., 1986. Cenozoic paleo-stress pattern in the Alpine Foreland and Structural interpretation in a plateform basin. Tectonophysics 132, 215-235.

[16] Dercourt, J., Zonenshain, L.P., Ricou, L.E., Kazmin, V.G., Le Pichon, X., Knipper, A.M., Grandjacquet, C., Sborshikov, J.M., Geyssaut, J., Lepvrier, C. Pechersky, D.H., Boulin, J., Sibuet, J.M., Savostin, L.A., Sorokhtin, O., Westphal, M., Bazhenov, M.L., Lauer, J.P., Biju-Duval, B., 1986. The geological evolution of the Téthys belt from Atlantic to Pamir since Liassic. Tectonophysics 123, 241-315.

[17] Martinez, C., Truillet, R., 1987. Evolution paléogéographique et structurale de la Tunisie. Mémoire de la Société Géologique d'Italie 38, 35-45.

[18] Melki, F., 1997. Tectonique de l'extrémité nord-est de la Tunisie (Bizerte-Menzel Bourguiba-Mateur). Evolution tectonique de blocs structuraux du Crétacé supérieur au Quaternaire. Thèse de Doctorat d'Université, Université de Tunis II, Faculté des Sciences de Tunis: 1-207.

[19] Tricart, P., Torelli, L., Zitellini, N., Bouhlèl, H., Creuzot, G., De Santis L., Morlotti, E., Ouali, J., Peis, D., 1990. La tectonique d'inversion récente dans le canal de Sardaigne : résultats de la compagne MATS 87. C.R. Acad. Sc. Paris, t. 310, série II, 1083-1088.

[20] Mascle, G.H., Tricart, P., Torelli, L., Bouillin, J.P., Compagnoni, R., Depardon, S., Mascle, J., Pecher, A., Peis, D., Rekhiss, F., Rolfo, F., Bellon, H., Brocard, G., Lapierre, H., Monié, P., Poupeau, G., 2004. Structure du canal de Sardaigne : réamincissement crustal et extension tardi-orogénique au sein de la chaîne Apennino-Maghrébienne ; résultats des campagnes de plongées Cyana SARCYA et SARTUCYA en Méditerranée occidentale. Bulletin de la Société Géologique de France, Nov. 2004, 175 (6), 607-627 ; DOI : 10.2113/175.6.607.

[21] El Euchi, H., Saidi, M., Fourati, L., El Mahersi, C., 2004. Northern Tunisia thrust belt: Deformation models and hydrocarbon systems. In: Swennen R., Roure F. and Granath, J. W. (Ed.), Deformation, fluid flow, and reservoir appraisal in foreland fold and thrust belts. American Association of Petroleum Geologists Hedberg Series (1), 371-390.

[22] Frizon de Lamotte, D., Michard, A., Saddiqi, O. 2006. Some recent developments on the geodynamics of the Maghreb. C. R. Geoscience 338, 1–10.

[23] Frizon de Lamotte, D., Leturmy, P., Missenard, Y., Khomsi, S., Ruiz, G., Saddiqi, O., Guillocheau, F., Michard, A., 2009. Mesozoic and Cenozoic vertical movements in the Atlas system (Algeria, Morocco, Tunisia): An overview. Tectonophysics 475, 9-28.

[24] Melki, F., Zouaghi, T., Ben Chelbi, M., Bédir, M., Zargouni, F., 2010. Tectono-sedimentary events and geodynamic evolution of the Mesozoic and Cenozoic basins of the Alpine Margin, Gulf of Tunis, north-easternTunisia offshore. C. R. Geoscience 342, 741–753.

[25] Solignac, M., 1927. Étude géologique de la Tunisie septentrionale, Dir. Gen. Trav. Publ., Tunis, Thèse d'État, Lyon, 1-756.

[26] Burollet, P.-F., 1951. Étude géologique des bassins mio-pliocènes du Nord-est de la Tunisie. Annales des Mines et de Géologie, Tunis (7).

[27] Castany, G., 1954. Les grands traits structuraux de la Tunisie. Bulletin de la Société géologique de France, 6, t. 4, (1-3), 151-173.

[28] Jauzein, A., 1967. Contribution à l'étude géologique des confins de la dorsale tunisienne (Tunisie Septentrionale). Thèse ès Sciences-Annales des Mines et de Géologie, Tunis (22).

[29] Auzende, J.M., Olivet, J.L., Bonnin, J., 1973. Le détroit sardano-tunisien et la zone de fracture nord-tunisienne. Tectonophysics, 21, 357-274.

[30] Caire, A., 1975. Les règles de la fracturation continentale dans l'évolution de l'écorce terrestre. Revue de Géographie physique et de Géologie dynamique (2), vol. XVII, Fasc. 4, 319-354.

[31] Zargouni, F., 1977. Etude structurale de la bande triasique de Baouala-Aroussia-El Mecherket (Zone de diapirs, Atlas tunisien). Bulletin de la Société des Sciences Naturelles, Tunisie 12, 79-82.

[32] Rouvier, H., 1977. Géologie de l'extrême Nord tunisien: tectonique et paléogéographie superposées à l'extrémité orientale de la chaîne nord maghrébine. Thèse ès Sciences, Université de Pierre et Marie Curie, Paris VI, France.

[33] Turki, M.M., 1988. Polycinématique et contrôle sédimentaire associés sur la cicatrice Zaghouan-Nabhana. Editée par le Centre des Sciences de la Terre, Institut National de Recherche Scientifique, Tunisie. Revue des Sciences de la Terre (7).

[34] Bishop, W.F., 1988. Petroleum Geology of East-Central Tunisia. The American Association of Petroleum Geologists Bulletin, 72 (9), 1033-1058.

[35] Tricart, P., Rorelli, L., Brancolini, G., Croce, M., De Santis, L., Peis, D., Zitellini, N., 1991. Dérives d'arcs insulaires et dynamique méditerranéenne suivant le transect Sardaigne-Afrique. C.R. Acad. Sc. Paris, t. 313, série II, 801-806.

[36] Alouani, R., Rais, J., Gaya, S., Tlig, S., 1991. Les structures en décrochement au Jurassique de la Tunisie du Nord : Témoins d'une marge transformante entre Afrique et Europe. C. R. Acad. Sci., Paris, pp. ???.

[37] Ben Ayed, N., 1993. Évolution tectonique de l'avant pays de la chaîne alpine de Tunisie du début Mésozoïque à l'Actuel. Annales des Mines et de Géologie, Tunisie (32).

[38] Dlala, M., 1995. Evolution géodynamique et tectonique superposée en Tunisie ; implication sur la tectonique récente et la sismicité. Thèse Doctorat ès-Sciences Géologique, Université de Tunis II, Faculté des Sciences de Tunis, 1-200.

[39] Chihi, L., 1995. Les fossés néogènes à quaternaires de la Tunisie et de la Mer pélagienne : étude structurale et leur sugnification dans le cadre géodynamique de la Méditerranée centrale. Thèse de Doctorat ès-Science, Faculté des Sciences de Tunis, Université Tunis II, 1-566.

[40] Boukadi, N., 1996. Un schéma structural nouveau pour le Nord de la Tunisie. The Tunisian Petroleum Exploration Conference, Tunis, Octobre 15-16, 91-100.

[41] Rigo, L., Garde, S., El Euchi, H., Bandt, K., Tiffert, J., 1996. Mesozoic fractured reservoirs in a compressional structural model for the north-eastern Tunisian atlasic zone. The Fifth Tunisian Petroleum Exploration Conference, Tunis, October 15-18, 233-255.

[42] Boutib, L., 1998. Tectonique de la région grand Tunis : évolution géométrique et cinématique des blocs structuraux du Mésozoïque à l'Actuel (Atlas nord oriental de Tunisie). Thèse d'Université, Université de Tunis II, Faculté des Sciences de Tunis, 1-151.

[43] Ben chelbi, M., 2007. Analyse tectonique des structures liées à la faille de Tunis Ellès. Thèse de Doctorat. Univ. Tunis El Manar, Fac. Sci. Tunis, 1-235.

[44] Rekhiss, F., 2007. Modèle d'évolution structurale et géodynamique à l'extrémité orientale de la chaîne alpine d'Afrique du Nord, Thèse es-sciences en Géol., Uni. de Tunis El Manar, 2007, 1-285.

[45] Talbi, F., Melki, F., Ben Ismail-Latrach, K., Alouani, R., Tlig S., 2008. Le Numidien de la Tunisie septentrionale: données stratigraphiques et interprétation géodynamique. Estudios Geológicos 64 (1), 31-44.

[46] Riahi, S., Soussi, M., Boukhalfa, K., Ben Ismail Lattrache, K., Dorrik, S., Khomsi, S., Bédir M., 2010. Stratigraphy, sedimentology and structure of the Numidian Flysch thrust belt in northern Tunisia. Journal of African Earth Sciences 57, 109–126.

[47] Melki, F., Z ouaghi, T., Harrab, S., Casas Sainz, A., Bédir, M., Zargouni, F., 2011. Structuring and evolution of Neogene transcurrent basins in the Tellian foreland domain, north-eastern Tunisia. Journal of Geodynamics 52 (2011) 57–69, doi:10.1016/j.jog2010.11.009.

[48] Vila, J.M., 1980. La chaîne Alpine d'Algérie orientale et des confins algéro-tunisiens. Thèse ès Sciences, Université Pierre et Marie Curie, Paris VI.

[49] Bracène, R., Frizon de Lamotte, D., 2002. The origin of intraplate deformation in the system of western and central Algeria: from Jurassic rifting to Cenozoic-Quaternary inversion. Tectonophysics 357, 207-226.

[50] Marmi, R., Guiraud, R., 2006. End Cretaceous to recent polyphased compressive tectonics along the "Mole Constantinois" and foreland (NE Algeria). Journal of African Earth Sciences 45, 123-136.

[51] Casero, P., Roure, F., 1994. Neogene deformation at the Sicilian-North African plate boundary. In Roure, F. (Ed), Peri-Tethyan platforms. Editions Technip, Paris, 189-200.

[52] Catalano, R., Franchino, A., Merlini, S., Sulli, A., 2000. Central western Sicily structural setting interpreted from seismic reflection profiles. Memorie della Società Geol. Italiana 55, 5-16.

[53] Sartori, R., Carrara, G., Torelli, L., Zitellini, N., 2001. Neogene evolution of the south-western Tyrrhenian Sea (Sardinia Basin and western Bathyal plain). Marine Geology 175, 47-66.

[54] Abbès, C., 2004. Structurations et evolutions tectono-sédimentaires mésozoïques et cénozoïques, associées aux accidents réghmatiques, à la jonction des marges téthysienne et Nord-africaine (Chaîne Nord-Sud-Tunisie central). Thèse Doctorat ès-Sciences, Université Tunis El Manar, 1-437.

[55] Zecchin, M., Massari, F., Mellere, D., Prosser, G., 2004. Anatomy and evolution of a Mediterranean-type fault bounded basin: the Lower Pliocene of the northern Crotone Basin (Southern Italy). Basin Research 16, 117-143.

[56] Corti, G., Cuffaro, M., Doglioni, C., Innocenti, F., Manetti, P., 2006. Coexisting geodynamic processes in the Sicily Channel in Dilek, Y., and Pavlides, S., eds., Postcollisional tectonics and magmatism in the Mediterranean region and Asia: Geolo. Soci. of America Special Paper 409, 83–96.

[57] Accaino, F., Catalano, R., Di Marzo, L., Giustiniani, M., Tinivella, U., Nicolich, R., Sulli, A., Valenti, V., Manetti, P., 2010. A crustal seismic profile across Sicily. Tectonophysics (2010), doi:10.1016/j.tecto.2010.07.011.

[58] Burollet, P.F., 1956. Contribution à l'étude stratigraphique de la Tunisie centrale. Ann. Mines et Géol., 18, Tunis, 22 pls., 1-352.

[59] Turki, M.M., 1980. La « faille de Zaghouan » est la résultante de structures superposées (Atlas tunisien central). Bulletin de la Société géologique de France (7), XXII, 321-325.

[60] Caire, A., 1970. Tectonique de la Méditerranée centrale. Annales de la Société géologique du Nord, 307-346.

[61] Caire, A., 1977. Interprétation tectonique unitaire de l'Atlas à fossés. Comptes Rendus de l'Académie des Sciences Paris, série D, t. 284, 403-406.

[62] Burollet, P.F., 1973. Importance des facteurs salifères dans la tectonique tunisienne. Livre Jubilaire M. Solignac, Annales des Mines et de la Géologie, Tunis, (26), 111-120.

[63] Perthuisot, V., 1978. Dynamique et pétrogenèse des extrusions triasiques en Tunisie septentrionale. Thèse ès Sciences, Ecole Normale Supérieure, ERA 604-CNRS, 1-321.

[64] Chikhaoui, M., Jallouli, C., Turki, M.M., Soussi, M., Braham, A., Zaghbib-Turkib, D., 2002. L'affleurement triasique du Debadib–Ben Gasseur (Nord-Ouest de la Tunisie) : diapir enraciné à épanchements latéraux dans la mer Albienne, replissé au cours des phases de compression tertiaires. C. R. Geoscience 334, 1129–1133.

[65] Vila, J.M., Ben Youssef, M., Charrière, A., Chikhaoui, M., Ghanmi, M., Kamoun, F., Peybernès, B., Saâdi, J., Souquet, P., Zarbout, M., 1994. Découverte en Tunisie, au SW du Kef, de matériel triasique inter stratifié dans l'Albien : extension du domaine à « glaciers de sel » sous-marins des confins algéro-tunisiens. Comptes Rendus de l'Académie des Sci. de Paris, 318, II (12), 1535-1543.

[66] Vila, J.M., Ben Youssef, M., Bouhlel, S., Ghanmi, M., Kassaâ, S., Miâadi, F., 1998. Tectonique en radeaux cénomano-turonien, au toit d'un « glacier de sel » albien de Tunisie du Nord-ouest : exemple du secteur minier de Gueurn Halfaya. Comptes Rendus de l'Académie des Sciences de Paris, 327, IIa (8), 563-570.

Role of the NE-SW Hercynian Master Fault Systems and Associated Lineaments on the Structuring and
Evolution of the Mesozoic and Cenozoic Basins of the Alpine Margin, Northern Tunisia
161

[67] Ghanmi, M., Vila, J.M., Ben Youssef, M., Jouirou M., Kechrid-Benkhrouf, F., 2000. Le matériel triasique interstratifié dans l'Albien de l'anticlinal autochtone atlasique du Jebel Takrouna (Tunisie) : Stratigraphie, arguments gravimétriques, signification dans la transversale N-S des confins algéro-tunisiens (Maghreb du Nord-est). Bull. de la Société de l'Histoire Naturelle de Toulouse, 136, 19-27.

[68] Ghanmi, M., Ben Youssef, M., Jouirou M., Zargouni, F., Vila, J.M., 2001. Halocinèse crétacé au Jebel Kebbouch (Nord-ouest tunisien) : mise en place à fleur d'eau et évolution d'un glacier de sel albien, comparaisons. Eclogea geol. Helv. 94, 153-160.

[69] Ben Chelbi, M., Melki, F., Zargouni, F., 2006. Mode de mise en place des corps salifères dans l'Atlas septentrional de Tunisie. Exemple de l'appareil de Bir Afou. C. R. Geoscience 338, 349-358.

[70] Bolze, J., 1954. Ascension et percée de diapirs au Crétacé moyen dans les monts de Teboursouk, C.R. somm. Soc. Géolo. France, 139-141.

[71] Crampon, N., 1971. Etude géologique de la bordure des Mogods, du Pays de Bizerte et du Nord des Hédil. Thèse es Sci., Uni. Nancy, 1-522.

[72] Bonnefous, J., 1972. Contribution à l'étude stratigraphique et micropaléontologique du Jurassique de Tunisie (Tunisie septentrionale et centrale, Sahel, zone des Chotts). Thèse d'État, université Paris-6, 1–397.

[73] Alouani, R., 1991. Le Jurassique du Nord de la Tunisie. Marqueurs géodynamiques d'une marge transformante. Turbidites, radiolarites, plissement et métamorphisme. Thèse de spécialité, Université de Tunis II, 1-200.

[74] Kammoun, F., Peybernès, B., Fauré P., 1999. Evolution paléogéographique de la Tunisie saharienne et atlasique du Jurassique. C. R. Académie des Sciences, Paris 328, 547-552.

[75] Soussi, M., 2000. Le Jurassique de la Tunisie atlasique : Stratigraphie, dynamique sédimentaire, paléogéographie et intérêt pétrolier. Thèse Doctorat ès Sciences, Université de Tunis-2, 1–661.

[76] Soussi, M., 2003. Nouvelle nomenclature lithostratigraphique « événementielle » pour le Jurassique de la Tunisie atlasique, Geobios, 36, 761-773.

[77] Boughdiri, M., Sallouhi, H., Maâlaoui, K., Soussi, M., Cordey, F., 2006. Calpionellid zonation of the Jurassic–Cretaceous transition in North-Atlasic Tunisia. Updated Upper Jurassic stratigraphy of the 'Tunisian trough' and regional correlations. C. R. Geoscience 338,1250–1259.

[78] Sekatni, N., 2009. Thèse d'université, Faculté des Sciences de Tunis.

[79] Rakus, M., Biely, A., 1971. Stratigraphie du Lias dans la Dorsale tunisienne. Notes Serv. Géol. Tunisie, n° 32, 45-63.

[80] Pini, S., 1971. Carte géologique de la Tunisie. Echelle 1/50 000. Feuille n° 13 : Ariana. Notice explicative, Dir. Mines et Géol. Tunis, 1-62.

[81] Ben Haj Ali, N., 2005. Les foraminifères planctoniques du Crétacé (Hautérivien à Turonien inférieur) de Tunisie : systématique, biozonation et précisions stratigraphiques. Thése d'Etat Es-Sciences géologiques, Université de Tunis El Manar, Faculté des Sciences de Tunis, 1-340.

[82] Maâmouri, A.L., Salaj, J., Maâmouri, M., Matmati, F., Zargouni, F., 1994. Le Crétacé inférieur du Jebel Oust (Tunisie nord-orientale) : microstratigraphie-biozonation-aperçu sédimentaire. Zemny plyn a nafta, 39, 1, 73-105. Hodonin, cerven 1994.

[83] Memmi, L., 1989. Le Crétacé inférieur Bérriasien-Aptien) de Tunisie: Biostratigraphie, Paléogéographie et Paléoenvironnement. Thèse d'Etat Es-Sciences géologqiues, Univ. Claude Bernard, Lyon 1, 1-158.

[84] Melki, F., Boutib, L., Zargouni, F., Alouani, R., 1999. Nouvelles données sur l'évolution structurale de l'extrémité nord-est de la Tunisie (région de Bizerte). Africa Geoscience Rev., 149-157.

[85] Ben Ismail-Lattrache, K,. Bobier, C., 1984. Sur l'évolution des paléo-environnements marins paléogènes des bordures occidentales du détroit siculo-tnisien et leurs rapports avec les fluctuations du paléo-océan mondial. Marine Geology, 55, 195-217.

[86] Ben Ismail-Lattrache, K., 2000. Précisions sur le passage Lutétien-Bartonien dans les dépôts éocènes moyens en Tunisie centrale et nord-orientale. Revue de Micropaléontologie 43 (1-2), 3-16.

[87] Boukhalfa, K., 2011. Sédimentologie et stratigraphie des séries oligo-miocènes de la Tunisie septentrionale. Thèse d'Université, Faculté des Sciences de Tunis, 1-327.

[88] Riahi, S., 2011. Sedimentologie, Stratigraphy, Provenance and Reservoir Potentiel of the Oligo-Miocene Numidian Flysch of Northern Tunisia. Thèse d'Université, Fac. Sci. de Tunis, 1-389.

[89] Yaïch, Ch., Hooyberghs, H.J.F., Durlet, Ch., Renard, M., 2000. Corrélation stratigraphique entre les unités oligo-miocènes de Tunisie centrale et le Numidien. Comptes Rendus de l'Académie des Sciences de Paris 331, 499-506.

[90] El Mehressi, C., 1991. Dynamique de dépôt du flysch numidien de la Tunisie (Oligo-Miocène). Thèse d'Université, Ecole des Mines de Paris. (15), 1-244.

[91] El Euch, N., Ferry, S., Suc, J.P., Clauzon, G., Carmen, M., Dobrinescu, M., Gorini, Ch, Safra, A., Zargouni, F., 2009. Messinian deposits and erosion in northern Tunisia: inferences on Strait of Sicily during the Mesinian Salinity Crisis. Terra Nova, Vol. 21, No. 1, 41-48.

[92] Ben Salem, H., 1992. Contribution à la connaissance de la géologie du Cap Bon: Stratigraphie, tectonique et sédimentologie. Thèse de 3ème cycle, Université de Tunis 2, Faculté des Sciences de Tunis.

[93] Bédir, M., Tlig, S., Bobier, Cl., Aissaoui, N., 1996. Sequence Stratigraphy, Basin Dynamics, and Petroleum Geology of the Miocene Eastern Tunisia. American Association of Petroleum Geologists Bulletin 80 (1), 63-81.

[94] Mannaï-Tayech, B., 2009. The lithostratigraphy of Miocene series from Tunisia, Revisited. Journal of African Earth Sciences 54, 53–61.

[95] Mannaï-Tayech, B., 2006. Les series silicoclastiques miocènes du Nord-Est au Sud-Ouest de la Tunisie: une mise au point, Geobios, 39, 71-84.

[96] Ben Ayed, N., Gueddiche, M., El Ghali, A., Amar, H., Boughdiri, M., 1996. Évolution géodynamique et néotectonique du bassin molassique de Kechabta. The Fifth Tunisian Petroleum Exploration Conference, Tunis, October 15-18, 83-89.

[97] Oueslati, A., 1989. Les côtes de la Tunisie. Recherche Géomorphologique. Thèse de Doctorat d'Etat en Géographie Physique, Uni. Tunis I, Fac. Sc. Hum. Soc., 1-682.

[98] Zargouni, F., 1978. Analyse structurale de la chaîne de Lansarine (zone des diapirs, Atlas tunisien). Bulletin de la Société des Sciences Naturelles, Tunisie, t. 13, 97-104.

[99] Ben Haj Ali, M., Chihi, L., Rabhi, M., 1998. Le Néogène de Tunisie: Analyse des séries stratigraphiques et le rôle de la tectonique et de l'halocinèse dans la genèse des bassins. Proceedings of the 6th Tunisian petroleum exploration and production conference, ETAP, Memoir (12), 59-76.

[100] Castany, G., 1951. Etude géologique de l'Atlas tunisien oriental. Annales des Mines et de Géologie, Tunis, 8.

[101] Caire, A., 1978. The central mediterranean mountains chains in the Alpine orogenic environment. The ocean basins and margins, Vol. 4B, Edited by Alan E.M. Nairm, William H. Kanes and Francis G. Stehli (Plenum Publishing Corporation), 201-256.

[102] Morgan, M.A., Grocott, J., Moody, R.T.J., 1998. The structural evolution of the Zaghouan-Ressas Belt, northern Tunisia. Geological Society, London, Special Publication, 132, 405-422.

[103] Boutib, L., Melki, F., Zargouni, F., 2000. Tectonique synsédimentaire d'âge crétacé supérieur en Tunisie nord orientale: blocs basculés et inversion de subsidence. Bull. Soc. Géol. France 171 (4), 431-440.

[104] Ben Chelbi, M., Melki, F., Zargouni, F., 2008. Précision sur l'évolution structurale de l'Atlas septentrional de Tunisie depuis le Crétacé (Bassin de Bir M'Cherga). Echos d'une évolution polyphasée de la marge tunisienne dans son cadre méditerranéen. African Geoscience Review, 15, (3 & 4), 229-246.

[105] El Ouardi, H., Turki, M.M., 1995. Tectonique salifère polyphasée dans la région de Mejez el Bab–Testour (zone des dômes, Tunisie septentrionale): contrôle de la sédimentation méso-cénozoïque. Geol. Mediterr. 22 (2) (1995) 73–84.

[106] Hamdi-Nasr, I., Inoubli, M.H., Ben Salem, A., Tlig, S., Mansouri, A., 2009. Gravity contributions to the understanding of salt tectonics from the Jebel Cheid area (dome zone, Northern Tunisia). *Geophysical Prospecting,* 2009, 57, 719–728.

[107] Ben Slama, M.M., Masrouhi, A., Ghanmi, M., Ben Youssef, M., Zargouni, F., 2009. Albian extrusion evidences of the Triassic salt and clues of the beginning of the Eocene atlasic phase from the example of the Chitana-Ed Djebs structure (N.Tunisia): Implication in the North African Tethyan margin recorded events, comparisons. C. R. Geoscience 341 (2009) 547–556.

[108] Perthuisot, V., Jauzein, A., 1974. L'accident El Alia-Tebessa dans la région de Téboursouk. Notes Service Géologique, Tunis (9), 57-63.

[109] Marzougui, W., Melki, F., Zargouni, F, 2008. Modèle évolutif de mise en place de la structure d'Aouana-Bazina (extrémité orientale de la chaîne alpine de la Tunisie septentrionale) : apport du filtrage des anomalies gravimétriques. 22nd Colloquium of African Geology, 13th Conference of the Geological Society of Africa, November 04-06-2008, Abstract, 289-290.

[110] Ben Ayed, N., 1994. Les décrochements-chevauchements EW et NS convergents de la Tunisie septentrionale : géométrie et essai de reconstruction des conditions de

déformations. Proceeding of the 4th Tunisian Petroleum Exploration Conference, Mémoire n°7, ETAP, 25-37.

[111] Melki, F., Alouani, R., Talbi, F., Zargouni, F., 1996. Evolution géodynamique de la région de Bizerte. Entraînement et blocage des structures à partir du Paléogène. The Fifth Tunisian Petroleum Exploration Conference, Tunis, October 15-18, 413-421.

[112] Melki, F., Zargouni, F., 1991. Tectonique cassante post Jurassique de la mine de Hammam Zriba (Tunisie nord-orientale). Incidence sur la karstification et les concentrations de fluorine, barytine et célestine, d'environnement carbonaté. Bull. Soc. Géol. France 162, (5), 851-858.

[113] Dubourdieu, G., 1959. La déformation récente de l'Afrique du Nord. Extrait des comptes rendus de l'Académie des Sciences, t. 249, pp. 2799-2801, séance du 21 décembre 1959.

[114] Glangeaud, 1951. Interprétation tectono-physique des caractères structuraux et paléogéographiques de la Méditerranée occidentale. Bulletin de la Société géologique de France, t.6, 735-762.

[115] Chandoul, H., Burollet, P.F., Ben Ferjani, A., Memmi, L., 1993. Recueil des Coupes Types de Tunisie –I- Trias et Jurassique. Entreprise tunisienne d'activités pétrolières, Mémoire (4), 1-95.

[116] Zouaghi T., Guellala R., Lazzez M., Bédir M., Ben Youssef M., Inoubli M.H., Zargouni F., The Chotts Fold Belt of Southern Tunisia, North African Margin: Structural Pattern, Evolution, and Regional Geodynamic Implications. New Frontiers in Tectonic Research - At the Midst of Plate Convergence. Intech; 2011. p49-72.

[117] Alouani, R., Tlig, S., Zargouni, F., 1990. Découverte de radiolarites du Jurassique supérieur dans le « sillon tunisien ». Faciès et structures d'une marge SE de la Téthys maghrébine. C. R. Acad. Sci. Paris, t. 310, série II, 609-612.

[118] Soussi, M., Mangold, C., Enay, R., Boughdiri, M., Ben Ismail, M.H., 2000. Le Jurassique inférieur et moyen de la Tunisie septentrionale ; corrélations avec l'axe Nord-Sud et paléogéographie. GEOBIOS, 33, 4: 437-446.

[119] Alouani, R., Rais, J., Gaya, S., Tlig, S., 1992. Les structures en décrochement au Jurassique de la Tunisie du Nord : Témoins d'une marge transformante entre Afrique et Europe. C. R. Acad. Sci. Paris, t. 315, série II, 717-724.

[120] Rabhi, M., 1999. Contribution à l'étude stratigraphique et analyse de l'évolution géodynamique de l'axe Nord-Sud et des structures avoisinantes (Tunisie centrale). Thèse d'Université, Faculté des Sciences de Tunis, Université de Tunis II, 1-217.

[121] Chikhaoui, M., Turki, M.M., Delteil, J., 1991. Témoignages de la structurogenèse de la marge téthysienne en Tunisie, au Jurassique terminal-Crétacé (Région du Kef, Tunisie septentrionale). Géologie Méditerranéenne, XVIII (3), 125-133.

[122] El Ouardi, H., 2002. Origine des variations latérales des dépôts yprésiens dans la zone des dômes en Tunisie septentrionale, C. R. Geoscience 334 (2002) 141–146.

[123] Bouaziz, S., Barrier, E., Soussi, M., Turki, M.-M., Zouari, H., 2002. Tectonic evolution of the northern African margin in Tunisia from paleostress data and sedimentary record. Tectonophysics 357, 227-253.

[124] Rais, J., Gaya, S., Alouani, R., Mouguina, M., Tlig, S., 1991. Le sillon tunisien : structuration synsédimentaire jurassique. Rift avorté et cicatrisé au Dogger supérieur–Malm, à la marge SE de la Téthys maghrébide, C. R. Acad. Sci. Paris, Ser. II 312 (1991) 1169–1175.

[125] Sekatni, N., Fauré, P., Alouani, R., Zargouni, F., 2008. Le passage Lias–Dogger de la Dorsale de Tunisie septentrionale. Nouveaux apports biostratigraphiques. Âge Toarcien supérieur de la distension téthysienne. C. R. Palevol 7 (2008) 185–194.

[126] Bobier, C., Viguier, C., Chaâri A., Chine, A., 1991. The post-Triassic sedimentary cover of Tunisia: seismic sequences and structure. Tectonophysics, 195, 371-410.

[127] El Ouardi, H., 1996. Halocinèse et rôle des décrochements dans l'évolution géodynamique de la partie médiane de la zone des dômes, Thèse d'Université, Faculté des Sciences de Tunis, Université de Tunis II, 1- 242.

[128] Martinez, C., Chikhaoui, M., Truillet, R., Ouali, J., Creuzot, G., 1991. Le contexte géodynamique de la distension albo-aptienne en Tunisie septentrionale et centrale : structuration éocrétacée de l'Atlas tunisien. Eclogae geol. Helv., 84, 1, 61-82.

[129] Salaj, J., 1980. Microbiostratigraphie du Crétacé et du Paléogène de Tunisie septentrionale et Orientale (hypostratotypes tunisiens). Thèse d'Etat, Institut de Dionyzhir, 238 p., Bratislava.

[130] Boukadi, N., Bédir, M., 1996. L'halocinèse en Tunisie : contexte tectonique et chronologique des évènements. Comptes Rendus de l'Académie des Sciences, Paris, 322, série IIa, 587-594.

[131] Laâridhi-Ouazaâ, N., Bédir, M., 2004. Les migrations tectono-magmatiques du Trias au Miocène sur la marge orientale de la Tunisie. African Geoscience Review 11 (3), 179-196.

[132] Talbi, F., Jaafari, M., Tlig, S., 2005. Magmatisme néogène de la Tunisie septentrionale : Pétrogenèse et évènements géodynamiques. Revista de la Sociedad Geológica de España, 18(3-4).

[133] Bishop, W.F., 1980. Eocene and Upper Cretaceous reservoirs in the East Central Tunisia. Oil and Gas Journal, 137-142.

[134] Er-Raioui, L., 1994. Environnements sédimentaires et Géochimie des séries de l'Eocène du Nord-Est de la Tunisie. Thèse 3ème Cycle, Faculté des Sciences de Tunis, 1-244.

[135] Zaïer, A., Beji-Sassi, A., Sassi, S., Moody, R.T.J., 1998. Basin evolution and deposition during the Early Paleogene in Tunisia. In: Macgrfgor D.S., Moody R.T.J. & Clarck-Lowis D.D. (eds) 1998. Petroleum Geology of North Africa, Geological Society, London, Special Publication, (132), 375-393.

[136] Tlig, S., Sahli, S., Er-Raioui, L., Alouani, R., Mzoughi, M., 2010. Depositional environment controls on petroleum potential of the Eocene in the North of Tunisia. Journal of Petroleum Science and Engineering 71 (2010), 91–105. doi:10.1016/j.petrol.2010.01.009.

[137] Masrouhi, A., Ghanmi, M., Ben Slama, M.M., Ben Youssef, M., Vila, J.M., Zargouni, F., 2008. New tectono-sedimentary evidence constraining the timing of the positive tectonic inversion and the Eocence Atlasic phase in northern Tunisia: Implication for the North Africa paleo-margin evolution. C. R. Geoscience 340, 771-778.

[138] Batik, P., 1977. Faciès de l'Eocène dans la région des Hedil. Notes Service géologique de Tunisie, (43), 63-72.

[139] Letouzey, J., Trémolière, 1980. Paleo-stress around the Mediterranean since the Mesozoic from microtectonics : comparison with plate tectonics data. Rock Mech., 9, 173-192.

[140] Gharbi, R.A., Chihi, L., Hammami, M., Soumaya, A., Kadri, A., 2005. Manifestations tectono-diapiriques synsédimentaires et polyphasées d'âge Crétacé supérieur-Quaternaire dans la région de Zag Et Tir (Tunisie, centre-nord), C. R., Geoscience 337, 1293–1300.

[141] Ben Mehrez, F., Kacem, J., Dlala. M., 2009. Late Cretaceous synsedimentary diapirism of Bazina-Sidi Bou Krime Triassic evaporates (Northern Tunisia): Geodynamic implication. C. R. Geoscience 341, 78-84.

[142] El Ghali, A., Ben Ayed, N., Bobier, C., Zargouni, F., Krima, A., 2003. Les manifestations tectoniques synsédimentaires associées à la compression éocène en Tunisie, implications paléogéographiques et structurales sur la marge Nord-Africaine. C. R. Geoscience 335 (2003) 763–771.

[143] Meulenkamp, J.E., Sissingh, W., 2003. Tertiary palaeogeography and tectonostratigraphic evolution of the Northern and Southern Peri-Tethys platforms and the intermediate domains of the African-Eurasian convergent plate boundary zone. Palaeogeography, Palaeoclimatology, Palaeoecology 196 (2003) 209-228.

[144] Khomsi, S., Bédir, M., Soussi M., Ben Jemia, M.G., Ben Ismail-Lattrache, K., 2006. Mise en évidence en subsurface d'événements compressifs Éocène moyen–supérieur en Tunisie orientale (Sahel) : généralité de la phase atlasique en Afrique du Nord. C. R. Geoscience 338, 41–49.

[145] Benaouali-Mebarek, N., Frizon de Lamotte, D., Roca, E., Bracène, R., Faure, J.-L., Sassi, W., Roure, F., 2006. Post-Cretaceous kinematics of the Atlas and Tell systems in central Algeria: Early foreland folding and subduction-related deformation. C. R. Geoscience 338, 115-125

[146] De Jong, K.A., 1975. Gravity tectonics or plate tectonics : example of the Numidian flysch, Tunisia. Geol. Mag., 112 (4), 373-381.

[147] Er-Raioui, L., Alouani, R., Tlig, S., 1995. Les bassins de l'Oligo-Miocène inférieur de la Tunisie nord-orientale : structures et remplissage de fossés intracontinentaux. Notes du Service Géologique de Tunisie, (61), 43-61.

[148] Rebaï, S., 1992. Sismotectonique et champ de contrainte dans les chaînes alpines et dans les plate-formes de l'Europe, d'Afrique du Nord et du Moyen-Orient. Thèse de Doctorat, Académie de Montpellier, Université Montpellier II, Sciences et Techniques du Languedoc, 1-210.

[149] Tlig, S., Erraoui, L., Ben Aissa, L., Alouani, R., Tagourti, A., 1991. Tectogenèse alpine atlasique: Des évènements distincts de l'histoire géologique de la Tunisie. Corrélations avec les évènements clés de la Méditerranée. Comptes Rendus de l'Académie des Sciences de Paris 312, 295-301.

[150] Ben Ayed, N., Viguier, C., Bobier, C., 1983. Les éléments structuraux récents essentiels de la Tunisie nord-orientale. Notes Service géologique de Tunisie (47), 5-19.

Role of the NE-SW Hercynian Master Fault Systems and Associated Lineaments on the Structuring and
Evolution of the Mesozoic and Cenozoic Basins of the Alpine Margin, Northern Tunisia

167

[151] Gueddiche, M., Harjono, H., Ben Ayed, N., Hfaiedh, M., Diament, M., Dubois, J., 1992. Analyse de la séismicité et mise en évidence d'accidents actifs dans le Nord de la Tunisie. Bulletin de la Société géologique de France, (4), 415-425.

[152] Hfaiedh, M., Ben Ayed, N., Dorel, J., 1985. Etude néotectonique et séismotectonique de la Tunisie nord-orientale. Notes du Service Géologique de Tunisie, n° 51, 41-55.

[153] Gueddiche, M., Ben Ayed, N., Mohammadioun, G., El Ghali, A., Chekhma, H., Diament, M., Dubois, J., 1998. Etude sismotectonique de la Tunisie nord-orientale. Bulletin de la Société géologique de France, t. 169, (6), 789-796.

[154] Zouaghi, T., Bédir, M., Melki, F., Gabteni, H., Gharsalli, R., Bessioud, A., Zargouni, F., 2010. Neogene sediment deformations and tectonic features of northeastern Tunisia: evidence for paleoseismicity. Arabian Journal of Geosciences, DOI 10.1007/s12517-010-0225-z.

[155] Kazi-Tani, N., 1986. Evolution géodynamique de la bordure Nord africaine: Le domaine intraplaque nord-algérien. Approche mégaséquentielle. Thèse ès Sc. Uni. Pau, Bordeaux, 1-871.

[156] Guardia, P., 1975. Géodynamique de la marge alpine du continent africain d'après l'étude de l'Oranie nord-occidentale. Thèse ès Sciences, Inst., Polytech. Médit., Nice.

[157] Laville, E., 1985. Evolutions sédimentaire, tectonique et magmatique du bassin jurassique du Haut Atlas (Maroc). Modèle en relais multiples de décrochement. Thèse ès Science.

[158] Adil, S., 1993. Dynamique du Trias dans le Nord de la Tunisie: Bassin en relais multiples de décrochement, magmatisme et implications minières. Thèse de 3ème cycle, Université de Tunis 2, 1-248.

[159] Alvaro, M., Capote, R., Vegas, R., 1979. Un modela de evolucion geotectonica para la cadena celtiberica. Acta. Geol. Hispanica, 172-177.

[160] Uchupi, E., 1988. The Mesozoic-Cenozoic evolution of Iberia. A tectonic link between Africa and Europe. Rev. Soc. Geol. Espana, 1, 257-294.

[161] Raoult, J.-F., 1974. Géologie du centre de la chaîne numidienne (Nord du Constantinois, Algérie). Thèse ès Sciences, Paris, 1-163.

[162] Melki, F., Zouaghi, T., Ben Chelbi, M., Bédir, M., Zargouni, F., 2009. Mesozoic and Cenozoic tectono-sedimentary events and geodynamic evolution of the Alpine margin in the North-Eastern Tunisia offshore (Gulf of Tunis) using the subsurface data. 4ème Congrès Maghrébin de Géophysique Appliquée, Hammamet (Tunisie), 26-27-28 mars 2009, Abstract, 79.

[163] Castany, G., 1956. Essai de synthèse géologique du territoire Tunisie-Sicile. Annales des Mines et de Géologie (16).

[164] Philip, H., 1987. Plio-Quaternary evolution of the stress field in Mediterranean zones of subduction and collision. Annales Geophysicae 5 (3), 301-319.

[165] Dlala, M., Rebaï, S., 1994. Relation compression-extension Miocène supérieur à Quaternaire en Tunisie : implication sismotectonique. C. R. Académie des Sciences, Paris 319, 945-950.

[166] Chihi, L., Philip, H., 1998. Les fossés de l'extrémité orientale du Maghreb (Tunisie et Algérie orientale): tectonique mio-plio-quaternaire et implication dans l'évolution

géodynamique récente de la Méditerranée occidentale, Notes du Service Géologique de Tunisie (64), 103-116.

[167] Kacem, J., 2004. Étude sismotectonique et évaluation de l'aléa sismique régional du Nord-Est de la Tunisie: apport de la sismique réflexion dans l'identification des sources sismogéniques. Thèse de Doctorat d'Université, Université de Tunis El Manar, Faculté des Sciences de Tunis.

[168] Le Pichon, X., Bergerat, F., Roulet, M.J., 1988. Plate kinematics and tectonics leading to the Alpine belt formation; a new analysis. Geol. Soc. Am., sp. Paper, 218, 111-131.

[169] Bédir, M., Zargouni, F., Tlig, S., Bobier, C., 1992. Subsurface Geodynamics and Petroleum of Transform Margin Basins in the Sahel of Mahdia and El Jem (Eastern Tunisia). American Association of Petroleum Geologists Bulletin 76 (9), 1417-1442.

Neotectonics

Paleoseismological Three Dimensional Virtual Photography Method; A Case Study: Bağlarkayası-2010 Trench, Tuz Gölü Fault Zone, Central Anatolia, Turkey

Akın Kürçer and Yaşar Ergun Gökten

Additional information is available at the end of the chapter

1. Introduction

In order to make earthquake risk analysis in reliable manner in the regions carrying seismic risks, earthquake behavior of the active faults present in that region must be well determined. An active fault can be defined as a fault that produced earthquakes in Quaternary time associated with surface rupturing or deformation and has the potential to produce earthquakes in the future. The most important method which is accepted and commonly used at the present time to reveal earthquake behavior of active faults is paleoseismology. Paleoseismology is a method that tries to obtain information on the location, nature and date of historical earthquakes making use of geological and geomorphological data (McCalpin and Nelson, 2009).

Paleoseismic trenching is one of the methods that is frequently applied and provides noteworthy data in paleoseismology (Pantosti and Yeats, 1993; Demirtaş, 1997). In this method, geological evaluations are made according to the principles of sedimentology, stratigraphy and structural geology within trenches excavated perpendicular or parallel to the active fault trace depending on the faulting type. Appropriate samples collected from sediments are enable to date the historical earthquakes by suitable radiometric dating techniques.

Excavation types and dimensions of paleoseismological trenches can show variations according to the properties of the studied fault (faulting type, annual slip rate, earthquake recurrence interval, amount of displacement occurred in each earthquake, the elapse time from the last known historical earthquake until today, etc.), and the physical parameters of the trench site (groundwater condition, stability of trench sediments, etc.). The paleoseismological excavations which are conducted in faults having relatively lower slip rate (< 1 mm / year) and in which dip slip is dominant, are made deeper and wider. In areas where the groundwater level is shallow or the trench walls are not stable, trenches are excavated benched and/or sloped for the sake of safety.

After the trench walls are cleaned, gridding is performed taking into consideration detailed geological features within the trench. In gridding, a grid interval of 1 m² is generally accepted as standard (McCalpin, 2009). After gridding the opposite walls in same scale the stratigraphic levels and displacement figures on the walls are visually recorded and photographed. Logging can be conducted based on two principles: subjective and objective (Hatheway and Leighton, 1979). In subjective logging, the logger reflects his personal point of view in the trench log. As for the objective logging, in that, all sorts of details observed on trench walls are tried to be reflected in the trench log rather than the personal point of view. Although objective logging is possible in theory, in practice its application is difficult and most of the time not fit for purpose. For this reason, the logger must transfer paleoseismological details on the trench walls with an objective point of view adding some subjective interpretation. Three methods are used in trench logging. These are: manual trench logging, electronic trench logging and photomosaic trench logging (McCalpin, 2009). These methods have numerous advantages and disadvantages compared to each other (for further detail, see McCalpin, 2009).

Trench logs are, after all, two dimensional figures that reflect the personal interpretation of the logger about that trench. In recent years, photomosaic trench logging technique started to be frequently applied in paleoseismology. Photomosaic trench logging technique is a fast and inexpensive method which minimizes the possible errors may be rised from drawing. However, this method remains inefficient in deep and benched trench excavations since it causes some image losts on photographs during stitching them together.

In this article, a new photographic method is suggested to be used for paleoseismological trench works and this method is named 'Paleoseismological Three Dimensional Virtual Photography Method'. This method has been applied for the first time in the paleoseismological study conducted on the Tuz Gölü Fault Zone (Central Anatolia, Turkey).

2. Tuz Gölü Fault Zone – Akhisar-Kılıç segment

The Tuz Gölü Fault Zone (TGFZ) is one of the most important intracontinental active fault zones of the Central Anatolian Region (Şaroğlu et al., 1992; Dirik and Göncüoğlu, 1996; Çemen et al., 1999; Koçyiğit, 2000) (Figure 1). When taken into account its morphotectonic features and the distribution of the epicenters of the earthquakes with magnitudes reaching up to 5, it can be said that TGFZ is still active (Dirik and Erol, 2000). TGFZ is a NW-SE trending, approximately 200 km long, oblique-slip normal fault zone with a right lateral strike-slip component. It extends from Tuz Gölü to the northwest and Kemerhisar (Niğde) to the southeast. This fault zone is at the same time a transition zone that separates the region of normal component strike slip neotectonic regime to the east from the region of extensional neotectonic regime to the west (Koçyiğit and Özacar, 2003).

TGFZ is composed of fault segments parallel or subparallel to each other and with lengths varying between 4 and 33 km. Because of its geological fault length and morphotectonic features, the Akhisar-Kılıç Segment (AKS) constitutes one of the most important fault segments of TGFZ. AKS is a 25 km long, N25^0 - 30^{0}W trending fault. It is located in the middle section of TGFZ and it extends between Akhisar Village and Hasandağ (Figure 2).

In Figure 3, a geologic map illustrates the region along the Akhisar-Kılıç Segment. According to this map, the Akhisar-Kılıç Segment cuts Lower Pliocene aged Kızılkaya ignimbrites in the vicinity of Akhisar village to the northwest. Between Akhisar and Yuva villages, it forms the boundary between Oligocene aged Yassıpur formation and Quaternary aged alluvial fan deposits and in places it cuts these alluvial fan deposits. The Akhisar-Kılıç Segment cuts Hasandağ volcanites starting from the northwest of Yuva village. The fault segment which borders from the northeast the NW-SE trending area of depression with an ellipsoidal geometry, at the same time cuts Late Pleistocene-Holocene deposits in places. It again cuts Hasandağ volcanites in the section starting from the southeast of Helvadere to Kılıç Ridge and comes to an end at Kılıç Ridge locality. The Akhisar-Kılıç Segment is characterized by the alluvial fans arranged parallel to each other and the linear fault scarps between Akhisar and Yuva villages. A number of cold and brackish water springs are present along the fault trace in this area.

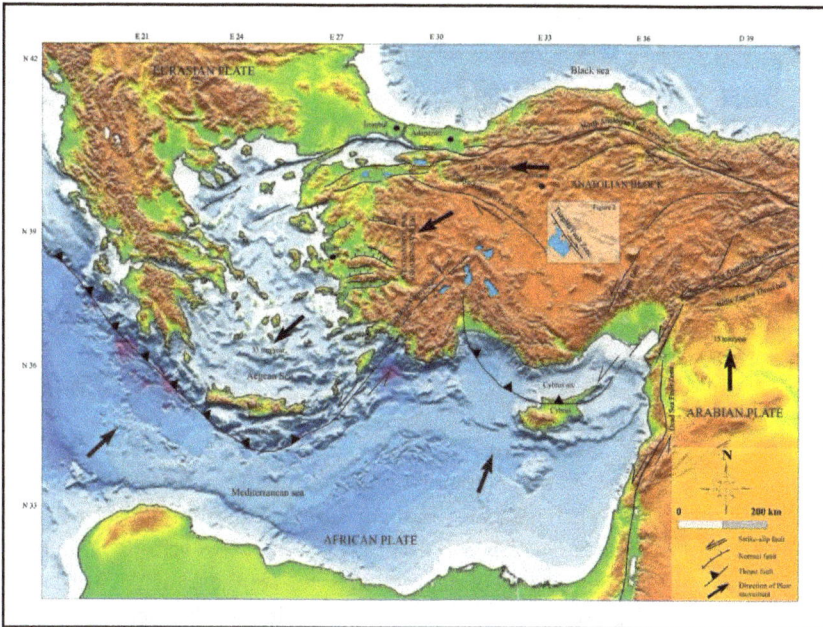

Figure 1. Main neotectonic elements of Eastern Mediterranean Region and Location of Tuz Gölü Fault Zone (Modified from Okay et al., 2000). Black arrovs and corresponding number show GPS-derived plate velocities (mm-year) relative to Eurasia (Reilinger at al., 2006).

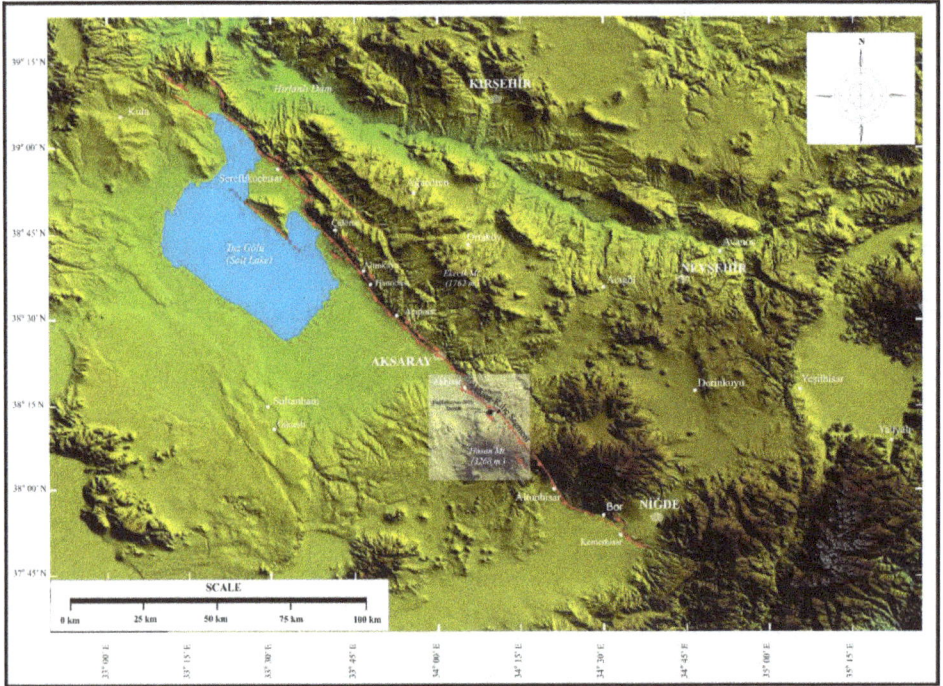

Figure 2. Segments of Tuz Gölü Fault Zone and setting of Akhisar-Kılıç Segment and Bağlarkayası-2010 Trench on the Digital Elevation Model of the region (Shuttle Radar Topography Mission-SRTM data were used for the Digital Elevation Model)

3. Bağlarkayası-2010 trench

Bağlarkayası-2010 Trench excavation is the first paleoseismological study carried out on the Tuz Gölü Fault Zone. The Trench is located in the middle part of the Akhisar-Kılıç Segment (See Figure 3, GPS Coordinates: 34.178505^0 E – 38.235561^0 N). The trench site is situated within the fault-controlled Quaternary area of depression to the southeast of Yuva village. In this depression area, Hasandağ volcanites were cut by the fault and obliquely displaced. Since the normal component of the fault was dominant, the hanging wall was tilted towards the fault and owing to this back-tilting a topographical saddle came into existence. Bağlarkayası-2010 Trench was excavated on this topographical saddle (Figure 4).

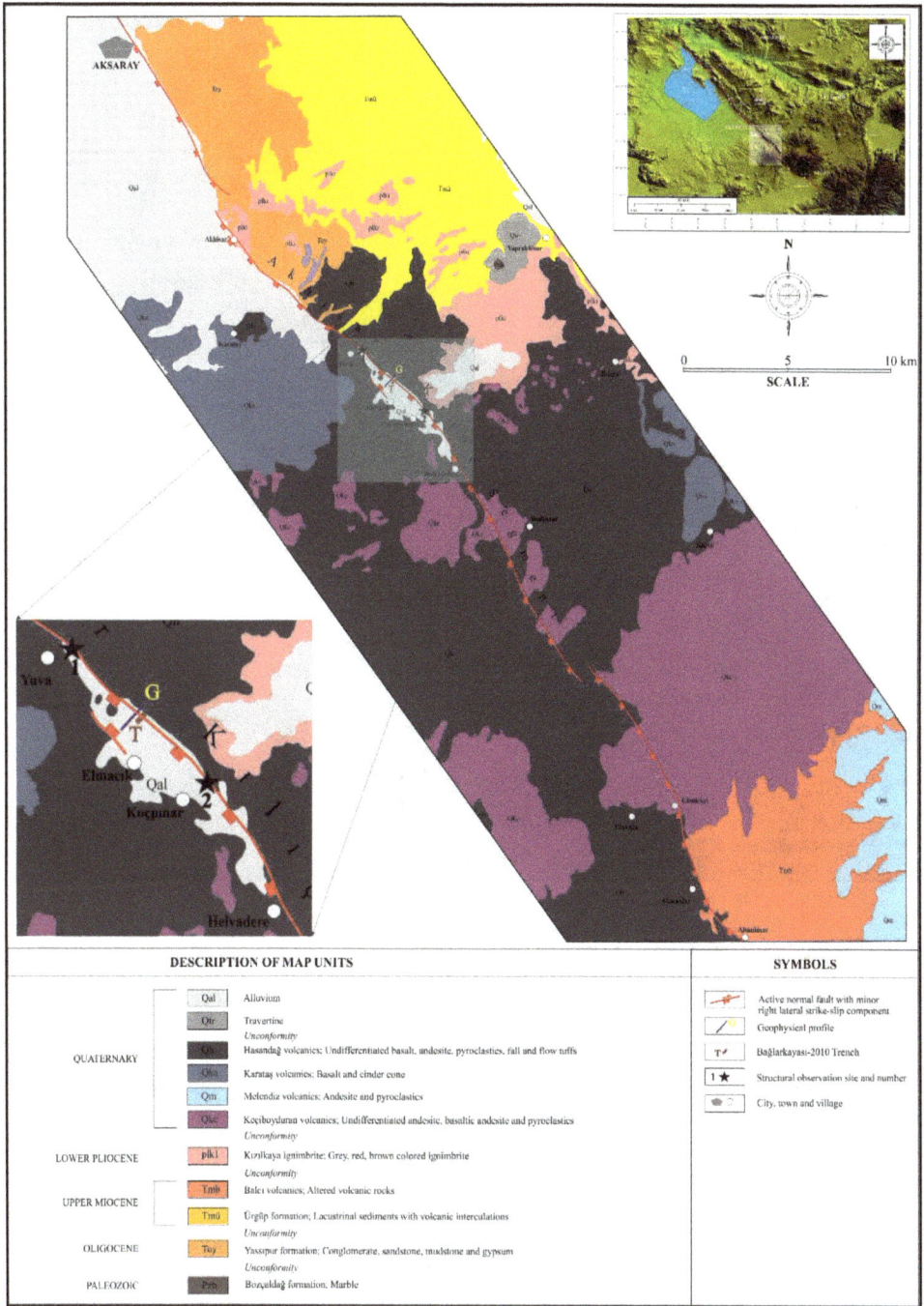

Figure 3. Geological map of Akhisar-Kılıç Segment and its vicinity (Modified from Dönmez et al., 2005).

Figure 4. Panoramic view of Bağlarkayası-2010 Trench area (view from NE to SE)

In the selection of trench site, the fault plane outcrops existing in the near north and south of the trench area were made use of (See, points 1 and 2 in Figure 3 and Figure 5). In addition to this, 8 vertical electric soundings with intervals of 250 meters, along a line perpendicular to the fault trace were carried out in the area which was considered for the trench site. In these vertical electric soundings, data were gathered from a depth of approximately 2000 meters. A geoelectric cross section was generated making use of these data. On the generated geoelectric cross section, a fault that reached up to the surface was detected (Figure 6). The site of the trench was selected by integrating the fault data determined from the geoelectric cross section and structural observations with the geomorphology. The 1/500 scale microtopographical map of the Bağlarkayası-2010 Trench area is presented in Figure 7.

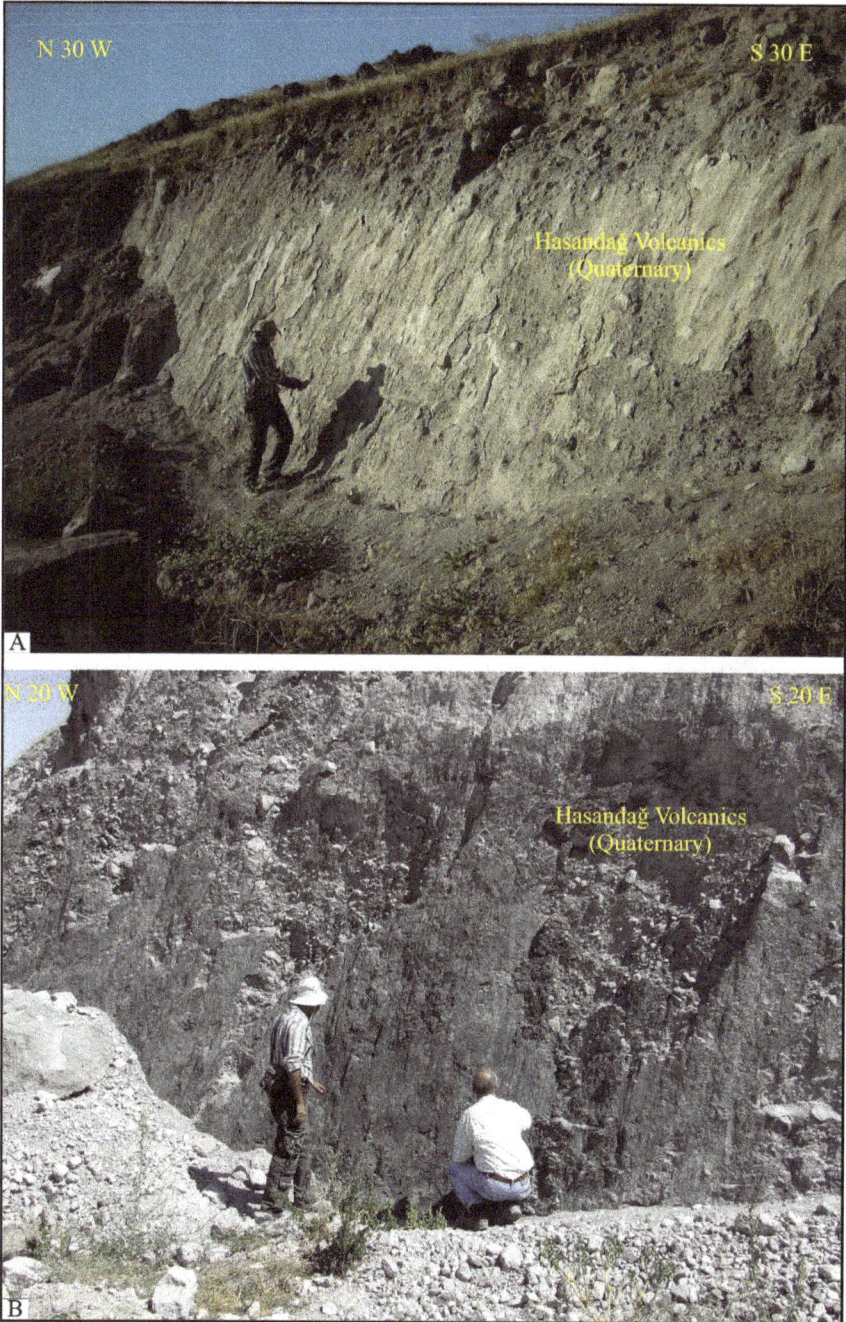

Figure 5. Fault planes belonging to Akhisar-Kılıç Segment that cuts Hasandağ volcanites A. SE of Yuva Village, B. NE of Koçpınar Village (see Figure 3 for Locations)

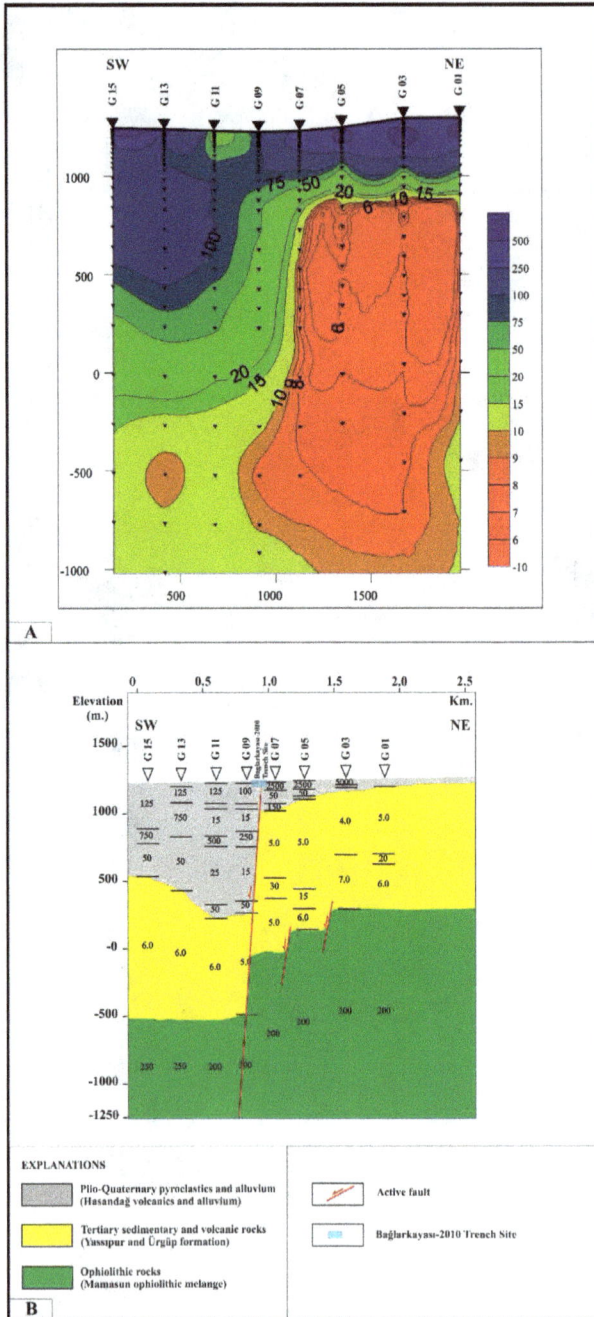

Figure 6. A. Apparent iso-resistivity cross section, B. Geoelectric cross section, generated from Vertical Electric Soundings (see Figure 3 for Profile Location)

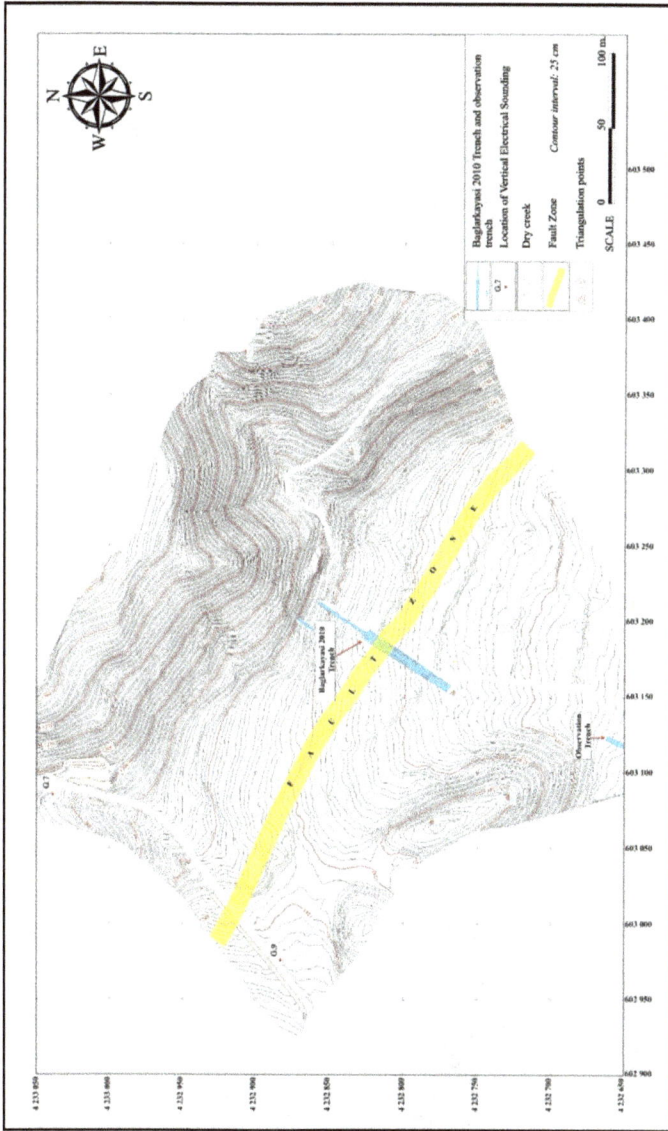

Figure 7. Microtopographic map of the Bağlarkayası 2010 Trench site

The Bağlarkayası-2010 Trench was excavated perpendicular to the fault. It has a length of 94 meters, a width of 5 meters and an average depth of 5 meters (maximum depth: 8.5 meters). While the northern 40-meter section of the trench was excavated as a single slot trench, the southern part of it was excavated as multi-bench trench (Figure 8). A gridding of 1 m² was applied in the trench. Manual trench logging method was applied for the entire trench, at a scale of 1/20. In Figure 9, the stages of the trench work are presented.

Figure 8. Aerial panoramic view of Bağlarkayası-2010 Trench, Photograph was taken from an altitude of 25 meters at inclined angle, from the fire tower (view from SW to NE)

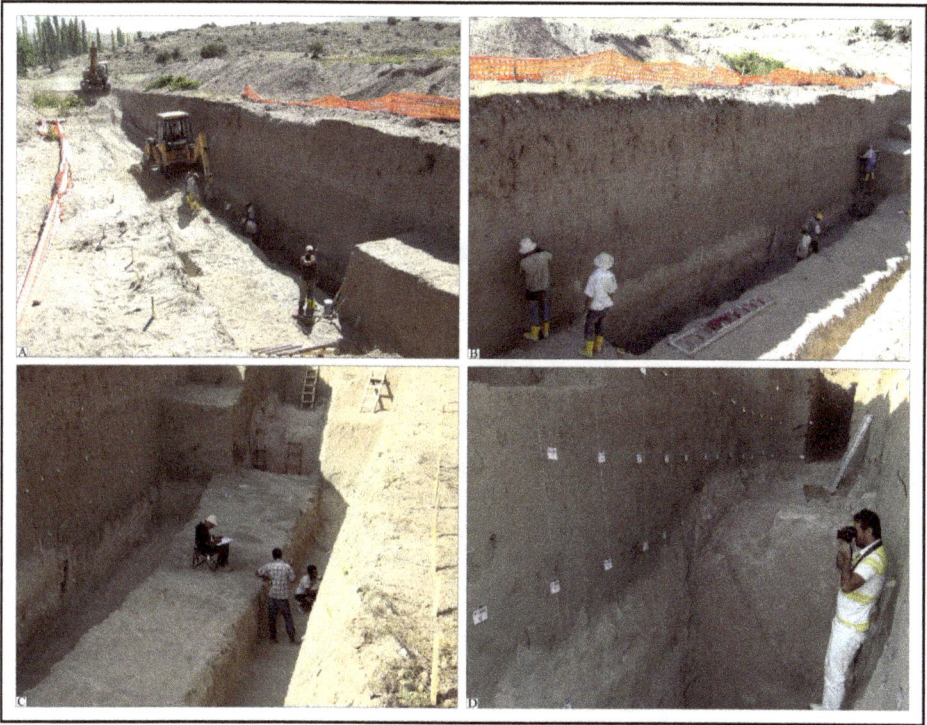

Figure 9. The stages of the trench work A. Excavation and cleaning, B. Gridding, C. Manual logging D. Photographing

In Figures 10 and 11, wall logs belonging to Bağlarbaşı-2010 Trench are presented. Seven different microstratigraphic levels were identified within Bağlarbaşı-2010 Trench. The first two of these levels that are relatively older were interpreted as ash and block flows of Hasandağ volcanism. And the relatively younger units are deposits associated with plinian activity and fluvial processes. Three deformation zones were defined within the trench. These zones are main fault zone, synthetic faulting zone and antithetic faulting region. The photomosaic of the main fault zone on the NW wall of the trench is seen in Figure 12.

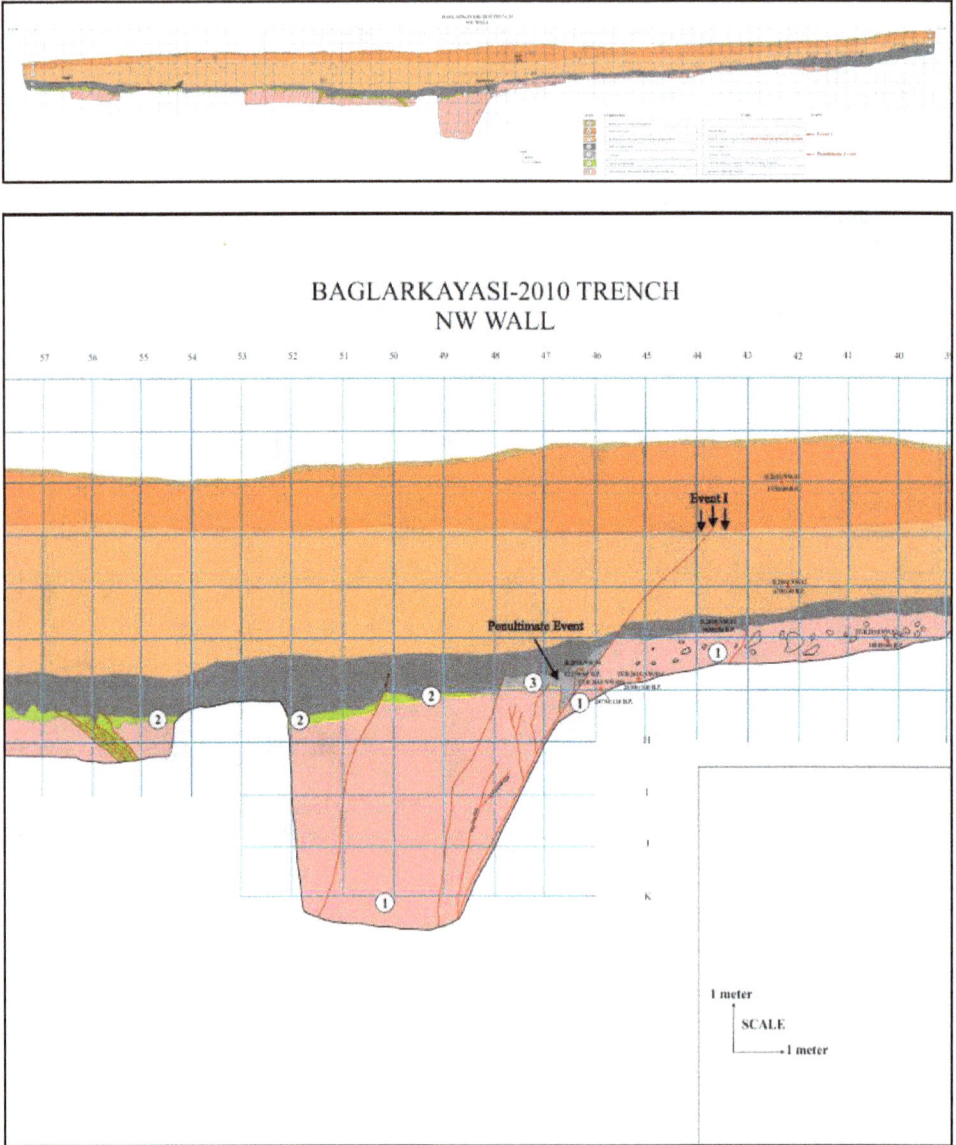

Figure 10. NW Wall Log of Bağlarkayası-2010 Trench (close shot of general log and main fault zone)

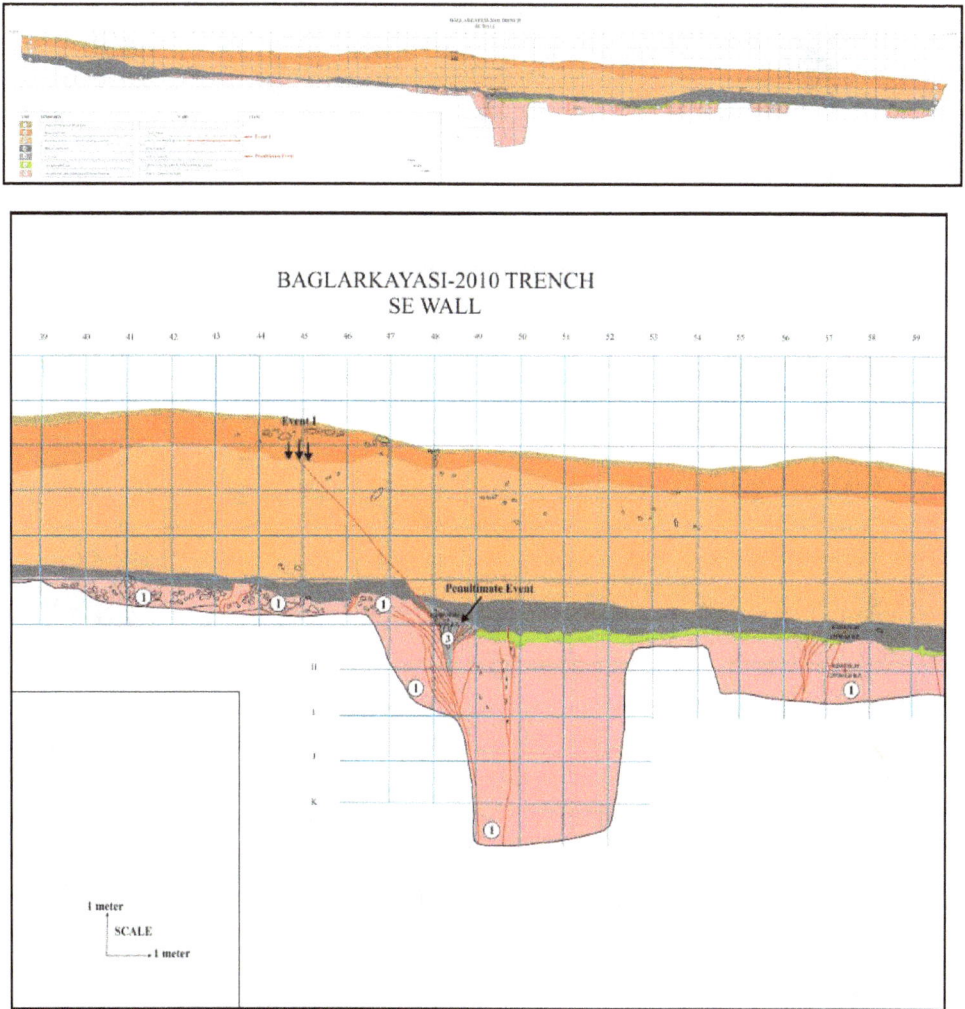

Figure 11. SE Wall Log of Bağlarkayası-2010 Trench (close shot of general log and main fault zone)

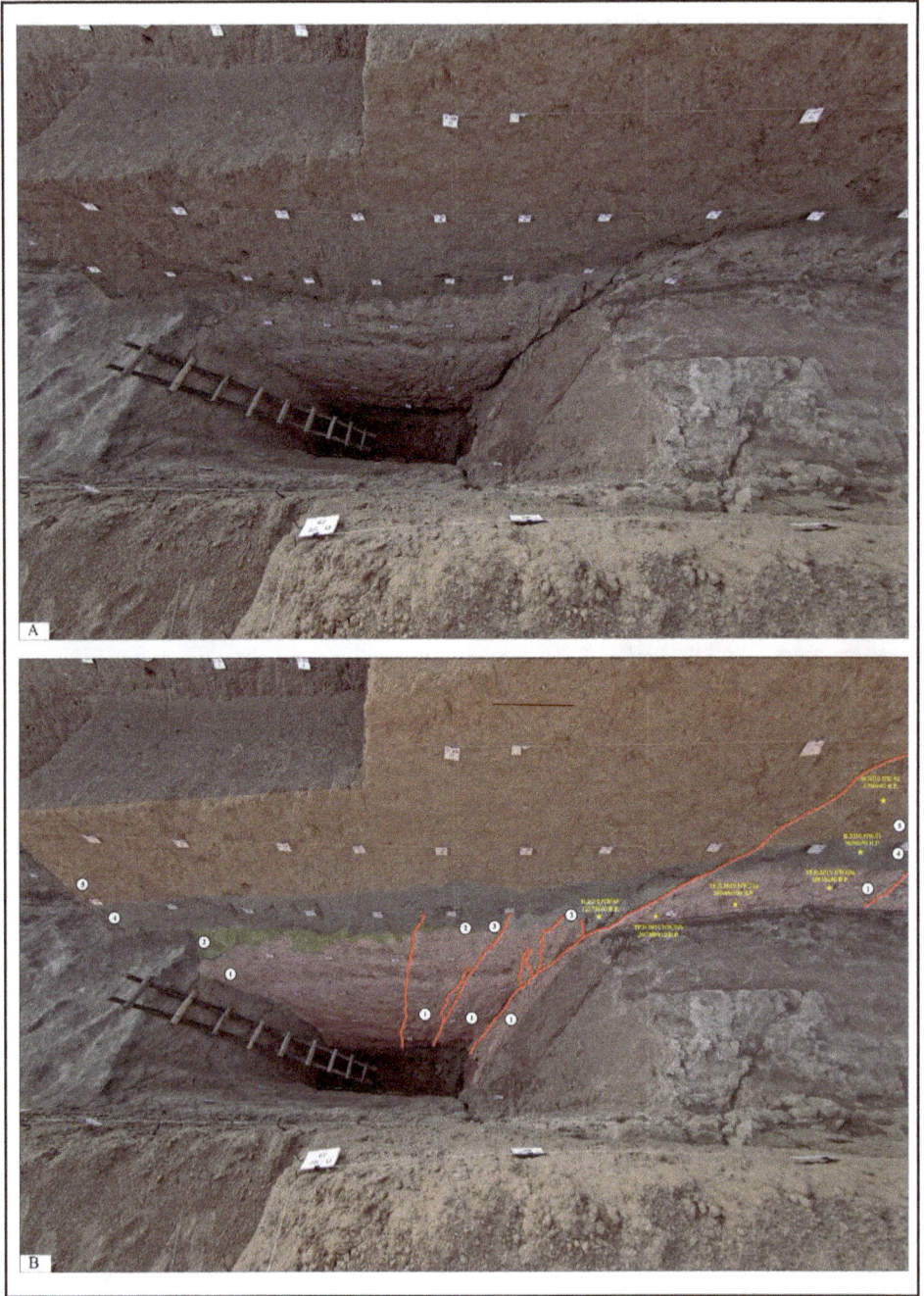

Figure 12. The photomosaic of the main fault zone on the NW wall of the trench; A : Uninterpreted B : Interpreted

Since carbon-rich material is limited within Bağlarkayası-2010 Trench, an supplementary trench was excavated approximately 100 meters SW of the trench (see Figure 4 and 7). In this trench, at a depth of 2 meters from the surface, some bones were encountered within a level that can be correlated with the unit 5 of Bağlarkayası-2010 Trench (Figure 13). According to the determination of Anthropologist Dr. Gerçek Saraç, these bones belong to a human.

Figure 13. Human bones detected within the supplementary trench

18 samples from Bağlarkayası-2010 Trench and 3 samples (2 of which were bones) from the Supplementary Trench were collected. The radiometric age determination (^{14}C -AMS) of the total 21 samples was carried out at BETA Analytical Lab (Table 1). The ^{14}C ages determined by BETA Analytical Laboratory were afterwards evaluated making use of Oxcal v3.10 calibration program developed by Ramsey (2001) (Figure 14).

Number of Sample	Unit	Number of Beta Lab.	Material	Measured age (B.P.)	13C/12C	2 Sigma Calibration
TF.B. 2010.NW.01a	1	291116	(organic sediment): acid washes	20300 +/- 100 BP	-20.1 o/oo	
TF.B. 2010.NW.01b	1	291118	(organic sediment): acid washes	20730 +/- 110 BP	-21.2 o/oo	
TF.B. 2010.NW.02a	1	291117	(organic sediment): acid washes	18810 +/- 80 BP	-22.2 o/oo	Cal BC 20560 to 20320 (Cal BP 22510 to 22270)
B.2010.NW.01	4	291093	(organic sediment): acid washes	9650 +/- 50 BP	-22.9 o/oo	Cal BC 9260 to 9120 (Cal BP 11210 to 11070), Cal BC 9010 to 8910 (Cal BP 10960 to 10860), Cal BC 8900 to 8860 (Cal BP 10840 to 10810)
B.2010.NW.02	5	291094	(organic sediment): acid washes	6790 +/- 40 BP	-23.5 o/oo	Cal BC 5740 to 5640 (Cal BP 7690 to 7580)
B.2010.NW.03	6	291095	(organic sediment): acid washes	1370 +/- 40 BP	-23.8 o/oo	Cal AD 600 to 680 (Cal BP 1350 to 1270)
B.2010.NW.04	3	291096	(organic sediment): acid washes	12270 +/- 60 BP	-22.8 o/oo	Cal BC 12470 to 12090 (Cal BP 14420 to 14040)
B.2010.NW.06	2	291098	(organic sediment): acid washes	17440 +/- 80 BP	-23.0 o/oo	Cal BC 18840 to 18470 (Cal BP 20790 to 20420)
B.2010.NW.07	2	291099	(organic sediment): acid washes	15060 +/- 70 BP	-23.5 o/oo	Cal BC 16690 to 16160 (Cal BP 18640 to 18100)
B.2010.NW.08	5	291100	(organic sediment): acid washes	8420 +/- 50 BP	-25.9 o/oo	Cal BC 7570 to 7420 (Cal BP 9520 to 9380), Cal BC 7420 to 7360 (Cal BP 9370 to 9310)
B.2010.NW.10	6	291101	(organic sediment): acid washes	480 +/- 30 BP	-25.3 o/oo	Cal AD 1410 to 1450 (Cal BP 540 to 500)
B.2010.NW.11	2	291102	(organic sediment): acid washes	15660 +/- 130 BP	-24.1 o/oo	Cal BC 17120 to 16810 (Cal BP 19070 to 18760)
B.2010.NW.15	5	291104	(organic sediment): acid washes	7360 +/- 40 BP	-23.3 o/oo	Cal BC 6380 to 6210 (Cal BP 8330 to 8160)
B.2010.SE.04	3	291105	(organic sediment): acid washes	10410 +/- 50 BP	-22.5 o/oo	Cal BC 10690 to 10190 (Cal BP 12640 to 12140)
B.2010.SE.06	2	291107	(organic sediment): acid washes	13330 +/- 60 BP	-23.2 o/oo	Cal BC 14070 to 13730 (Cal BP 16020 to 15680)
B.2010.SE.07	2	291108	(organic sediment): acid washes	12420 +/- 60 BP	-23.5 o/oo	Cal BC 12890 to 12220 (Cal BP 14840 to 14170)
B.2010.SE.10	5	291109	(organic sediment): acid washes	6870 +/- 40 BP	-23.7 o/oo	Cal BC 5850 to 5710 (Cal BP 7800 to 7660)
B.2010.SE.13	4	291111	(organic sediment): acid washes	8640 +/- 50 BP	-23.8 o/oo	Cal BC 7760 to 7580 (Cal BP 9710 to 9540)
BK.2010.SE.02	2	291113	(organic sediment): acid washes	15270 +/- 70 BP	-23.1 o/oo	Cal BC 16830 to 16650 (Cal BP 18780 to 18600)
BK.2010.SE.03	5	291114	(bone collagen): collagen extraction: with alkali	5590 +/- 40 BP	-18.8 o/oo	Cal BC 4610 to 4450 (Cal BP 6560 to 6400)
BK.2010.SE.04	5	291115	(bone collagen): collagen extraction: with alkali	5560 +/- 40 BP	-18.9 o/oo	Cal BC 4560 to 4440 (Cal BP 6500 to 6390), Cal BC 4420 to 4390 (Cal BP 6370 to 6340)

Table 1. Results of [14]C Age Determination for the samples collected from Bağlarkayası-2010 Trench and Bağlarkayası Observation Trench

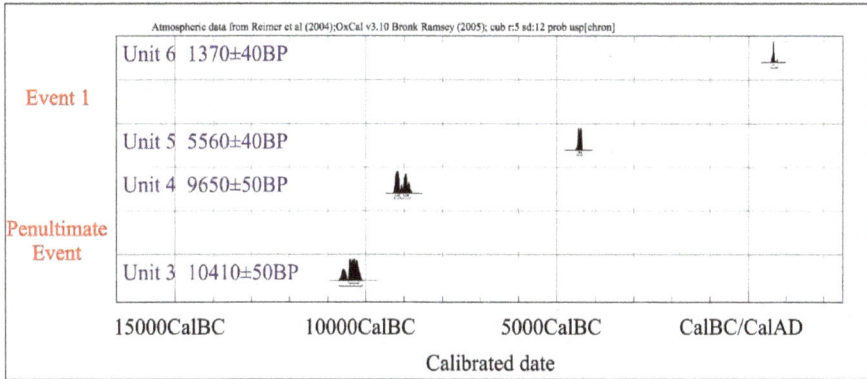

Figure 14. Radiocarbon ages calibrated making use of Oxcal v3.10 calibration program (Ramsey, 2001) of the samples collected from Bağlarkayası-2010 Trench and Bağlarkayası Observation Trench

As a result of the evaluation carried out making use of paleoseismological criteria such as trench microstratigraphy, geometry of fault colluvial wedge, upward termination of fault strands, and ^{14}C age data, two paleoseismic events were described within the Bağlarkayası-2010 Trench, which occurred during the last 10 500 years. The first of these earthquakes (penultimate event) occurred after the deposition of unit number 2 and as a result of it the colluvial wedge number 3 was developed (see Figures 10, 11 and 12). The youngest age obtained from the unit number 2 is 12 420 ± 60 B.P. years. And the base age obtained from the unit number 4 that covers the colluvial wedge number 3 is 9650 ± 50 B.P. years. In addition to these, the ages obtained from the colluvial wedge number 3 are 12 270 ± 60 B.P. and 10 410 ± 50 B.P. years, respectively. Colluvial wedges can be directly used in paleoseismological studies as event horizons in dating of earthquakes. In the light of these data, the age of the first earthquake (penultimate event) defined in the Bağlarkayası-2010 Trench was determined between the years 10 410 ± 50 B.P. and B.P. 9650 410 ± 50 B.P.

The second earthquake (Event 1) defined in the trench occurred after the deposition of the unit number 5. Faulting traces belonging to this earthquake can be traced within the unit number 5, but the faults do not cut the unit number 6 (see Figures 10, 11 and 12). The youngest age obtained from the unit number 5 of the trench is 6790 ± 40 B.P. years. In addition, the human bones discovered within the observation trench were collected from the unit number 5. The ages obtained from these bones were determined as 5590 ± 40 B.P. and 5560 ± 40 B.P. years, respectively. And the base age determined from the unit number 6 is 1370 ± 40 B.P. years. According to these age data the last earthquake took place between the years 5560 ± 40 B.P. and B.P. 1370 ± 40 B.P.

The amount of displacement occurred during the last earthquake (Event 1) was measured as 25 cm by taking the top of the unit 4 as reference plane (see Figure 12). Along the fault, the amount of displacement in the older units is 50 cm. This shows that these two earthquakes are of similar magnitude and an average displacement of 25 cm occurred in each earthquake. This data demonstrates that the Akhisar-Kılıç segment displayed characteristic earthquake behavior in Holocene.

On the other hand, it was determined that Kızılkaya Ignimbrites was displaced 268 meters in the vertical direction by Tuz Gölü Fault Zone in the vicinity of Akhisar village (Figure 15). The age of Kızılkaya Ignimbrites was determined by various researchers using different methods. For example, employing K/ AR method, while Innocenti et al. (1975) obtained 5.4 ± 1.1 million years, Besang et al. (1977) 5.4 ± 1.1 million years, and Schumacher and Schumacher (1996) 4.3 ± 0.2 and 4.5 ± 0.2 million years of age for Kızılkaya Ignimbrites. On the other hand, Aydar et al. (2012) obtained 5.19 ± 0.07 and 5.11 ± 0.37 million years of age employing the method of Ar/Ar. According to these age data, the age of the Kızılkaya Ignimbrites is approximately 5 million years. Taking into account the radiometric age (5 million years) determined in previous studies and the measured total displacement (268 meter), the annual slip rate of Akhisar-Kılıç Segment of Tuz Gölü Fault Zone was calculated as 0.0536 mm starting from Early Pliocene.

Through the evaluation of the annual slip rate (0.0536 mm/ year) of the Akhisar-Kılıç Segment together with the amount of vertical displacement (25 cm), the average earthquake recurrence interval of the Akhisar-Kılıç Segment has been found as 4664 years. The age of the first earthquake (penultimate event) that occurred within the Bağlarkayası-2010 Trench was determined between 10 410 ± 50 B.P. and 9650 410 ± 50 B.P. years which can be accepted as 10000 years BP in average. After 4664 years from this earthquake, which is the average earthquake recurrence interval, the last earthquake (Event 1) must have been occurred. The age of this earthquake is approximately 5336 B.P. years. Consequently, the elapse time from the last earthquake to the present time is 5336 years. Since this value is greater than the average earthquake recurrence interval that is 4664 years, the Akhisar-Kılıç Segment could produce an earthquake at any moment. By taking into consider the annual slip rate and the elapse time from the last earthquake to the present time, the amount of average vertical displacement accumulated on the Akhisar-Kılıç Segment was calculated as 28.76 cm.

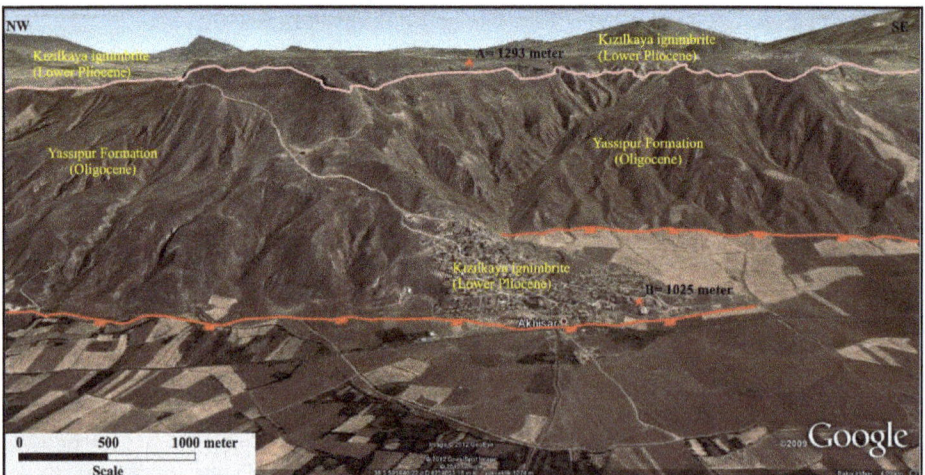

Figure 15. Google earth image of Kızılkaya Ignimbrites displaced by Tuz Gölü Fault, in the vicinity of Akhisar Village; the amount of vertical displacement between the points A and B is 268 meters.

Supposing that the whole of the Akhisar-Kılıç Segment, which has a length of 27 km, of Tuz Gölü Fault Zone was broken, the maximum earthquake magnitude that this segment can produce was calculated making use of the empirical equations proposed by Wells and Coppersmith (1994).

According to these equations, the magnitude of the maximum earthquake that the Akhisar-Kılıç segment could produce (M), the amount of maximum displacement (MD) and the amount of average displacement (AD) were computed as follows:

M = a + b x log (SRL)
a = 4.86
b = 1.32
SRL = 27 km
M = 4.86 + 1.32 x log (27)
and M was computed as M = 6.74 .
log (MD) = a + b x M
M = 6.74
a = −5.90
b= 0.89
log (MD) = 0.0986
and MD was computed as MD = 1.25 meters.
log (AD) = a + b x M
M = 6.74
a = −4.45
b = 0.63
log (AD) = − 0.2038
and MD was computed as MD = 0.62 meters.

The amount of the displacement measured for an earthquake in the Bağlarkayası- 2010 Trench is 25 cm. This value is smaller than the amount of average displacement computed employing the empirical equations proposed by Wells and Coppersmith (1994). There might be several reasons for this: 1) Since Akhisar- Kılıç Segment has a minor right lateral compo-nent, the amount of the measured normal displacement is smaller than the computed one 2) or the segment that has a length of 27 km was not entirely broken when the earthquake(s), during which a displacement of 25 cm occurred, took place.

By means of the evaluation made taking into account the second possibility, the magnitude (M) of the earthquake that should occur in order that an average displacement of 25 cm could take place and the length of the surface rupture (SRL) required for an earthquake with that magnitude to take place were computed as follows making use of the empirical equations proposed by Wells and Coppersmith (1994) :

log (AD) = a + b x M
AD = 0.25 cm
a = −4.45
b = 0.63

$-0.60205991 = -4.45 + 0.63$ M

and M was computed as M = 6.10

log (SRL) = a + b x M

M = 6.10

a = -2.01

b = 0.50

log (SRL) = 1.04

and SRL was computed as SRL = 10.96 km.

4. Paleoseismological three dimensional virtual photography method; Bağlarkayası-2010 trench application

In recent years there have been significant developments in the applications related to the cameras that provide panoramic viewing, especially depending on the developments of image processing programs. Such cameras started to be widely used in various fields such as security, teleconference, publicity, virtual tour and robot navigation (Baştanlar and Yardımcı, 2005; Ergün and Şahin, 2009). Since the angle of sight of cameras is always smaller than that of human beings and the image of large objects cannot be pictured within a single photograph, a demand for the creation of a panorama arose just at the beginning of the photography (Parian, 2007). The efforts to obtain a photographic panorama were realized at the end of 1800s with joining together several photographs taken from different directions to form a full panorama (Baştanlar, 2005). 'Panorama' is a word that was formed by joining two Greek words together. 'Pan' means 'all', and 'horama' means 'sight' (Baştanlar, 2005). The difference between traditional photograph and panoramic photograph is similar to the difference between looking at a city from the window of a small office and from the roof of it (Kwiatek, 2005). One of the main objectives of Photogrammetry is, by carrying the real image into the virtual environment, to arouse the impression of 'really being there' in viewers and readers. In recent years, with the development of Internet and multimedia technologies, important developments were recorded in photogammetry and the Three Dimensional Virtual Photography Technique started to be used in various fields. Although the method is frequently used in different sectors such as architecture, tourism and museum publicity; it has no application either in earth sciences or in paleoseismology. Paleoseismological Three Dimensional Virtual Photography Method has been applied for the first time in this study, in the paleoseismological studies conducted on the Tuz Gölü Fault Zone (Central Anatolia, Turkey).

The application area of the Earth Sciences is the Nature. The presentation of the studies, which are carried out in the Nature, to the readers without any loss of data and with high resolution is important. Paleoseismological trench works are comprehensive and high cost works. Trench works should be completed within a specified period of time and with specific standards. The specific physical conditions of the trench area (the state of groundwater, instability of trench sediments, etc.) in some cases complicate the trench works may sometimes and cause loss of life and property due to collapse and cave in. In order to eliminate such drawbacks, trenches should be excavated as deep and wide as possible and the work should be completed in a manner as fast and qualified as possible.

Paleoseismological Three Dimensional Virtual Photography Method makes it possible the transfer of all details within the trench to the reader without any loss of data and with high resolution. In addition, as the reader can directly see the trench wall image, the method gives the reader the right to interpret. In the section that follows, the application of this method to a paleoseismological trench study (Bağlarkayası-2010 Trench) conducted in the middle part of the Tuz Gölü Fault Zone is explained.

Paleoseismological Three Dimensional Virtual Photography Method is mainly composed of four stages. These are :

1. Planning
2. Photographing
3. Stitching the photographs and forming scenes
4. Connecting the scenes and forming virtual tour

4.1. Planning

In order a trench work to be photographed in three dimensions, first of all, the number and locations of the points at which photographs are to be taken should be planned. In planning the number of points,

-   Paleoseismological detail within the trench and
-   Dimensions and excavation type of the trench

should be taken into consideration. In the determination of the locations of the point at which photographs are to be taken, care should be taken not to leave any blind spots. A virtual tour of the entire trench can be obtained only under these conditions.

Bağlarkayası-2010 Trench was excavated as 94 meters long, 5 meters wide and 5 meters deep on the average (Maximum depth: 8.5 meters). While the northern 40-meter section of the trench was excavated as a single slot trench, the southern part of it was excavated as multi-bench trench (Figure 8). A gridding of 1 m² was applied on the walls of the trench. Three deformation zones are present within the trench, which are main fault zone, synthetic faulting zone and antithetic faulting zone.

Before starting paleoseismological three dimensional virtual photographing, a general photograph should be taken which covers the entire trench and shows the points where the photographs are to be taken. This photograph can be taken, if possible, from above, at an inclined angle. The general trench photograph is to be used as the index photograph in the formation of the virtual tour in the final stage. In this study, the general photograph of the Bağlarkayası-2010 Trench was taken from the fire tower, at an elevation of 25 meters and at an inclined angle (Figure 16).

As a result of the evaluation made taking into account the form of the trench and the detailed paleoseismological features on the trench walls, it was decided to take photographs at 15 points in the Bağlarkayası-2010 Trench (Figure 16). The points where photographs were taken were selected near the parts of the trench walls that contained fault. In order to trace the continuation of the fault on one wall of the trench, on the other wall, the floor of the trench should also be cleared and the fault traces should be made distinct at the floor as well.

Figure 16. The general image of the Bağlarkayası-2010 Trench taken from the air at an inclined angle. Yellow circles represent the points where photographs were taken (look from SW to NE)

In the virtual tour to be obtained as a result of this method, the image should be as clear as possible. That's why, one day before photographing, the time and the angle at which sunlight comes to the trench walls should be controlled and recorded. Thus, it would be planned at which point, at which hour photographs be taken. Since 8 photographs are to be taken, an average of 15 minutes can be envisaged for taking photographs at each point.

4.2. Photographing

The hardware and software required for Paleoseismological Three Dimensional Virtual Photography are presented in Table 2.

Software	The aim of use	Hardware	The aim of use
Adope Photoshop CS5	Correction of color and light	Tripod	Photographing
PTGui Pro 9.0	Stitching	Digital camera	Photographing
Cubic Converter 2.2.1	Stitching and creating of stages	Pano Head	Photographing
PTLens 1.5.3	Correction of distortion	Fisheye Lens	Photographing
Panotour Pro 1.6.0	Cretating of Virtual tour	Computer	Office work
Web Browser	Internet Explorer, Google Chrome, Mozilla Firefox		

Table 2. The hardware and software required for Paleoseismological Three Dimensional Virtual Photography

In this study, Manfrotto 055x probe tripod, Nikon D200 Model and 10.2 megapixel Digital photo camera, Manfrotto sph 303 panoramic head and Nikkor fish-eye objective having 180^0 angle of sight were used.

At the stage of photographing, 8 photographs are taken at each point. Six of these photographs are taken at the horizontal plane using an angle of 60^0 and thus a horizontal image of 360^0 of that point is obtained. Afterwards, the sky and floor image of the same point are taken as two separate photographs and the image of that point at the vertical plane is obtained. In this manner, a total of 8 photographs, 6 of which at horizontal plane and 2 at vertical plane, of each point are taken.

4.3. Stitching photographs together and forming scenes

Paleoseismological Three Dimensional Virtual Photography Method is a method which is most of the time applied employing some computer programs. However, from time to time it requires some manual intervention. In order that the virtual tour to be produced from the photographs can be as realistic as possible, some corrections are needed to be made on the raw photographs.

At this stage, first of all, color and light corrections of 8 photographs taken at each point are made using Adobe Photoshop CS5 program. Following this, employing PTGui Pro 9.0

program, photographs are connected by stitching. At the first stage of the connection, 6 photographs taken at horizontal plane and the sky image taken at the vertical plane are used. During this process, from time to time manual corrections may be required when common parts of the photographs are overlapped. At the end of the first stage of the connection, all photographs, except the floor image, become stitched together.

These 7 stitched photographs are then transferred to Cubic Converter 2.2.1 program. Cubic Converter 2.2.1 program creates partial photographs by dividing the image composed of 7 stitched photographs into 6 parts in the form of cubes. One of these partial photographs is the floor photograph yet to be completed.

To complete the floor photograph, of the raw photographs taken at first, the one belonging to the floor is opened in the program PT lens 1.5.3. The image, which is obtained after its distortion correction is made in this program, is assembled with the partial photograph belonging to the floor. This assembled photograph is again transferred to the program Cubic Converter 2.2.1 and thus the floor photograph becomes completed. Lastly, the image of the whole scene is exported in TIFF format from the Cubic Converter 2.2.1 program. In Figure 17, the stitched and connected scene image belonging to the point 8 (see SE second bench in Figure 16) of Bağlarkayası-2010 Trench is presented.

By repeating these processes for all of the scenes, a total of 15 files were obtained.

Figure 17. The scene image obtained as a result of stitching and connecting the photographs belonging to the point 8 (see SE second bench in Figure 16) of Bağlarkayası-2010 Trench.

4.4. Connecting the scenes and forming the virtual tour

All of the scenes that are connected and recorded are opened in Panotour 1.6.0 program. Transition links are formed for the transition of the scenes with each other to be made. Afterwards, index photograph is imported into the program and transition links are added to the index photograph as well. In this manner, the access to a desired scene is made possible both over the virtual tour and over the index photograph.

The completed virtual tour is recorded as MTML file. This file not only makes it possible the visual presentation of a trench study to the reader but also makes significant contributions especially at the stage of interpretation of the trench data

You can reach the virtual tour produced employing Paleoseismological Three Dimensional Virtual Photography Method developed in this study over the following link:

https://hotfile.com/dl/155062161/c15656d/webquality.rar.html

5. Results

In this study, a new photography method for paleoseismological studies was developed and named 'Paleoseismological Three Dimensional Virtual Photography Method'.

It was observed that this method could be successfully applied especially in deep and benched paleoseismological trench excavations.

The most important advantage of this method is that all the geologic detail within the trench can be transferred to the reader without any loss of data and with high resolution. In this manner, the reader is given the right to interpret.

Paleoseismological Three Dimensional Virtual Photography Method is mainly composed of four stages which are planning, photographing, forming scenes by stitching the photographs together and forming the virtual tour by connecting the scenes.

The virtual tour obtained as a result of this method enables the reader to make a tour, feeling as if he were really there.

The method has been applied for the first time in this study on the Akhisar-Kılıç Segment of the Tuz Gölü Fault zone which is one of the most important active fault zones of the Central Anatolia Region.

Tuz Gölü Fault Zone (TGFZ) is a NW-SE trending, dipping towards SW, active, right lateral strike slip component normal fault zone which extends between north of Tuz Gölü and Kemerhisar (Niğde) and has a length of 200 km.

TGFZ is composed of fault segments that have lengths varying between 4 and 33 km. The Akhisar-Kılıç Segment is one of the most important segments of this zone owing to its geological fault length and morphotectonic features.

The total amount of vertical displacement of the Akhisar-Kılıç Segment is 268 meters according to Lower Pliocene aged Kızılkaya Ignimbrites.

According to the calculation made taking into account the total amount of vertical displacement (268 meters) and the age of Kızılkaya Ignimbrites (5 million years), the slip rate of the Akhisar-Kılıç Segment during the period from Lower Pliocene to the present day is 0.0536 mm/year.

In this study, a paleoseismological trench study has been conducted on the Akhisar-Kılıç Segment of TGFZ for the first time and the trench has been named 'Bağlarkayası-2010 Trench'.

According to the paleoseismological findings obtained from Bağlarkayası-2010 Trench, two earthquakes that resulted in surface faulting occurred on the Akhisar-Kılıç Segment during the last 10 500 years. The first of these earthquakes occurred approximately in the year 10 000 B.P. and the second one took place in the year 5336 B.P. Consequently, the elapse time from the last earthquake up to the present is approximately 5336 years.

The amount of the average vertical displacement accumulated on the Akhisar-Kılıç Segment from the last earthquake to the present time was calculated making use of empirical equations (Wells and Coppersmith, 1994) as 28.76 cm.

An average displacement of 25 cm was measured in the earthquakes that were defined and dated within Bağlarkayası-2010 Trench. According to the evaluation made taking into consideration this value together with the slip rate (0.0536 mm/year), the earthquake recurrence interval for the Akhisar-Kılıç Segment was calculated as 4664 years.

The fact that the average earthquake recurrence interval (4664 years) is smaller than the elapse time from the last earthquake to the present day indicates that the Akhisar-Kılıç Segment has completed its average earthquake recurrence interval and could produce earthquake(s) at any moment.

The geologic fault length of the Akhisar-Kılıç Segment is 27 km. The moment magnitude of the maximum earthquake that this segment could produce in case the whole of it were to be broken was calculated as M = 6.74 making use of the empirical equations proposed by Wells and Coppersmith (1994).

According to the computations made using the other equations of the same study, the amount of the maximum displacement which could occur in an earthquake with the magnitude M = 6.74 is MD = 1.25 meters and the amount of the average displacement is AD = 0.62 meters.

A displacement of 25 cm was measured in both of the earthquakes defined by means of the paleoseismological excavation studies conducted in the trench. This value is smaller than the amount of the average displacement that a pure normal fault with a length of 27 km can produce. The reason for this difference may be the fact that the fault has a minor strike slip component. Or the whole of the segment that is 27 km long is not broken. In this case, in order for a displacement of 25 cm to occur, the moment magnitude of the fault is expected to be M = 6.10 and the required surface rupture length for an earthquake with the magnitude M =6.10 to occur is expected to be SRL = 10.96 km.

Author details

Akın Kürçer
General Directorate of Mineral Research and Exploration, Department of Geology, Çankaya, Ankara, Turkey

Yaşar Ergun Gökten
*Ankara University, Faculty of Engineering, Department of Geological Engineering,
Tectonic Research Group, Ankara, Turkey*

Acknowledgement

This study has been carried out within the scope of the project (2010-30-14-02-3) titled 'The
Neotectonic Properties and Paleoseismology of the Tuz Gölü Fault Zone, Central Anatolia,
Turkey'conducted by the Department of Geological Researches of the General Directorate of
Mineral Research and Exploration (MTA). We would like to express our thanks to the
General Directorate of Mineral Research and Exploration (MTA). In addition, our thanks
extend to Dr. Ömer Emre (MTA), Dr. Tamer Yiğit Duman (MTA) and Dr. Selim Özalp
(MTA) for their useful critiques of paleoseismological interpretations, Geophysical Engineer
Hayrettin Karzaoğlu (MTA) for his contributions to the geophysical studies and Dr. Gerçek
Saraç for his contributions to the determination of the human bones.

6. References

Aydar, E, Schmitt, A.K.., Çubukçu, H.E., Akin, L., Ersoy, O., Sen, E., Duncan, R.A., ve Atici,
    G., 2012. Correlation of ignimbrites in the central Anatolian volcanic province using
    zircon and plagioclase ages and zircon compositions. Journal of Volcanology and
    Geothermal Research 213-214 (2012) 83–97.

Baştanlar, Y., 2005. Parameter extraction and image enhancement for catadioptric
    omnidirectional cameras. Msc. Thesis, METU Informatic Institute, Ankara

Baştanlar, Y. and Yardımcı, Y., 2005. Hiperbolik Aynalı Katadioptrik Tüm Yönlü Kameralar
    için Parametre Çıkarımı, IEEE Xplore,
    http://ieeexplore.ieee.org/stamp/stamp.jsp?arnum ber=01567669 (2009.04.20).

Besang, C., Eckhardth, F.J., Harre, W., Kreuzer, H. and Müller, P., 1977. Radiometrische
    Altersbestimmungen an neogenen

Bronk Ramsey, C., 2001. Development of the radiocarbon program OxCal: Radiocarbon, v.
    43, no. 2A, p. 355–363.

Çemen, İ., Göncüoğlu, M.C. and Dirik, K., 1999. Structural evolution of the Tuz Gölü basin
    in Central Anatolia, Turkey. Journal of Geology, 107, 693-706.

Demirtaş, R., 1997. Paleoseismology, General Directorate of Disaster of Turkey, Ankara,
    Turkey.

Dirik, K. and Göncüoğlu, M.C., 1996. Neotectonic characteristics of Central Anatolia, Int.
    Geology Review, 38, 807-817.

Dirik, K. and Erol, O., 2000. Tuz Gölü ve civarının tektonomorfolojik evrimi Orta Anadolu,
    Türkiye, Haymana-Tuz Gölü-Ulukışla Basenleri Uygulamalı Çalışma (Workshop),
    T.P.J.D. Bülteni, Özel sayı 5

Dönmez, M., Akçay, A.E., Kara, H., Türkecan, A., Yergök, A.F. and Esentürk, K., 2005.
    Geological maps of Turkey in scale of 1/100 000 Aksaray L 32 sheet. MTA publications,
    52, Ankara.

Ergün, B. and Şahin, C. 2009. Digital Spherical Photogrammetry Techniques Recently in Use. Harita Dergisi, 142, p. 40-50.

Hatheway, A. W., and Leighton, F. B., 197). Exploratory trenching. Geol. Soc. Am., Rev. Eng. Geol. 4, 169–195.

Innocenti, F., Mazzuoli, G., Pasquare, F., Radicati Di Brozola, F. and Villari, L., 1975. The Neogen calc-alcalin volcanism of Central Anatolia: geochronological data on Kayseri-Niğde area: Geol. Mag., 112 (4), 349-360.

Koçyiğit, 2000. General neotectonic characteristics and seismicity of Central Anatolia, Haymana-Tuz Gölü-Ulukışla basenlerinin uygulamalı çalışması (workshop). Abstracts, 1-26, Aksaray.

Koçyiğit, A. and Özacar, A., 2003. Extensional neotectonic regime through the NE edge of the Outer Isparta Angle, SW Turkey: New Field an Seismic Data. Turkish Journal of Earth Sciences, 12, 67-90.

Kwiatek, K., 2005. Generation of a virtual tour in the 3D space applying panoramas, exercised on the sites of Dresden and Cracow. Lisans Tezi, AGH University ofScience and Technology, Dresden.

McCalpin, J., P., 2009. Field Tecniques in Paleoseismology – Terrestrial Environments, Paleoseismology, edited by James P. McCalpin, second edition, International Geophysics Series v. 95, p. 29-118.

McCalpin, J., P. and Nelson, A., R., 2009. Introduction to Paleoseismology. Paleoseismology, edited by James P. McCalpin, second edition, International Geophysics Series v. 95, p. 1-25.

Okay, A.I., Tüysüz, O., Satır, M., Özkan-Altıner, S., Altıner, D., Sherlock, S. and Eren, R.H., 2006. Cretaceous and Triassic subduction-accretion, HP/LT metamorphism and continental growth in the Central Pontides, Turkey. Geological Society of America Bulletin, 118: 1247-1269.

Pantosti, D. and Yeats, R.S., 1993. Paleoseismology of great earthquakes of the late Holocene, Analidi Geofisica, v. XXXVI, n:3-4, p. 237-257.

Parian, J. A., 2007. Sensor Modeling, Calibration and Point positining with Terrestrial panoramic Cameras. Doctora Thesis, E.T.H. Zurich, Swisszerland.

Reilinger, R.E., Mcclusky, S.C., Vernant, P., Lawrence, S., Ergintav, S., C. akmak, R., Nadariya, M., Hahubia, G., Mahmoud, S., Sakr, K., Arrajehi, A., Paradissis, D., Al-Aydrus, A., Prilepin, M., Guseva, T., Evren, E., Dmitritsa, A., Filikov, S.V., Gomes, F., Al-Ghazzi, R., and Karam, G., 2006. GPS constraints on continental deformation in the Africa-Arabia-Eurasia continental collision zone and implications for the dynamics of plate interactions: Journal of Geophysical Research, v. 111, p. V05411, doi: 10.1029/2005JB004051.

Schumacher, R., Mues-Schumacher, U., 1996. The Kizilkaya ignimbrite—an unusual low-aspect-ratio ignimbrite from Cappadocia, Central Turkey. Journal of Volcanology and Geothermal Research 70, 107–121.

Şaroğlu, F., Emre, Ö. and Kuşçu, İ., 1992. Active fault map of Turkey, MTA, Ankara.

Wells, D.L. and Coppersmith, K.J., 1994. New Empirical Relationships among Magnitude, Rupture Length, Rupture Width, Rupture Area, and Surface Displacement. Bulletin of the Seismological Society of America, Vol. 84, No. 4, pp. 974-1002.

Chapter 8

An Assessment of the Earthquakes of Ancient Troy, NW Anatolia, Turkey

Akın Kürçer, Alexandros Chatzipetros, Salih Zeki Tutkun, Spyros Pavlides, Süha Özden, George Syrides, Kostas Vouvalidis, Emin Ulugergerli, Özkan Ateş and Yunus Levent Ekinci

Additional information is available at the end of the chapter

1. Introduction

Many destructive earthquakes occurred in Northwestern Anatolia during historical and instrumental periods and as a result of these earthquakes civilies were damaged. Approximately 30 km southwest of Çanakkale the ancient city of Troy is located, containing remains belonging to the period between B.C. 3000 and A.D. 400 (Figure 1).

According to the intermittent archeological excavations, carried out from 1871 up to the present, there exist nine different layers of settlements in Troy. Although there is some archeological evidence which indicates that some of these layers, especially Troy III (B.C. 2200-2050) and Troy VI (B.C. 1800-1275) have been damaged by one or more earthquakes, no multidisciplinary geoscientific research has been carried out so far on the active faults which could have caused these earthquakes.

Troy which once controlled the commercial crossing point between Asia and Europe over Dardanos (the Dardanelles) used to be one of the most important trade centers of that era. Because of this fact Troy, besides being one of the hundreds of ancient cities situated in Anatolia, was a city that played an important role in the development of Western Anatolian and Aegean cultures. As a result of Troy's dominant position, several European countries believe that their roots lie in Troy and Trojans. When the architecture of Troy, represented by 9 layers of settlements spanning the period between B.C. 3000 and A.D. 400, is examined, it is observed that passages between civilizations are not gradual; instead, there are radical changes in building styles and materials. This observation can lead to the assumption that the events causing passages between civilizations are natural phenomena such as earthquakes rather than wars, fires or epidemics. On the other hand, Professor Manfred Korfman, who meticulously presided over Troy excavations between 1968 and 2005, talks

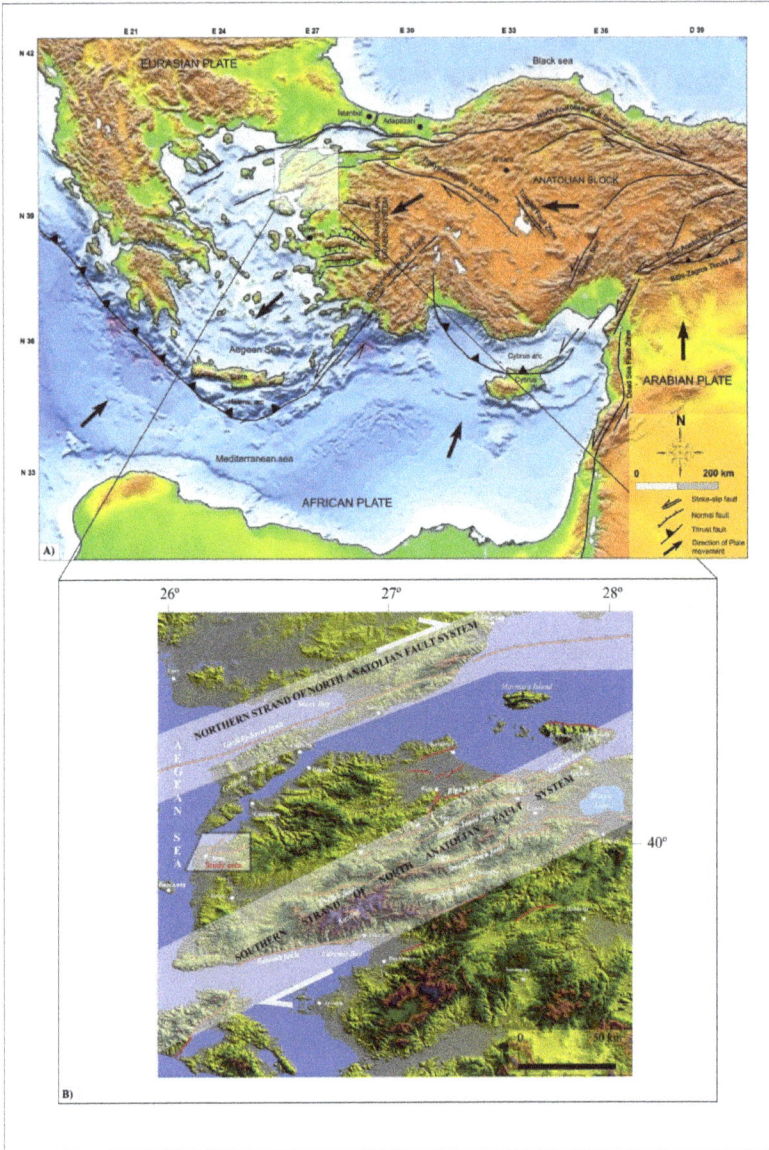

Figure 1. A. The major active tectonic structures of Eastern Mediterranean Region (Modified from Şengör et al., 1985), B. The locational relationship of Troy with the northern and southern branches of NAFS and the major active faults in the region (Modified from Şaroğlu et al., 1992).

about archaeological findings indicating that especially Troy VI Layer (B.C. 1800-1275) was damaged by one or more earthquakes. In the light of this information, some deformation structures thought to be of seismic origin have been observed, especially on the walls of Troy VI Layer. These deformation structures can be classified as systematic cracks, rotations and blocks tilting (Figure 2). In this article, the results of the research related to the earthquakes that caused the deformation structures observed in the remains of ancient Troy city are presented.

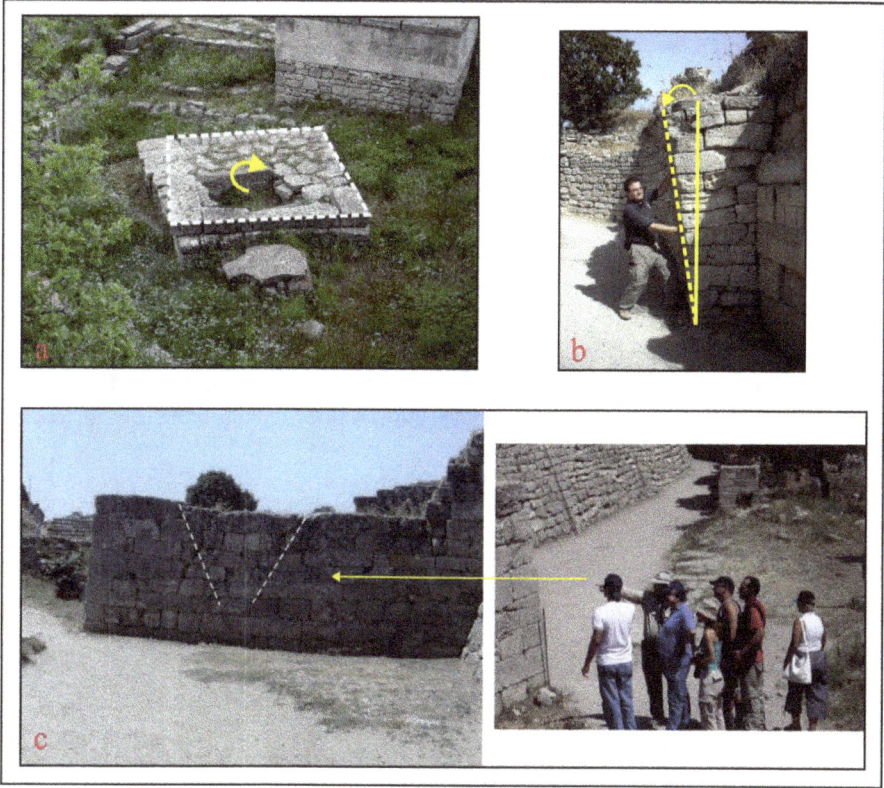

Figure 2. Deformation structures of seismic origin within Troy VI Layer a) clockwise rotation b) Tilting c) systematic cracks in the walls of Troy VI Layer that Prof. Manfred Korfman showed the project team

2. Regional geology and tectonic framework

In the study area, rock units having ages from Lower Cretaceous to present-day outcrop (Figure 3). The basement of the study area is constituted by Lower Cretaceous aged Denizgören ophiolites (Okay, 1987). Denizgören ophiolites are unconformably overlain by Çanakkale formation (Şentürk and Karaköse, 1987). Çanakkale formation, which outcrops at relatively high plateaus over a large part of the study area, is composed of a succession of pebblestones, sandstones, sandy limestones and limestones deposited in lagoonal, coastal and offshore environments. The age of the unit is Upper Miocene-Pliocene according to ostracoda and pelecyopoda fossils contained in it (Şentürk and Karaköse, 1987). All units outcropping in the study area are unconformably covered by Quaternary deposits. Quaternary deposits are represented by paleoterrace deposits of Dümrek river, flood plain deposits of Dümrek and Karamenderes rivers and alluvial fan deposits developed at the front of the Troy Fault.

The study area is located in the peninsula in the NW corner of Anatolia. This area, known as Biga Peninsula, is being deformed under the effect of the roughly N-S oriented Western Anatolian Extension System and the western extensions of the North Anatolian Fault System (NAFS). The study area, situated at the NW part of Biga Peninsula, is mostly under the influence of NAFS. NAFS, which is one of the most important active tectonic structures of the eastern Mediterranean Region, is divided into two branches as the north and south branches, starting from the east of Adapazarı. The northern branch that reaches the Sea of Marmara from Hersek delta, after traversing the Marmara Sea in approximately E-W direction is connected to the Ganos-Saroz Fault (GSF) on which Şarköy-Mürefte Earthquake (Mw=7.2) of 1912 occurred. GSF is oriented N70⁰E and cutting through Gallipoli Peninsula in this orientation from Saroz Gulf, reaches North Aegean Sea. The fault segments constituting the south branch of NAFS, by traversing Biga Peninsula in NE-SW direction reach North Aegean Sea. The most important ones of these NE-SW oriented faults are, from north to south, Edincik fault, Biga fault, Sarıköy-İnova fault, Yenice-Gönen fault, Evciler fault, Pazarköy fault and Edremit fault.

The ancient city of Troy is situated within a right lateral deformation zone bordered by the northern and southern branches of the North Anatolian Fault System (NAFS). The northern border of this deformation zone is formed by Gaziköy-Saroz Fault. And the southern border of the deformation zone is represented by the south branch of NAFS composed of faults which are parallel or subparallel to each other and extend between Kapıdağ Peninsula and Edremit Bay (Figure 1b). Troy Fault System, which is represented by E-W oriented normal faults, NE-SW oriented right lateral and NW- SE oriented left lateral strike slip faults, exists in the area within this deformation zone. When taken into account its geologic fault length and morphotectonic characteristics, 12-km long Troy Fault is the most important one of the E-W oriented faults within the Troy Fault System (Tutkun and Pavlides, 2005). And the most important of the NE-SW oriented right lateral faults is Kumkale Fault which possesses a total length of 15 km including its continuation in the sea.

Destructive earthquakes occurred on the active faults in Biga Peninsula (Troy is also located here) and in its vicinity, during historical and instrumental eras (Figure 4 and 5).

3. Material and method

This study is a multidisciplinary research in which geomorphological, geological and geophysical methods were applied in a specific order.

Within the framework of the geomorphological studies, a 1/25 000 scale digital elevation model of the study area was constructed and on this model, geomorphological features such as stream/valley drainage areas, mountain front sinuosity ratio and ratio of valley floor width to valley height were investigated.

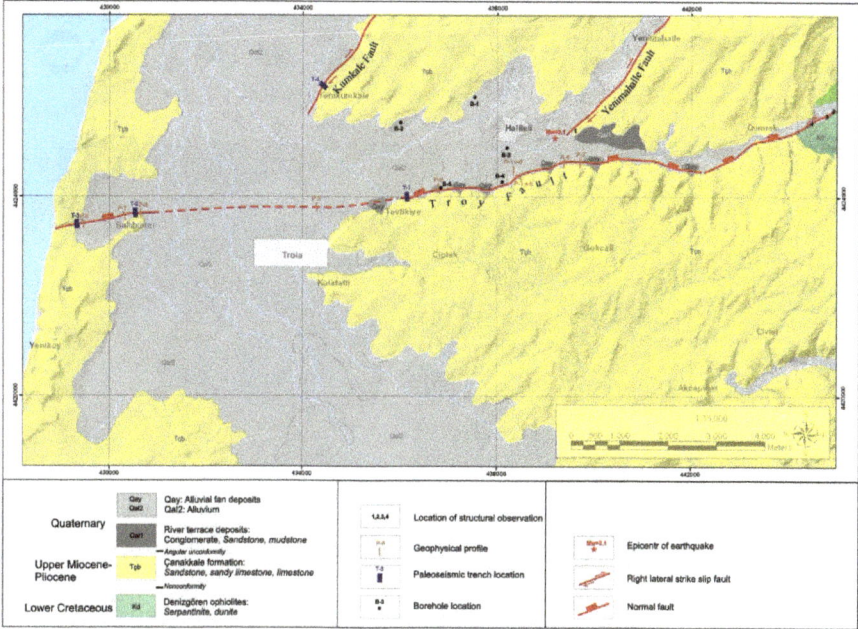

Figure 3. Geological map of Troy and its vicinity.

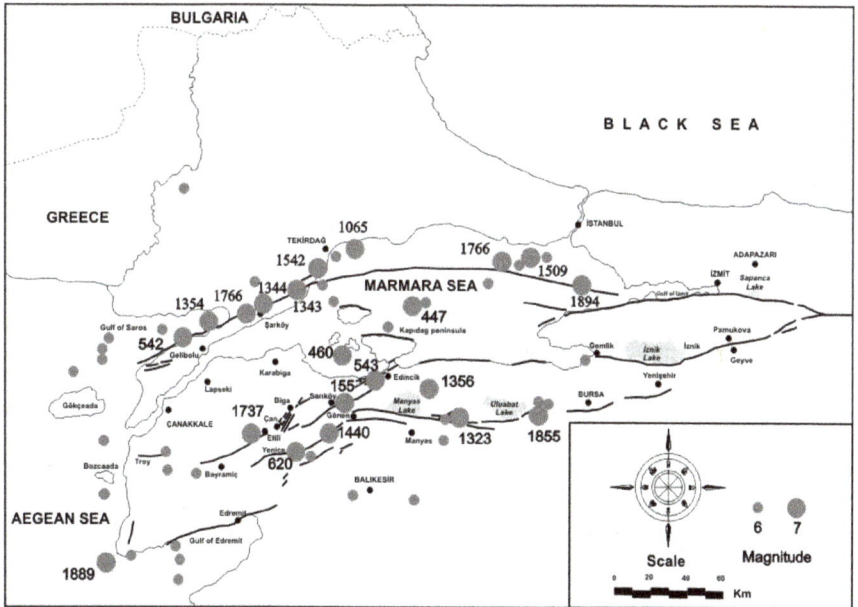

Figure 4. Western segments of NAFS in the study area and in its near vicinity (modified after Şaroğlu et al., 1992) and the historical earthquakes that occurred on these faults between A.D. 32 and A.D. 1900 (modified after Ambraseys and Finkel, 1991)

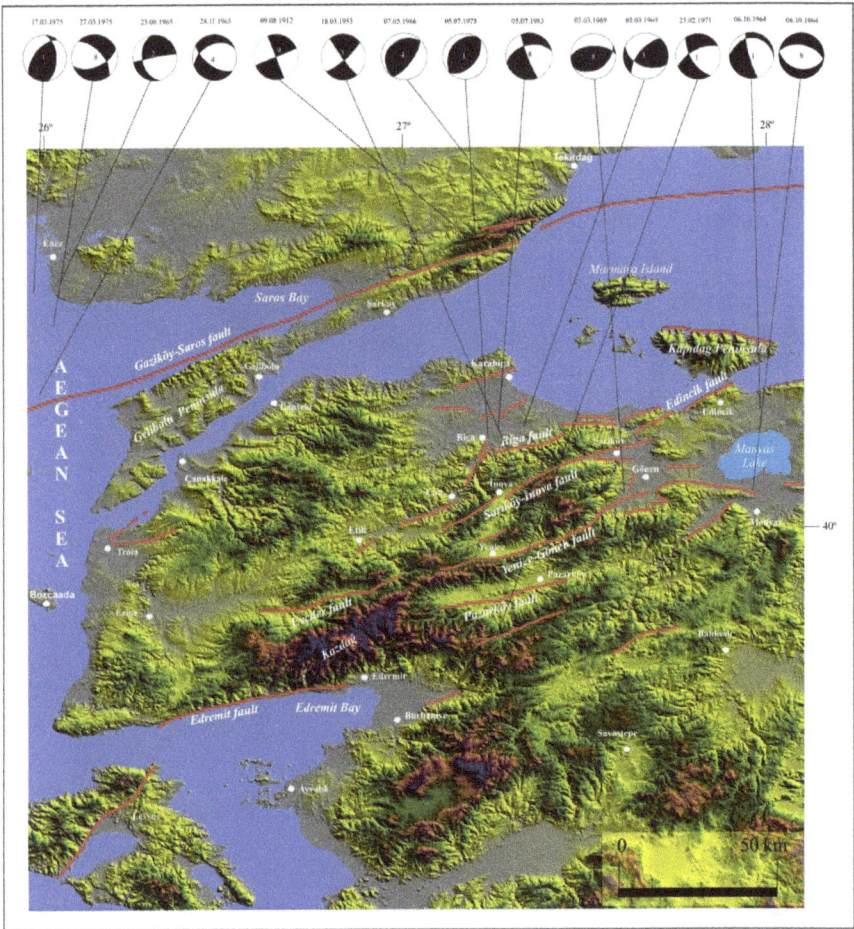

Figure 5. Extensions of North Anatolian Fault System (NAFS) in Biga Peninsula, instrumental period earthquakes related to these faults and focal mechanism analyses (Focal mechanism analyses were taken from 1: Kıyak, 1986; 2: Canıtez and Tokgöz, 1971; 3: McKenzie, 1972; 4: Kalafat, 1978; 5: Kalafat, 1988; 6: Jackson and McKenzie, 1984; 7: McKenzie, 1978; 8: Taymaz, 2001; 9: Aksoy et al., 2010)

Within the framework of the geological studies, a 1/25 000 scale geologic map of the region was prepared. Afterwards, structural observations were made on the faults within the Troy Fault System. The data obtained from these observations were evaluated in the kinematical analysis program developed by Carey (1979) and the stress regimes effective in the region at the present time were determined. In active tectonic researches, shallow (0-15meter) drilling works can be conducted in order to determine and document basin asymmetry originating from active faults. With the help of shallow core drillings, lateral and vertical variations of the Quaternary deposits in the basin can be determined. In this study, a total of 5 shallow drillholes with depths varying between 3 and 9 meters were opened in the area north of the Troy fault and where Quaternary deposits outcrop. The core samples compiled from the drillings were logged and interpreted. In this study, along the Troy fault, using Direct Current Resistivity method measurements were made along 8 profiles with an average length of 200 meters and perpendicular to the fault. And applying two-dimensional inverse solution, geoelectric cross- sections for the shallow depths (0-15 meters) of the earth were obtained. And lastly, in the study paleoseismological trench works were conducted in the areas determined according to geological observations and geophysical data. Within the scope of this work, a total of 4 trench works were carried out, 3 trenches on Troy fault and 1 on Kumkale fault. All of the trenches were excavated perpendicular to the fault, with an average length of 10-15 meters, 3-4m wide and 4-6m deep. A total of 20 carbon-containing soil samples were compiled from the trenches for C-14 dating. These samples were analyzed at Beta Analytical Laboratory in the United States of America.

4. Findings

4.1. Morphotectonic analysis of Troy Fault

The geomorphology of a region can be described based on specific measurements of its morphological characteristics. This method, defined as morphometry, is realized by digitally deriving information about geomorphological elements from the elevation values (DEM - Digital Elevation Model) belonging to the region and analyzing them. These values, obtained with the help of morphometry, can provide consistent and fast information both on the evolution of the drainage in the study area and on the degree, distribution and character of the structural/lithological control on this evolution (Keller and Pinter, 1996).

For this purpose, through the digitization of the 1/25 000 scale sheets of Ayvalık I 16 a2 and Ayvalık I 16 b1 belonging to the study area, the digital elevation model of the region was created (Figure 6). The geomorphological characteristics of the region were investigated on this model. The geologic map of the region was draped over the digital elevation model as a separate layer and a relief geologic map was obtained (Figure 3).

The purpose of the morphotectonic analysis is to digitally reveal the degree of influence of the erosional and tectonic processes effective in the morphological shaping of a region. In this study, such geomorphological indices as mountain front sinuosity ratio (Smf index), stream/valley drainage areas and profiles, and valley floor width to valley height ratio (Vf index) were calculated.

Mountain front sinuosity ratio (Smf) is an index that reflects the balance between erosional forces that tend to cut embayments into a mountain front and the tectonic forces that tend to produce a straight mountain front. Mountain fronts uplifted by active tectonism are straight and have low Smf values. And mountain fronts that move slowly or have lost their activity display irregular forms and high values because they are destroyed by erosional forces (for further information, see Keller and Pinter, 1996). If Smf values are between 1 and 2, the fault in question has high activity. If this value is greater than 2, the activity of the fault should be considered as doubtful. However, one must bear in mind that Smf index can also be affected by the strength properties of the rocks forming the mounting front involved in faulting and by erosional activities.

In the evaluation made by taking into account mountain front sinuosity ratios, Troy Fault is subdivided into three geometric segments (Figure 6). These are, from east to west, Dümrek, Halileli and Tevfikiye segments, respectively. According to their mountain front sinuosity ratios, Tevfikiye and Halileli segments in the west have similar properties and are more active (Smf=1.15 and 1.48, respectively), while Dümrek segment in the east has relatively lower activity (Smf=2.52).

Morphological cross-sections were prepared in five areas along Troy fault (Figure 7). While there is disharmony between Dümrek and Halileli segments with regard to normal fault scarp elevation, an agreement is observed between Halileli and Tevfikiye segments. Normal fault scarp values also indicate that Halileli and Tevfikiye geometric segments should be evaluated together.

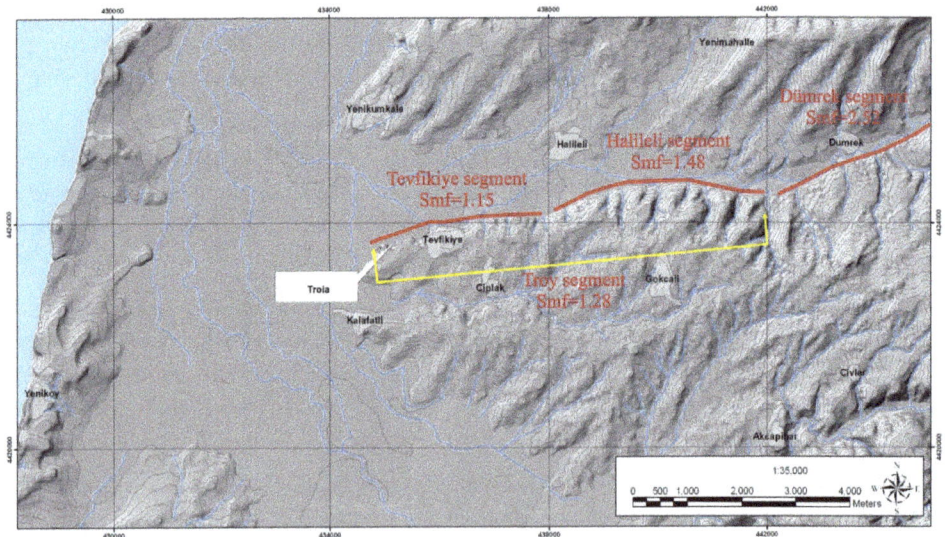

Figure 6. Digital elevation model of the study area and on this model, the segments of Troy fault according to mountain front sinuosity ratio.

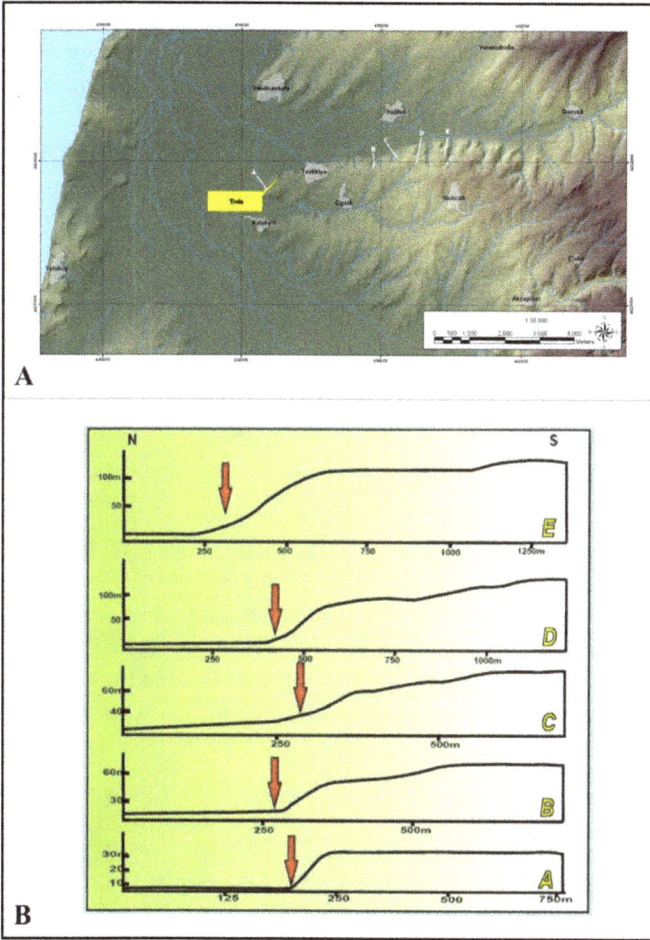

Figure 7. Locations of cross-sections along Troy fault

When the ratio of valley floor width to valley height (Vf) is calculated, the parameters in the formula are calculated at a certain distance from the mountain front for each valley. Higher Vf values show lower uplift speed and lower Vf values indicates actively uplifted areas.

Vf values showed that Tevfikiye and Halileli segments show similar properties and are more active; on the other hand, Dümrek segment has higher Vf values and hence is less active.

As a result of the morphotectonic analysis studies, since Tevfikiye and Halileli segments display similar morphotectonic characteristics, they were evaluated together and named the Troy segment. Since Troy segment is more active compared to Dümrek segment, the shallow geophysical and paleoseismological investigations related to the Troy fault were mostly conducted on this segment.

On the other hand, the expected maximum earthquake magnitude for Troy fault was calculated using the empirical equations proposed by Pavlides and Caputo (2004). The total length of the Troy fault is about 11-12km, hence, the expected maximum earthquake magnitude for the Troy fault was computed from the formula

$$Ms= 0.9\text{-log (SRL)} + 5.48 \text{ (Pavlides and Caputo, 2004)}$$

as Ms= 6.2-6.5

4.2. Kinematic analysis studies along the Troy fault

Troy fault system, located within the right lateral deformation zone between the northern and southern branches of NAFS, is represented by approximately E-W oriented normal faults, NE-SW oriented right lateral and NW- SE oriented left lateral strike slip faults. When taken into account its geologic fault length and morphotectonic characteristics, the Troy fault is the most important one of the approximately E-W oriented normal faults within the Troy Fault System.

Troy fault was defined for the first time by Tutkun and Pavlides (2005). It is a normal fault which is 11-12 km long, approximately E-W oriented, dipping 60^0 to the north (see Figure 3). Troy fault has highly pronounced normal fault morphology and is composed of two segments: Troy and Dümrek segments (See Figure 6).The Troy fault cuts Lower Cretaceous aged Denizgören ophiolites in the NE parts of the study area. To the east of Dümrek village, the fault makes a bend towards the west and continues in E-W direction, and it brings Neogene aged Çanakkale Formation and Quaternary deposits face to face in this area (Kürçer et al., 2006).

The fault enters the alluviums of Karamenderes Stream starting from the area where ancient Troy settlement is located, but its continuation from here towards the west is not clear. However, when its direction is followed, it can be seen that the fault trace again becomes evident between Ballıburun and Kesiktepe (See Figure 3). That's why, the sector of the Troy fault between ancient Troy settlement and Ballıburun was interpreted as probable active fault. In the drilling works we conducted in the Quaternary sediments in the hanging wall of the Troy fault, the top of the Çanakkale Formation was cut at 7-8 meter depth in the drillings B-4 and B-5 (See Figure 3) just north of the Troy fault. In the morphological cross-section (See Profile B in Figure 7) taken from an area between B-4 and B-5 drill locations, a normal dip slip of about 50 meter was measured. When the top of Çanakkale formation is taken as reference plane, cumulative dip slip in the Troy fault is around 60 meters including 7-8 meters obtained from the drillings.

In the study area, apart from the Troy fault, strike slip reverse and normal faults are observed, as well. In this context, within the Troy fault system, from 4 sites (for site locations, see Figure 3, Table 1) where outcrop conditions permit, a total of 67 fault planes were measured and calculated using numerical analysis method (Table 2). Site 1 was measured in Neogene aged Çanakkale formation. Thanks to this site, the kinematic condition of the final tectonic regime in the Troy region was determined.

Site	Longitude (N)	Latitude (E)	Age	Lithology
1	35 s 0435107	4423466	Neogene	Sandstone
2	35 s 0446204	4426875	Lower Cretaceous	Serpantinite
3	35 s 0446695	4427003	Lower Cretaceous	Serpantinite
4	35 s 0445035	4426292	Lower Cretaceous	Serpantinite

Table 1. Coordinates of the sites where fault planes or fault assemblages outcropping in the vicinity of Troy were measured; lithology and age of the measured geological units.

Site	N	σ_1 Az / dip	σ_2 Az / dip	σ_3 Az /dip	R	M.D.	S. D.
1	11	80 / 26	256 / 64	349 / 1	0.29	9.4	13.1
2	27	127 / 59	287 / 29	22 / 9	0.25	7.7	12.0
3	11	279 / 18	100 / 72	9 / 0	0.43	10.4	9.4
4	18	85 / 11	213 / 73	352 / 13	0.69	8.9	11.0
		$S_1 = 285 \pm 22° / 4°$		$S_3 = 15 \pm 21° / 6°$	Rm= 0.37		

Table 2. The conditions of the principal stress axes [(σ_1), (σ_2), (σ_3)] computed as a result of the evaluation of fault assemblages measured in the vicinity of Troy using Carey (1979) numerical analysis method; R ratio, number of measurements (N), fault measurement sites, mean (M.D.) and standard deviation (S.D.) values

The results of the faults (measured in the vicinity of Troy), obtained employing the method developed by Carey (1979) for kinematical analysis of fault assemblages are as follows: Although the sites where fault assemblages were measured (Figure 8) are composed of geological units of various ages, one site (Site 1) was measured in Neogene aged units. Accordingly, the tectonic regime effective at the present time implies transtensional strike slip faulting. In harmony with this strike slip faulting (compression) normal faulting (tension) regime developed, which represent the Troy fault as well.

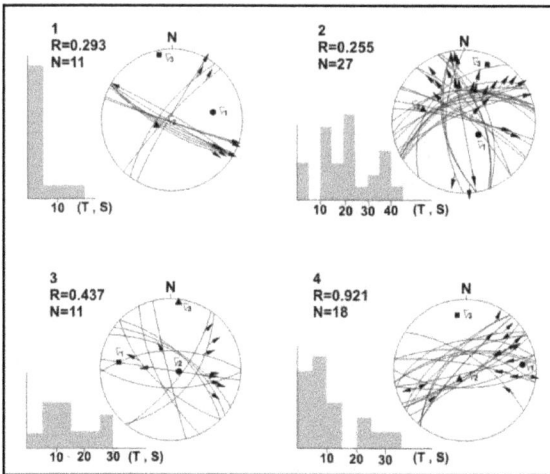

Figure 8. Presentation of the kinematical analyses carried out at locations belonging to faultings given in Table 2 on equal angle lower hemisphere (Wulf) (the distribution of the angle of deviation between the predicted slip vector (τ) and computed slip vector (s) is given in histograms).

For the purpose of correlating the results of the kinematical analyses of the fault assemblages with the focal mechanism inverse solutions of the recent earthquakes, focal mechanism inverse solution of the earthquake (Figure 3, Table 3) with a magnitude of 3.1 that occurred on 03.02.2008 was carried out (Figure 9). Focal mechanism inverse solution presents a faulting with strike slip normal component, developed under transtensional stress regime that is characterized by WNW-ESE oriented compression (P) and NNE-SSW oriented tension (T). This data is compatible with the stress conditions obtained from the kinematical analyses of the fault assemblages (Figure 10).

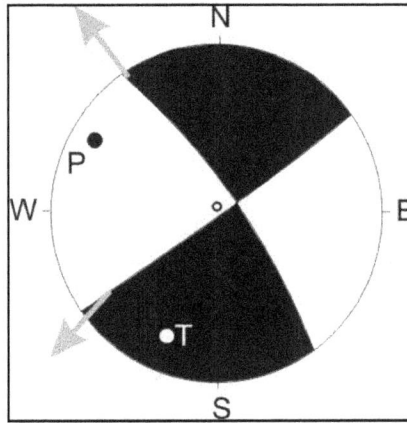

Figure 9. Focal mechanism inverse solution of the earthquake that occurred on 03.02.2008 (Özden et al., 2008)

Explanations	Values
Date	03.02.2008
Time	15 09
Latitude	39.94 N
Longitude	26.22 E
Magnitude (Mw)	3.1
Depth (km)	12
Plane 1	55°/88°/-165°
Plane 2	325°/75°/-3°
P axis	302°/10°
T axis	210°/10°
Source	BOUN-KOERI
Focal solution and reference	Özden et al., 2008

Table 3. Numerical values of the earthquake that occurred on 03.02.2008 (Özden et al., 2008)

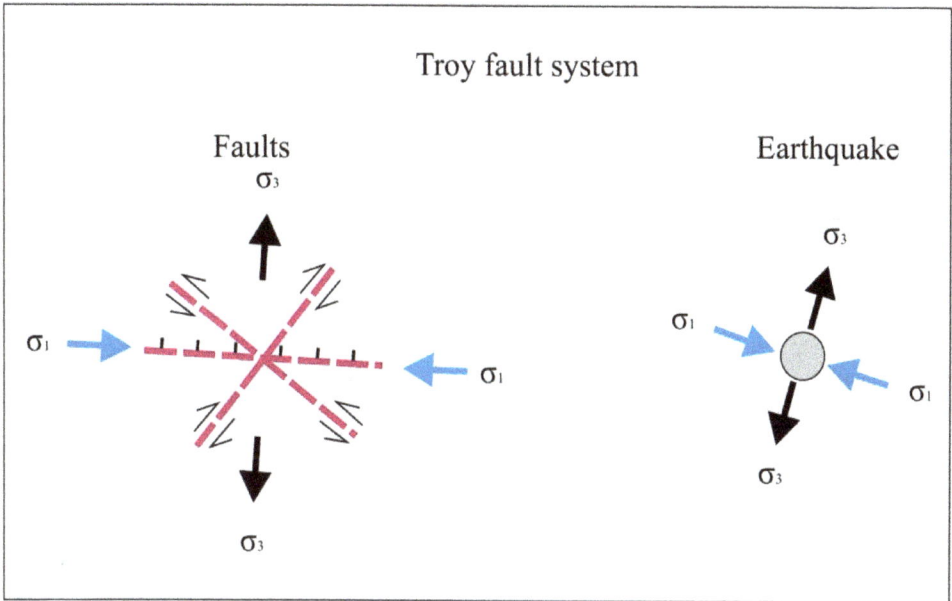

Figure 10. Comparative presentation of the regional stress conditions obtained from the kinematical analyses of the fault assemblages outcropping in the vicinity of Troy and the earthquake (black arrows represent effective regimes and blue ones represent non-effective regimes)

As a result; the Troy Fault System and stress regimes effective at the present time in the Troy region show that a NNW-SSE trending comppression and a NNE-SSW oriented tension is effective in this region and as a result of it, strike slip and normal faults developed under transtensional tectonic regime.

4.3. Shallow geophysical surveys along the Troy fault

Structural geology studies conducted along Troy fault aroused the suspicion that the fault scarp might have been degraded in the course of time. In order to determine the fault trace within Quaternary sediments, shallow geophysical researches employing direct current resistivity method were carried out. Within this context, shallow depths (0-20m) were investigated along a total of 8 profiles determined perpendicular to the fault trace along the Troy fault (for profile locations, see Figure 3). Two-dimensional inverse solutions of the obtained profiles were made and geoelectric cross-sections were constructed (Figure 11). According to the geoelectric cross-sections obtained by means of shallow geophysical studies, some discontinuities were encountered (Figure 11).

According to the Direct Current Resistivity studies conducted on the profiles perpendicular to the fault trace along Troy fault, it passes 30-100 meters north of morphologic escarpment bordering the Troy rise. This condition can be explained by the degradation of the Troy fault scarp in the course of time.

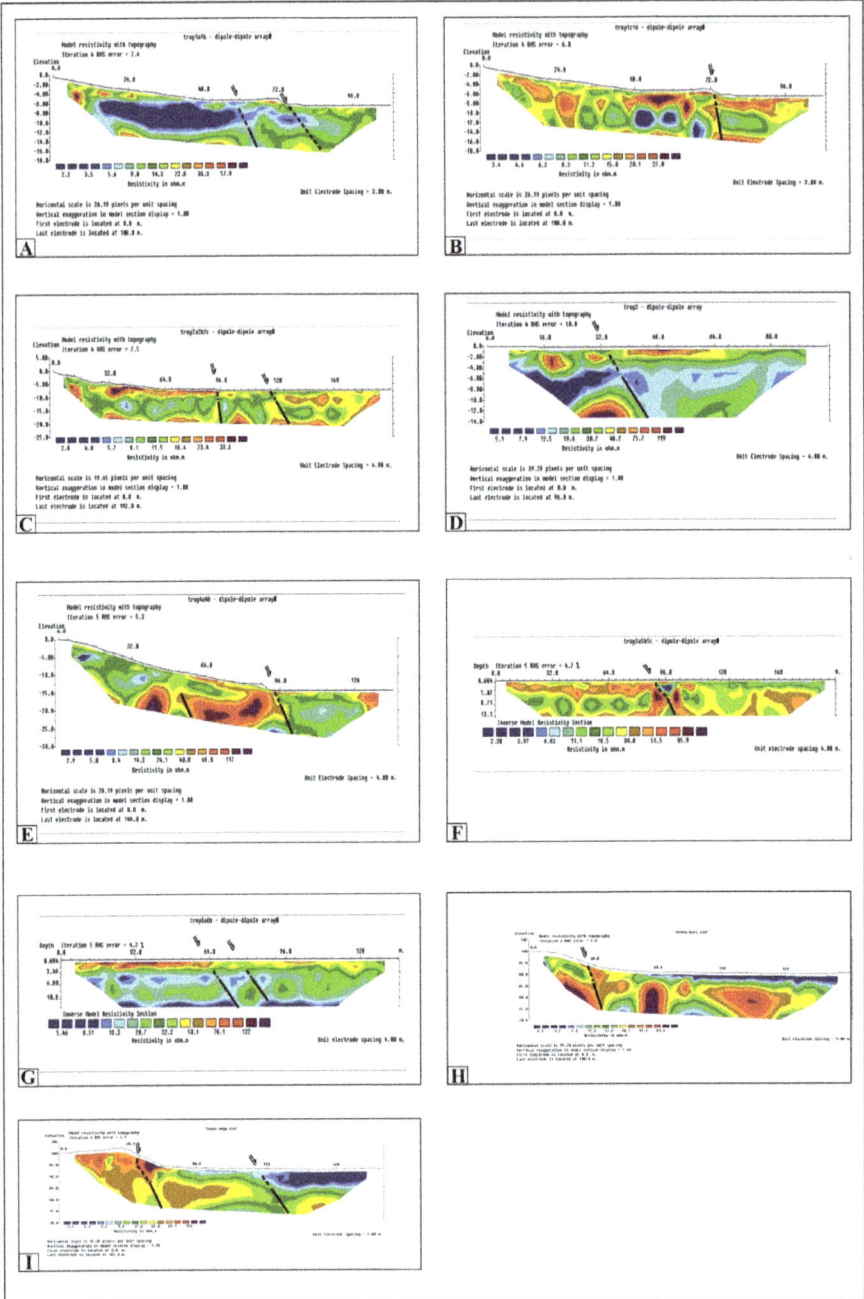

Figure 11. Geoelectric cross-sections measured by Direct Current Resistivity method along Troy fault and obtained employing two-dimensional inverse solutions. A: Profile 1a-b, B: Profile 1c-d, C: Profile 2, D: Profile 3, E: Profile 4, F: Profile 5, G: Profile 7, I: Profile 8 (for profile locations, see Figure 3).

4.4. Shallow drilling works on the hanging wall of the Troy fault

In active tectonic research, in order to determine and corroborate basin asymmetry, shallow (0-15 m.) drilling works can be carried out. Lateral and vertical variations of the Quaternary sediments within the basin can be determined with the help of shallow drillings. In this study, a total of 5 shallow drillholes having depths of 3-9m were opened in the area where Quaternary sediments outcrop, north of the Troy fault (for drill locations, see Figure 3).

The phases of the shallow drilling works are presented in Figure 12. In the drilling works, "Percussion drilling set for soils with gasoline powered percussion hammer-Cobra MK1" core drill machines were used. In the first phase, Quaternary sediments were drilled with the core drill machine (Figure 12A). Then with the help of a simple lever of the drill machine, the drill set was hoisted (Figure 12B). After labeling of the hoisted drill cores (Figure 12C), the drill cores were split into two along their long axes (Figure 12D). After these split cores were photographed (Figure 12E) and defined, the drill work was completed (Figure 12F). The well logs of the drillholes are given in Figure 13. According these well logs, Quaternary sediments become thicker towards the Troy fault.

When the information obtained from the shallow drillings were transferred onto the three-dimensional geologic cross-section of the region, it was observed that the plane constituting the boundary between Quaternary deposits and Neogene aged units are tilted towards south (towards the Troy fault) and consequently Quaternary sediments became thicker (Figure 14). In other words, the area north of the Troy fault is tilted back. This corroborates basin asymmetry and points to Quaternary activity of the Troy fault as well.

4.5. Paleoseismological trench works on Troy fault system

The tectonic regime type effective in Troy region at the present time is strike slip faulting with normal component (transtensional) developed under WNW-ESE oriented compression regime. The geological and geophysical investigations we conducted in Troy region showed that a NNE-SSW oriented tensional regime depended on this compression was effective in this region. The products of the transtensional tectonic regime effective in the region are E-W oriented normal faults, NE-SW oriented right lateral and NW-SE oriented left lateral strike slip conjugate faults. In this study, paleoseismological trench studies were conducted on the Troy normal fault, which is prominent by its geological fault length and morphotec-tonic properties, and on the right lateral strike slip Kumkale fault.

Within the scope of this work, a total of 4 trench works were carried out, 3 trenches along the Troy fault and 1 along Kumkale fault (for trench locations, see Figure 3). The trenches were excavated 15 meters long, 4 meters wide and 4 meters deep.

The existence of a fault, E-W oriented and dipping to the north, just north of the Troy rise was stated by various researchers (for example; Kayan, 2000; Tutkun and Pavlides, 2005; Kürçer et al., 2006). The fault named the Troy Fault (Tutkun and Pavlides, 2005) brings in contact the Neogene Çanakkale formation and Quaternary deposits outcropping in Dümrek plain (see Figure 3). There is an elevation difference of 50 meters on average between the

Figure 12. Phases of shallow drill works A: Drilling, B: Hoisting of drill set, C: Numeration, D: Splitting of drill cores, E: Photographing, F: Logging

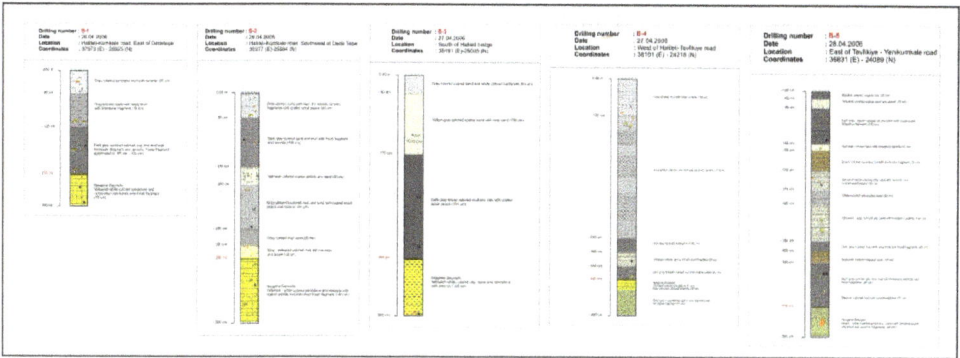

Figure 13. Well logs of shallow drillings conducted on the hanging wall of Troy fault

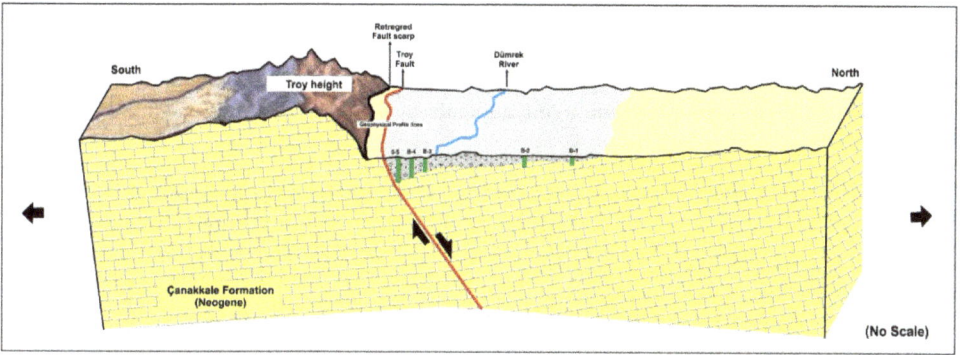

Figure 14. Three-dimensional schematic geologic cross-section of the Troy region constructed making use of surface geology and drilling data

floor of the plain and the Troy rise. In the shallow drilling works conducted by us, it was determined that Quaternary-Neogene boundary in sectors near the fault was at 8 meters. In the light of this information the cumulative dip slip on the Troy fault is thought to be approximately 60 meters.

When all the information about Troy fault, known and obtained by this study, is evaluated altogether, it is assumed that Troy fault was initially an E-W oriented normal fault developed within Western Anatolia Stress System (WASS). It gained some right lateral strike slip character owing to the western extensions of NAFS that affected the region as of Late Pliocene. The Troy fault that bears the traces of both WASS and NAFS has become one of the most significant morphologic elements in the region at the present time.

On the Troy fault, paleoseismological trench studies were conducted in three locations. These locations are: Tevfikiye (Figure 3, T1), Ballıburun (Figure 3, T2) and Kesiktepe (Figure 3, T3).

Tevfikiye Trench (T1)

Tevfikiye trench is located in an area east of the village road that connects Tevfikiye and Yenikumkale villages, very near to the drillhole B-5 and on the geophysical profile line number 2 (P-2) (see Figure 3, T-1). In this area, an excavation work was conducted that is 5-5.5 meters deep, 8 meters wide, 48 meters long and multibenched (Figure 15).

Tevfikiye trench was planned in S15E-N15W direction starting from Neogene aged deposits as to include the discontinuity indications on the geophysical profile line number 2 (see Figure 11-C). In figure 16, the composite trench cross-section belonging to the Tevfikiye trench location is presented. The composite trench cross-section was constructed making use of the excavation site belonging to Tevfikiye trench, the drilling information of drillhole B-5 and the observation of the small Tevfikiye trench excavated in this region.

As it can be seen from the trench section, Tevfikiye trench was excavated, at first, 48 meters long towards the north starting from the Neogene deposits. In the excavated 48- meter section, it was observed that Quaternary deposits unconformably overlie Neogene units with an erosional contact. The contact between Quaternary and Neogene can be distinctly traced over the 48-meter section. No trace of deformation that can be attributed to active tectonism was encountered in this section. However, numerous normal faults having dips not exceeding several centimeters and some paleoliquefaction structures in the type of fire structure were observed within the Neogene aged deposits at the basement, between 14th and 15th meters of the trench (Figure 17).

It was impossible to continue at the targeted depth towards north after 48th meter due to conditions of groundwater level. Thereupon, an additional trench with an average depth of 6 meters was excavated between 55th and 58th meters. No trace of deformation within Quaternary deposits at the upper parts of this trench was encountered, either. The Quaternary-Neogene contact was encountered at the depth of -6 meters (see Figure 16).

B-5 drillhole is located at the 74th meter of Tevfikiye excavation area (see Figure 16). In B-5 hole, Quaternary- Neogene contact was encountered at the depth of -8.10 meters (see Figures 13 and 16). When additional trench information and B-5 drilling data were transferred onto the trench cross-section belonging to Tevfikiye excavation area, it was seen that, at the 66th meter of the trench, Quaternary-Neogene contact is at different depths. This point corresponds to the discontinuity sign encountered at the 96th meter of the geophysical profile number 2 (see Figure 11-C). Geological observations in the Tevfikiye excavation area and shallow geophysical data point to the presence of a fault the south block of which was upthrown at the 66th meter of the excavation area. As can be seen from the composite trench cross-section prepared according to geological information, if a comparison is made taking Quaternary-Neogene contact as a basis, it is seen that the south block is uplifted about 2 meters compared to the north block. When evaluated under the regional tectonic framework, this fault is thought to be a normal fault dipping to the north (the Troy fault). No data, related to the continuation of the fault in the Quaternary deposits, was obtained from Tevfikiye trench. The age B.C.1190-1140 was obtained from the sample numbered TEV-E-04 compiled from Tevfikiye trench. TEV-E-04 sample was taken from the greenish yellow colored silty clay number 2. No trace of deformation was encountered within the unit number 2 in the Tevfikiye excavation area. This indicates that no earthquake occurred resulting from the Troy fault during the period from the deposition of the unit number 2 up to the present (B.C. 1190-Present-day), at least in Tevfikiye trench area.

Figure 15. Look at Tevfikiye Trench from the North

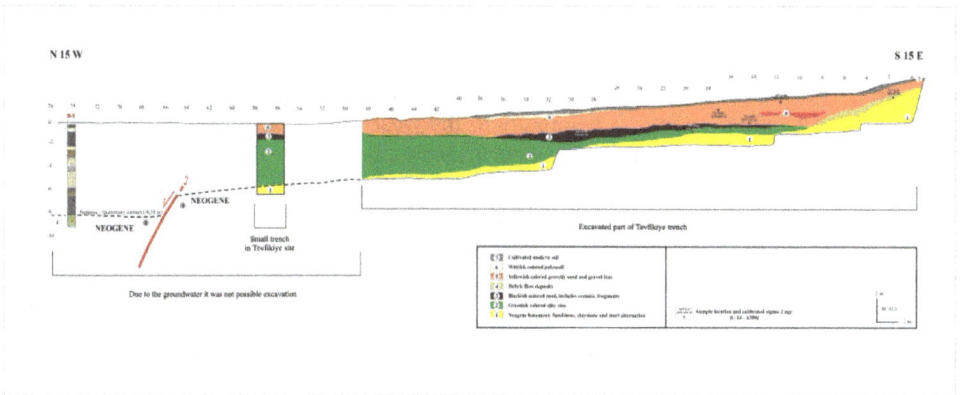

Figure 16. Composite section of the East Wall of Tevfikiye Trench (T1)

Figure 17. Small scale normal faults and paleoliquefaction structures observed between 14th and 15th meters of Tevfikiye Trench (Look from West to East)

Ballıburun Trench (T-2)

Ballıburun Trench is located between the electric resistivity profiles P-7 and P-8 at the Ballıburun location west of Tevfikiye (see Figure 3, T-2). The trench was excavated 22 meters long, 3 meters wide and 4 meters deep on average. Ballıburun trench was planned, starting from Neogene sediments, in S10W-N10E direction as to involve the discontinuity signs on the geophysical profile line number 8 (see Figure 11-I). In Figure 18, the trench cross section and the photograph belonging to the west wall of Ballıburun Trench are presented. As can be understood from the trench section, no sign of faulting was observed in Ballıburun trench. However, some paleoliquefaction structures that can be secondary proof of an earthquake were observed within the mud which is black colored and rich in organic matter, at a depth of -2 meters from the surface, between 12th and 14th meters of the trench (Figure 18B).

Three samples were collected and dated from Ballıburun trench. Two of these samples were collected from Unit 1 and the third one was taken from Unit 3 in which paleoliquefaction structures were observed. Since the age of the sample taken from Unit 3 is greater than the other two samples, it was considered as an allochthonous sample and was not counted in the interpretation. The ages of the remaining two samples support each other. As there is no usable age information from Unit 3, in which paleoliquefaction structures were observed, it can be said that the age of the earthquake, that might have caused these paleoliquefaction structures, is younger than the ages obtained from the samples BAL-5-13C and BAL-5-4C. Since the youngest age obtained from Unit 1 in the Ballıburun Trench is B.C. 760, the age of the earthquake that might have caused these paleoliquefaction structures is the time interval B.C. 760-Present-day.

Kesiktepe Trench (T-3)

The Troy fault enters the alluvium deposits of Karamenderes brook starting from the west of Tevfikiye village and loses its morphologic trace. It becomes again morphologically distinct in the sector between Ballıburun and the Aegean Sea (see Figure 3). In the highland that is located west of Ballıburun and composed of Neogene aged Çanakkale formation is known as "Kesik Tepe" (Truncated Hill) and this area is characterized by its "V" shaped morphology. It is thought that this morphologic anomaly was initially shaped by Troy fault and later modified by human activity. For the purpose of checking the anomaly, electrical resistivity measurements were carried out along P-5 profile from Kesiktepe location and geoelectric cross-section was obtained (see Figure 11-F). In the geoelectric cross-section, a discontinuity was determined in the region where Kesik Tepe is located and for the purpose of testing this discontinuity, Kesiktepe Trench was excavated, which is 6 meters long, 2 meters wide and 4.5 meters deep on average (Figure 3, T-3).The trench cross section of Kesiktepe trench is presented in Figure 19.

No sign of faulting was encountered in Kesiktepe trench. However, Kayan (2000) mentions that Neogene deposits were cut at different depths in two drillholes very near to each other, drilled in the south and north of Kesiktepe. According to the researcher, while the Neogene deposits were cut at -2 meter in the south, they were cut at -8meter just in the north. This indicates that the north block was relatively downthrown. As to support this, some signs of discontinuity were encountered at the 96th meter on the geophysical profile number 5 (see Figure 11-F). On the other hand, it is known that, during Troy era, in Kumkale plain some drainage canals were excavated in order to dry the swamp that caused malaria (personal communication with Dr. Rüstem Aslan. 2005; Figure 20). And one of these drainage canals is situated just on the Troy fault trace in Kesiktepe location. By means of the discontinuity in the geoelectric cross-section, archeological information and geomorphological approach, it is thought that the Troy fault reaches the Aegean Sea over Kesiktepe. However, since the ca-nal, excavated as drying canal, was filled with recent sediments during the period from its last activity to the present time, no trace of active faulting was encountered at the first 4 meters from the surface (from B.C. 1500 up to the present).

Kumkale Trench (T-4)

In the Troy region, there are also NE-SW oriented, right lateral strike slip faults that were developed within NAFS. The best examples for these faults are Kumkale and Yenimahalle faults (see figure 3). Kumkale fault is a $N25^0E$ oriented, dipping 75^0 to NW right lateral strike slip fault that borders from the north the ridge on which Kumkale village is situated. Yaltırak et al. (2000), in their shallow sea seismic studies conducted in Dardanelles and its

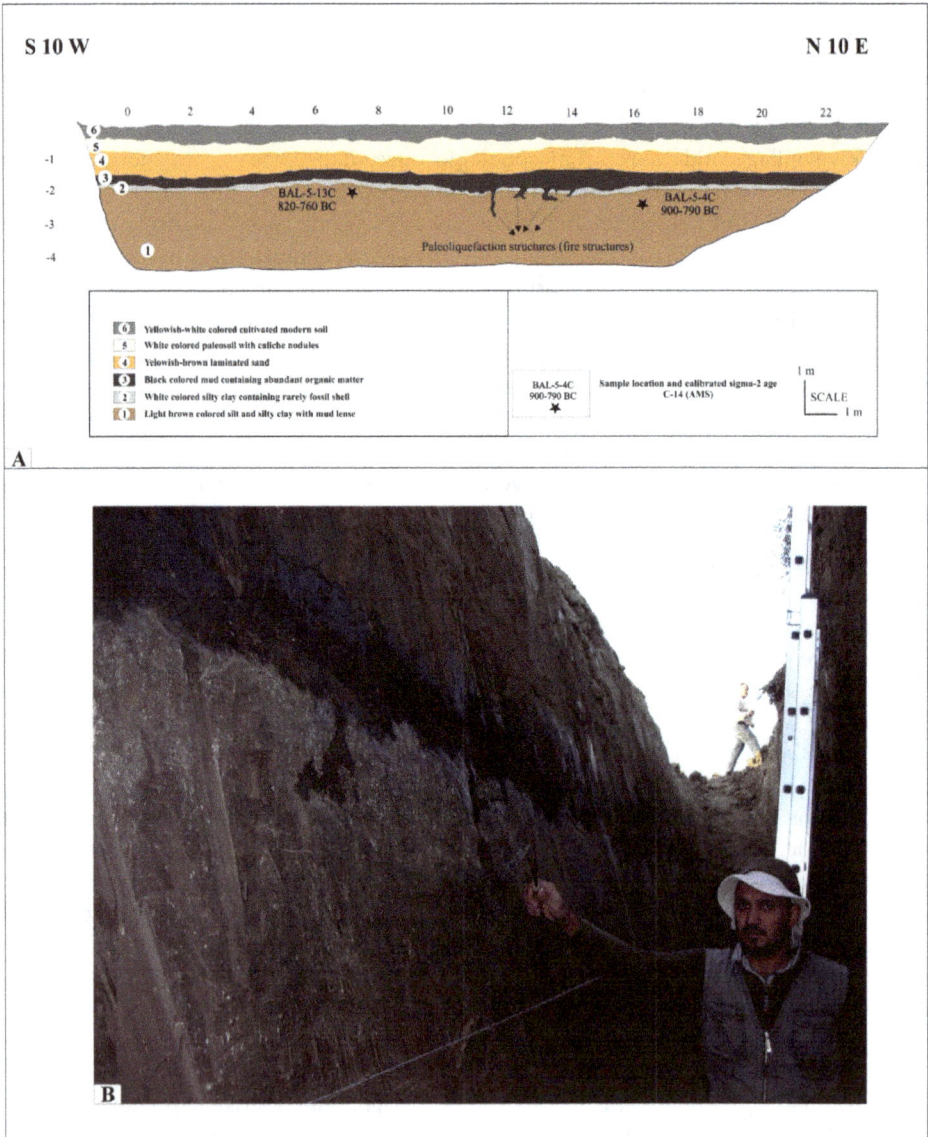

Figure 18. A: West wall cross-section of Ballıburun trench (T-2) B: Paleoliquefaction structures

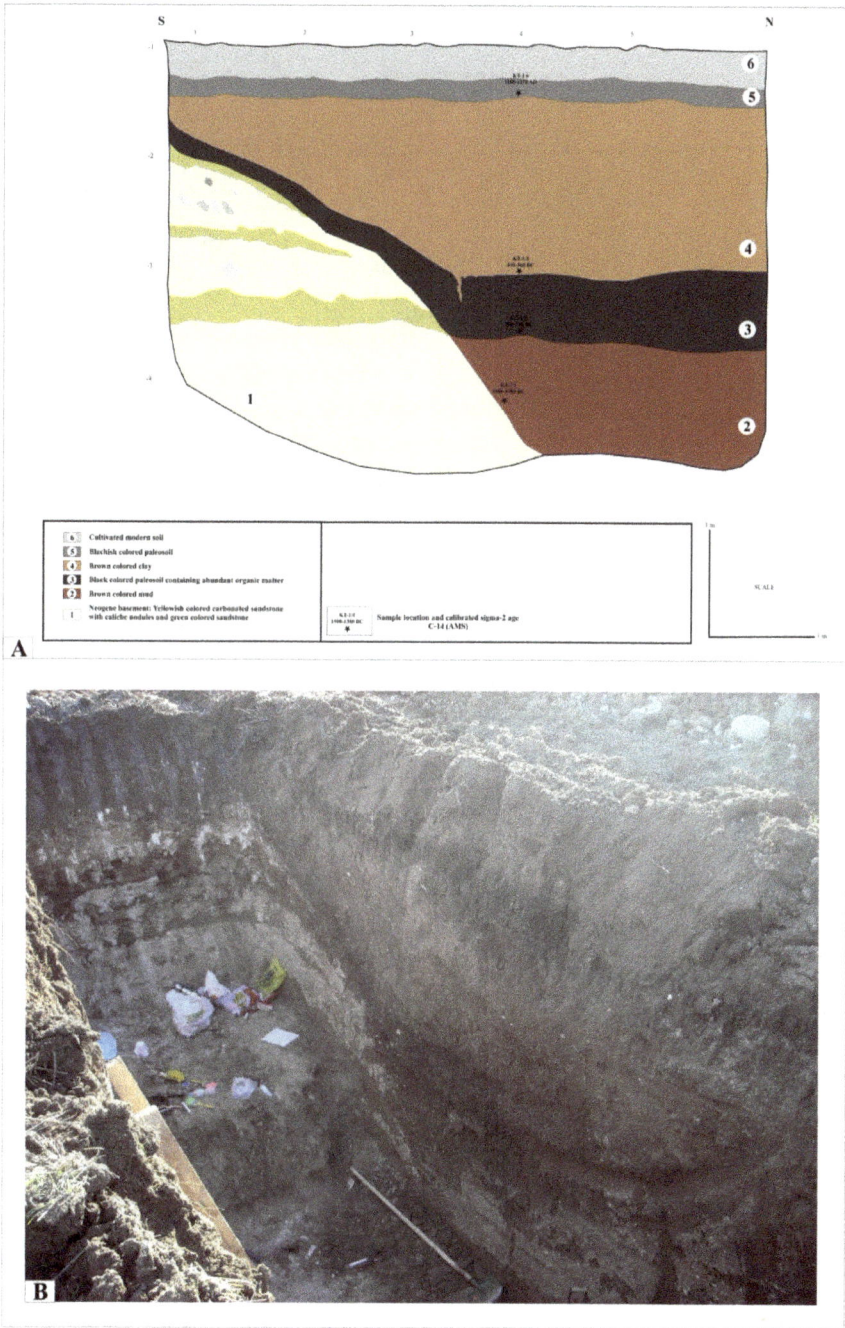

Figure 19. A: West wall cross section of Kesiktepe Trench (T-3), B: Photograph of West Wall of Kesiktepe Trench

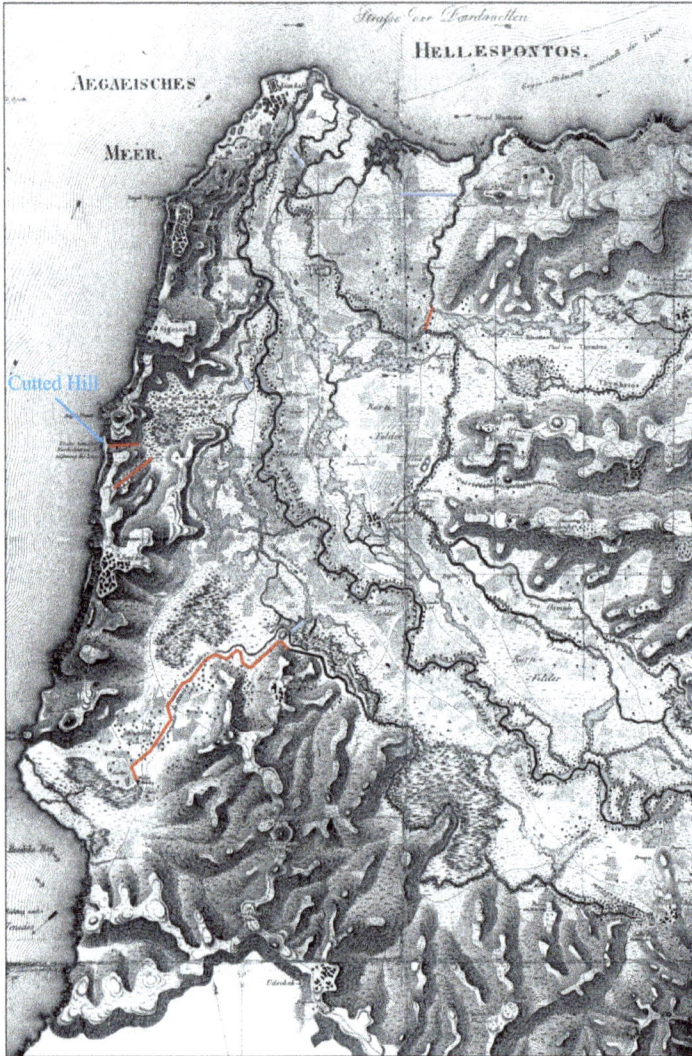

Figure 20. Relief geomorphological map of Troy region (Forchammer, 1850) Red lines show drainage canals opened in order to dry the swamp

vicinity, documented NW-dipping faults as the continuation of Kumkale fault. These faults are the continuation of Kumkale fault in the sea. The total length of Kumkale fault is 15 km including its continuation in the sea. Kumkale fault is thought to be one of the primary synthetic strike slip faults developed within NAFS.

Kumkale fault brings Neogene units and Quaternary deposits face to face (Figure 21). This relationship could be observed in the open outcrop in Figure 21 B. Afterwards; this observation point was leveled with the help of a working machine and transformed into a trench (see Figure 3, T-5). In Figure 22, the photomosaic and trench cross section are presented.

Four samples were taken from T-5 Trench and ages could be obtained from three of them.

Through the evaluation of trench microstratigraphy and C-14 age data together, two earthquakes were defined in T-5 Trench. The first of them is associated with the faulting that cuts Unit C and is covered by Unit B. Consequently, the event horizon for this fault (Event II) is between Units C and B. The age obtained from Unit C is between B.C. 40 and A.D. 130, and the age obtained from Unit B is between A.D. 780 and 1000 (see Figure 24). According to these age data, the first earthquake occurred between A.D. 130 and 780. The second fault (Event I) is associated with the faulting that cuts Unit B and is covered by Unit A. The event horizon for this earthquake is between Units B and A. The age obtained from Unit B is A.D. 780-1000. And the age obtained from Unit A is between A.D. 1300 and 1430 (see Figure 22). According to these age data, the second fault (Event I) occurred between A.D. 1000 and 1300.

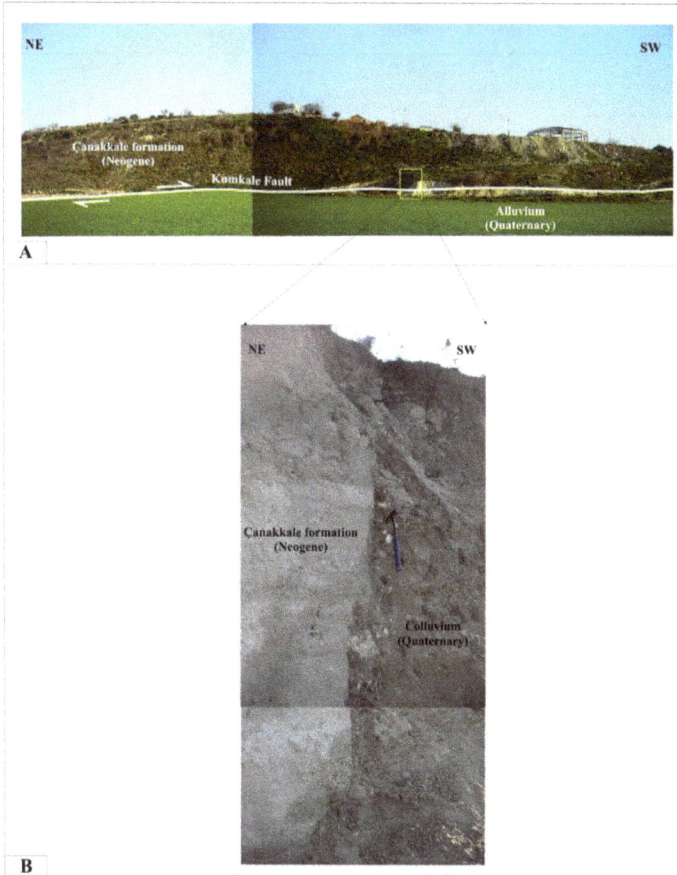

Figure 21. A: Panoramic sight of Kumkale fault, B: Sight of Kumkale fault in the outcrop

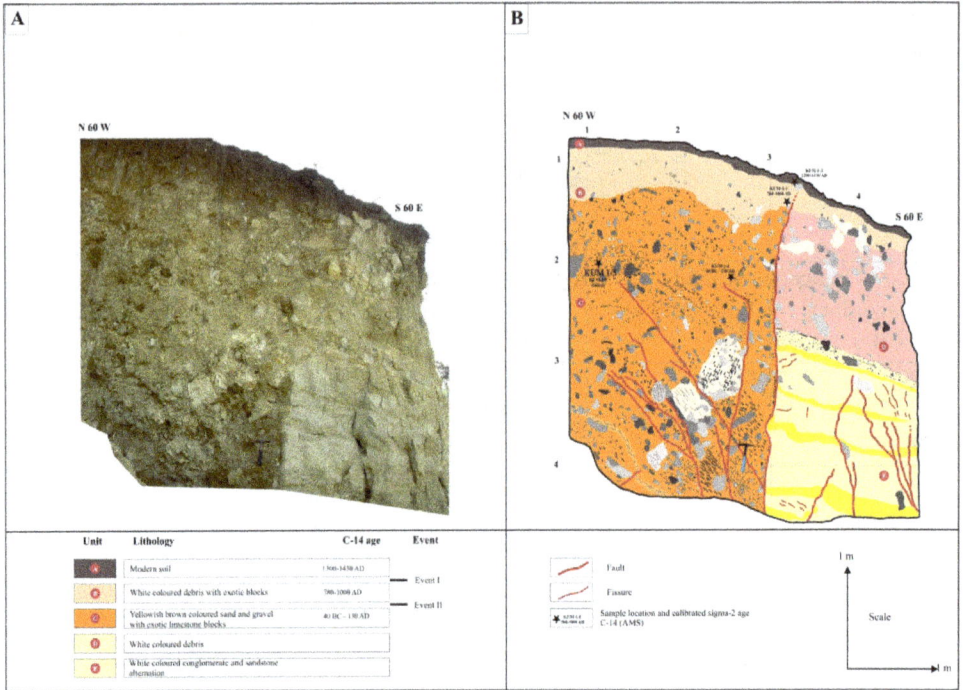

Figure 22. Photomosaic (A) and cross-section (B) of the south wall of Kumkale Trench (T-5)

5. Results and discussion

As a result of this study;

According to the morphotectonic studies conducted on the Troy fault, the Troy fault is divided into three segments, from west to east, as Tevfikiye, Halileli and Dümrek segments. Since their mountain front sinuosity ratio (Smf) and valley floor width to valley height ratio (Vf) are similar, Tevfikiye and Halileli segments were evaluated together and given the name "the Troy segment".

The mountain front sinuosity ratio of the Troy segment was computed as Smf= 1.28 and that of Dümrek segment was computed as Smf=2.52. Taking into account these values, the Troy segment was determined to be more active and therefore, subsequent geological and geophysical studies were concentrated on the Troy segment.

In the calculations carried out using the empirical equations suggested in the studies that investigate the relation between the geological fault length and the greatest earthquake (for example, Pavlides and Caputo, 2004), the greatest earthquake magnitude that can result from 11-12 km long Troy fault was computed as M= 6.2-6.5.

Both the results of the kinematic analysis studies of the fault assemblages and the focal mechanism inverse solution of the earthquake having the magnitude of 3.1 that occurred on the Troy fault on 03.02.2008, indicate a strike slip with normal component faulting developed under transtensional stress regime characterized by WNW-ESE oriented compression (P) and NNE-SSW oriented tension (T) in the Troy region at the present time. The fault systems in the Troy region and the stress regimes which are effective at the present time show that strike-slip faults and concordant normal faulting in active in the Troy region.

According to the Direct Current Resistivity profile studies conducted on profile lines perpendicular to the fault trace along the Troy fault, it was understood that the Troy fault passes 30-100 meters north of the morphological escarpment which borders the Troy rise from the north. This situation can be explained by means of the degradation of the Troy fault scarp in the course of time. According to the geoelectric cross-sections obtained from the shallow geophysical studies, some discontinuity signs were encountered. However, it was observed that these discontinuities do not reach the surface but come to an end at the depth of about -10 meters from the surface.

In the Quaternary deposition area north of the Troy fault, in order to examine the lateral thickness variation of the Quaternary deposits towards the fault, 5 shallow drilling works with depths varying between 3 and 9 meters were conducted at 5 points along a direction perpendicular to the Troy fault. When the information obtained from these drilling works were transformed onto the three dimensional schematic geological cross-section, it was seen that the plane which constitutes the boundary between the Quaternary deposits and the Neogene aged units gets thicker towards the south (towards the Troy fault). In other words, this boundary is tilted towards the Troy fault. This corroborates the asymmetry of the basin and at the same time points to the Quaternary activity of the Troy fault. From the paleo-seismological trench studies it was understood that no earthquake resulting from the Troy fault occurred in the last 3000 years from B.C. 1190 up to the present, in Tevfikiye trench. Since the thickness of the sediments accumulated on the fault during the period from the last earthquake which resulted from the Troy fault up to the present was greater than the depth that can be reached in paleoseismological trench works, it was not possible to reach traces of faulting in the trench excavations.

In Ballıburun trench, some paleoliquefaction structures, developed during the time from B.C. 760 up to the present as a result of an earthquake that was originated from the nearby faults, were observed.

Two earthquakes were dated in the paleoseismological studies conducted on the Kumkale fault. The first of these faults occurred between A.D. 130 and 780 and the second one occurred between A.D. 1000 and 1300.

The paleoliquefaction structures observed in Ballıburun trench might have resulted from the Kumkale fault.

In this study, it was understood that the earthquakes which destroyed Troy III and Troy VI layers had not developed depending on the Troy fault system. Although their source is not clearly known, these earthquakes might have originated from the active faults in Biga Peninsula or from Gaziköy-Saroz fault which represents the north branch of NAFS. The isoseist maps of the large earthquakes that caused surface faulting on Gaziköy-Saroz, Yenice-Gönen and Biga faults in this region, in the last century are presented in Figure 23.

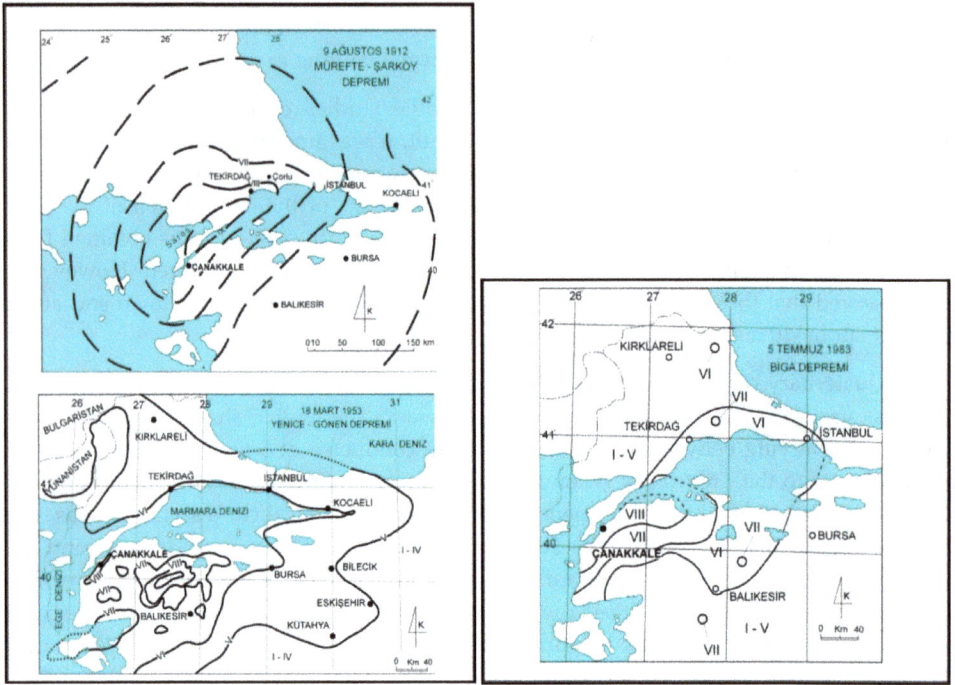

Figure 23. Isoseist maps of Mürefte-Şarköy (1912, Yenice-Gönen (1953), and Biga earthquakes (1983) (Kalafat et al., 2007)

As can be clearly seen in Figure 24, the region where Troy is located was affected at an intensity of VIII-IX by all three earthquakes. When such properties as local soil conditions and building style are taken into account, it is possible to state that historical earthquakes resulted from these faults affected Troy region. Rockwell et al., (2001) conducted paleoseismological studies on Gaziköy-Saroz segment of NAF in the vicinity of Kavakköy and documented 5 earthquakes from the present-day backwards. Kürçer et al., (2008), in their paleoseismological studies carried out on Yenice-Gönen fault, documented two earthquakes that they dated to A.D. 620 and A.D. 1440. In the light of all this information, it is the possible that the earthquakes that affected the Troy region might have resulted from the other active faults in the region.

Author details

Akın Kürçer
General Directorate of Mineral Research and Exploration,
Department of Geology, Çankaya, Ankara, Turkey

Alexandros Chatzipetros, Spyros Pavlides,
George Syrides, Kostas Vouvalidis and Özkan Ateş
Aristotle University of Thessaloniki, Faculty of Sciences,
Department of Geology, Thessaloniki, Greece

Salih Zeki Tutkun and Süha Özden
Çanakkale Onsekiz Mart University, Department of Geological Engineering, Çanakkale, Turkey

Emin Ulugergerli and Yunus Levent Ekinci
Çanakkale Onsekiz Mart University, Department of Geophysical Engineering, Çanakkale, Turkey

Acknowledgement

This study was supported by a bilateral cooperation project between TUBİTAK (The Scientific and Technological Research Council of Turkey) and GSRT (General Secretariat for Research and Tecnology). Grant: TUBİTAK - 105Y360. We would like to warmly thank the following colleagues that assisted in the fieldwork during various field periods: A. Michailidou, S. Sboras, S. Valkaniotis, A. Zervopoulou (from Aristotle Universityof Thessaloniki, Department of Geology) and Gündoğdu, Y (from Ankara University, Department of Geophysical Engineering).

6. References

Aksoy, M., E., Meghroui, M., Vallee, M. and Çakır, Z., 2010. Rupture characteristics of the is A.D. 1912 Mürefte (Ganos) earthquake segment of the North Anatolian fault (western Turkey). Geology, November 2010; v. 38; no. 11; p. 991-994; doi: 10.1130/G31447.1.

Ambraseys, N.,N. and Finkel, C.,F., 1991. Long-term seismicity of İstanbul and of the Marmara region, Engin. Seis. Earthq. Engin. Report, 91/8, İmperial College.

Canıtez, N. and Toksöz, M.N. 1971. Focal mechanism and source depth of earthquakes from body and surfacewave data. *Bulletin of Seismological Society of America* 61, 1369–79.

Carey, E.; 1976. Analyse numérique d'un modéle mécanique élémentaire appliqué a l'étude d'une population de failles : calcul d'un tenseur moyen des contraintes a partir des stries de glissement. Thése de 3º cycle, Université de Paris-Sud, Orsay, 138 p.

Carey, E., 1979. Recherche des directions principales de contraintes associées au jeu d'une population de failles, Revue Geological Dynamic and Géography physic., 21, 57-66

Jackson J., McKenzie D., 1984. Active tectonics of the Alpine–Himalayan Belt between Western Turkey and Pakistan, Geophy. J. Royal Astr. Soc. 7 (1984) (1984) 185–264.

Kalafat, D. 1988. Active tectonics and seismicity of SW Anatolia and its vicinity. Bulletin of Earthquake Research, 63, 5–98. Ankara (in Turkish with English abstract).

Kalafat, D., Güneş, Y., Kara, M., Deniz, P., Kekovalı, K., Kuleli, S., Gülen, L., Yılmazer, M., Özel, N., 2007. A revised and extended earthquake cataloque for Turkey since 1900 (M≥ 4,0). Boğaziçi University, Kandilli Observatory and Earthquake Research Institute, İstanbul.

Keller, A. E., and Pinter, N., 1996; Active Tectonics, Earthquakes, Uplift and Landscape. Prentice Hall (ISBN 0-02-304601-5) N. Jersey -pp. 377. (Second Edition 2002).

Kürçer, A. and Tutkun, S.Z., 2008. The role of the geological and tectonic processes on geomorphological evolotuion of the Biga Peninsula, NW Turkey. National Geomorphology Symposium, Abstracts, 213-214. Çanakkale, 2008.

Kürçer, A., 2006. Neotectonic Features of the vicinity of Yenice – Gönen and Paleoseismology of March 18, 1953 (Mw:7,2) Yenice-Gönen Earthquake Fault (NW Turkey) (Çanakkale Onsekiz Mart University Natural and Applied Sciences Institue - 2006). p. 170 (in Turkish with English abstract).

Kürçer, A., ,Tutkun, S. Z., Pavlides, S., Chatzipetros, A., Ateş, Ö., Özden, S., , Ulugergerli, E., Gündoğdu, Y., Bekler, T., Syrides, G., Vouvalidis, K., Valkaniotis, S., Zervopoulou, A., Şengül, E., Ekinci, Y. L.,Köse K., Demirci, A. and Elbek, Ş., 2006, Morphotectonical Features of Troia Fault and Preliminary Paleoseismological Studies, NW Turkey. 10*th* Meeting of the Active Tectonic Research Group 10*th*. Proceedings; pp: 60., 2-4 November 2006, Dokuz Eylül University, Department of Geology, İzmir.

Kürçer, A., Chatzipetros, A., Tutkun, S.Z., Pavlides, S., Ateş, Ö. and Valkaniotis, S. 2008. The Yenice-Gönen Active Fault (NW Turkey): Active Tectonics and Palaeoseismology, Tectonophysics, 453, 263-275.

McKenzie, D.P., 1972. Active tectonics of the Mediterranean region, Geophys. J. R. Astr. Soc., 30 (2), 109-185.

McKenzie, D.P., 1978. Active tectonics of the Alpine-Himalayan belt: The Aegean sea and surrounding regions (tectonics of aegean region), Geophys. J. R. Astr. Soc., 55, 217-254.

Okay, I.A, (1987); Geology and tectonics of the western part of Biga peninsula. TPAO, Report Number: 2374.

Özden, S.,Bekler, T., Tutkun, S. Z., Kürçer, A., Ateş, Ö., Bekler, F., Kalafat, D.,Gündoğdu, E., Bircan, F., Çınar, S., Çağlayan,Ö., Gürgen, M., İşler, H. and Yalçınöz, A., 2008, Seismotectonics of Biga Peninsula and South of Marmara Sea. Active Tectonic Research Group 12*th*. Abstracts; pp: 48-49, Akçakoca, MTA, Turkey.

Pavlides, S. and Caputo, R., 2004. Magnitude versus faults' surface parameters:quantitative relationships from the Aegean. Tectonophysics, 380 (3–4), 159–188.

Rockwell, T., Barka, A., Dawson, T., Akyüz, S. and Thorup K. 2001. Paleoseismology of the Gazikoy-Saros segment of the North Anatolia fault, northwestern Turkey: Comparison of the historical and paleoseismic records, implications of regional seismic hazard, and models of earthquake recurrence. Journal of Seismology, 5: 433-448.

Şaroğlu,F., Emre, Ö. and Kuşcu, İ., (1992), Active fault map of Turkey, MTA, Ankara.

Şengör, A.M.C., Görür, N. and Şaroğlu, F., 1985. Strike-slip faulting and related basin formation in zones tectonic escape: Turkey as a case study, in strike-slip deformation, Basin formation and Sedimentation, edited by Biddle, K.T. and Christie-Block, N., Soc. Econ. Paleontol. Mineral. Spec. Publ., 37, 227-264.

Şentürk, K. and Karaköse, C, 1987, Geology of Çanakkale strait and its vicinity, MTA Report Number: 9333 (unpublished).

Taymaz, T. 2001. Active tectonics of the North and Central Aegean Sea. Symposia on Seismotectonics of the North-Western Anatolia-Aegean and Recent Turkish Earthquakes, XI-XIX. Istanbul: ITU, 113 pp.

Tutkun, S.Z. and Pavlides, S.B. 2005. The Troy Fault, Bulletin of the Geological Society of Greece, Vol. XXXVII, 194-200.

Yaltırak, C., Alpar, B., Sakinc, M., Yüce, H., 2000. Origin of the Strait of Çanakkale (Dardanelles): regional tectonics and the Mediterranean–Marmara incursion. Mar. Geol. 164, 139– 156 (with Erratum 167, 189–190).

Chapter 9

Structural Geological Analysis of the High Atlas (Morocco): Evidences of a Transpressional Fold-Thrust Belt

Alessandro Ellero, Giuseppe Ottria, Marco G. Malusà and Hassan Ouanaimi

Additional information is available at the end of the chapter

1. Introduction

The High Atlas of Morocco, representing the southernmost element of the Perimediterranean Alpine belts, is a typical example of intracontinental belt (Mattauer et al., 1977). It was formed within the North African plate during convergence of the African and European plates during the Cenozoic (Dewey et al., 1989; Gomez et al., 2000). Like other intracontinental mountain belts, the High Atlas shows a double sense of vergence and a complex evolution of timing and sequence of thrusting.

Many studies have emphasized the role of the inversion tectonics in the evolution of the High Atlas system (Proust, 1973; Jacobshagen et al., 1988; Giese and Jacobshagen, 1992; Laville and Piquè, 1992; Beauchamp et al., 1996, 1999; Mustaphi et al., 1997; Hafid, 2000). Thrust and fold structures would resulted from the reactivation (inversion), caused by the Cenozoic compressional events, of the preexisting extensional faults associated with the Triassic-Liassic Atlasic rifting.

In this context several authors considered strike-slip faulting as an important component of the Alpine evolution of the High Atlas belt (Mattauer et al., 1977; Fraissinet et al., 1988; Froitzheim et al., 1988; Laville and Piquè, 1991, 1992; Morel et al., 2000; Piquè et al., 2002). However, more recent studies (Frizon de Lamotte et al., 2000; Teixell et al., 2003, 2005; Arboleya et al., 2004; Ayarza et al., 2005; Missenard et al., 2006) have been aimed to the definition of the geodynamic model for the Alpine evolution of the High Atlas belt. As a consequence structural studies have been performed within a regional scale geodynamic pure compressional framework, neglecting the kinematic meaning of data. Therefore, despite the fact that the geometries of deformation structures are locally well known, the details of Alpine structural evolution of the High Atlas belt, particularly kinematic information, are still poorly known.

In this contribution we present the results of a structural and kinematic study we have carried out in Morocco along a transect crossing the Western High Atlas and along the southern margin of the Central High Atlas in the Tinerhir-Boumalne area, two key sectors of the High Atlas belt. A main objective of this paper is to establish the possible relationships between the described structures to better understand the general processes of intracontinental mountain building that have constructed the High Atlas belt.

Our structural and kinematic study, suggesting main transpressional imbricate systems, indicates that strike-slip movements played an important role in the deformation evolution of the High Atlas fold and thrust belt. We propose that the main Alpine deformation of High Atlas belt was transpressional in character, with efficient kinematic strain partitioning focussing the strike-slip component along pre-existing major faults. The potential influence of pre-existing tectonic boundaries on weakness and evolving tectonic fabrics on kinematic strain partitioning can be therefore emphasized.

2. Regional geology

The High Atlas represents the highest mountain belt of Morocco, with peaks of over 4000 m a.s.l. (Mt. Toubkal, 4165 m), crossing the country along the SW-NE direction from the Atlantic Ocean to Algeria for a length of about 2000 km (800 km in Morocco) and a width ranging from about 50 km to 100 km, framed between the Meseta domains (Morocco Meseta and Oran Meseta) to the north, and the northern boundary of the West African Craton (Anti Atlas belt) to the south (Figure 1).

The High Atlas fold and thrust belt is formed by a Precambrian and Paleozoic basement and a Mesozoic-Cenozoic succession. The pre-Mesozoic basement is exposed in several inliers of the High Atlas, forming the most elevated areas of the Western High Atlas. The Mesozoic successions, mostly Jurassic in age, crop out almost exclusively in the Central High Atlas and the Atlantic basin south of Essaouira (Figure 1). The Cenozoic deposits, substantially absent along the belt axis, characterize the High Atlas boundaries with the plains where the Neogene formations deposited.

In the Western High Atlas, the Precambrian basement crops out mostly in the Ouzellarh Block, being composed by metamorphic rocks and granitoids topped by late Precambrian volcanics. The Paleozoic succession ranges from Lower Cambrian to Carboniferous and it is mostly characterized by clastic rocks deformed during the Variscan orogeny. In particular, the Late Visean-Early Westphalian tectonic event (Main Variscan Phase; Michard et al. 2008 and references) produced tight folds associated with metamorphism and granite intrusions. These folds show NE-SW to N-S axes and sub-vertical to generally E-dipping axial planes (western vergence) and developed a pervasive axial plane cleavage. The P-T conditions of metamorphism did not exceeded low-grade greenschist-facies conditions during the main Variscan folding, except in the regions close to the granite intrusions (Michard et al., 2008).

However the Variscan structural pattern is dominated by ENE-WSW to NE-SW major dextral fault zones (e.g. Tizi n'Test, Medinat, Erdouz) which broke up the Western High Atlas into several structural blocks (Proust et al., 1977; Ouanaimi and Petit, 1992; Houari and Hoepffner, 2003; Hoepffner et al., 2005). A recent paper (Dias et al., 2011) emphasises the occurrence of WNW-ESE sinistral shear zones, already documented by Fraissinet et al. (1988), in the framework of conjugated fault systems with the ENE-WSW dextral shear zones developed during the main Variscan phase.

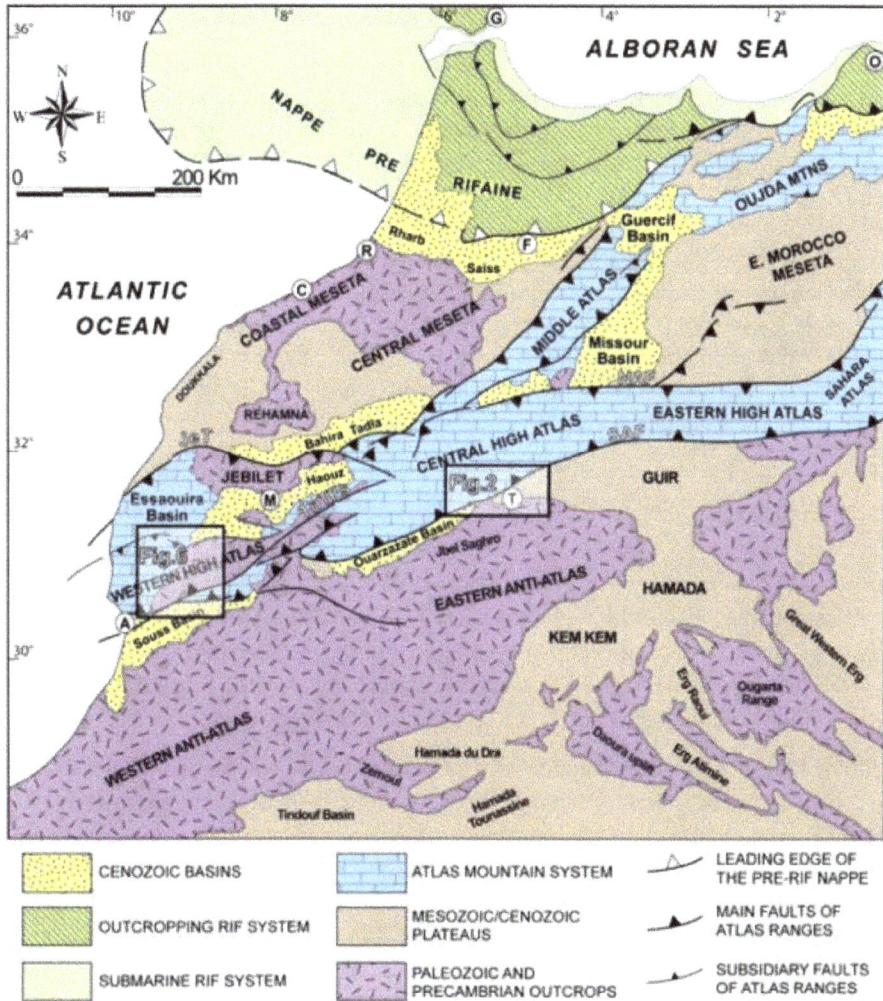

Figure 1. Schematic structural map of Morocco with location of Figures 2 and 6 (modified after Hafid et al., 2006 and Michard et al., 2008). SAF: South Atlas Fault; NAF: North Atlas Front; TnTFs: Tizi n'Test Fault system; JeT: Jebilet Thrust; A: Agadir; C: Casablanca; F: Fes; G: Gibraltar; M: Marrakech; O: Oujda; R: Rabat; T: Tinerhir.

The Mesozoic succession of the High Atlas belt started with the Late Permian-Triassic red beds (conglomerates, sandstones, siltstones and mudstones), unconformably resting on the Lower Paleozoic rocks or on the Precambrian basement. These continental deposits (Fabuel-Perez et al., 2009) represent the detrital infilling of basins developed during the Late Permian-Triassic Atlasic pre-rifting phase when the Variscan shear zones were reactivated as normal and strike-slip faults. The pre-rift deposits are capped by tholeiitic basalt flows of the CAMP (Central Atlantic Magmatic Province) that provide absolute ages of about 200 Ma corresponding to the Triassic-Jurassic transition (Fiechtner et al., 1992; Knight et al, 2004; Marzoli et al, 2004).

The overlying limestones and dolomites represent the transgressive Lower Liassic platform. Within the Liassic the transition from massive carbonates to a layered sequence of marls and limestones indicates a platform-basin boundary documenting the progressive disruption and drowning of the Liassic platform (Jossen and Filali-Moutei, 1992). The Upper Liassic-Lower Dogger (from Toarcian to Bajocian) are varicolored marls and reefal limestones underlying Bathonian red sandstones and silty shales pointing to a continental sedimentation (e.g. Ellouz et al., 2003). The Cretaceous is characterized by red sandstones and conglomerates ("Infracenomanian"; Gauthier, 1957) evolving to platform white limestones of Cenomanian-Turonian age which mark a global transgression to the scale of the entire Atlas domain (Ettachfini and Andreu, 2004). During the Upper Cretaceous-Paleogene the sedimentation is mainly continental and lacustrine with minor marine bioclastic limestones of Eocene age (Marzoqi and Pascal, 2000).

The Neogene continental deposits, occurring above a regional unconformity, resulted essentially from the syndeformation erosion of the mountain belt (Miocene-Pliocene molasses).

The post-Jurassic deposits which probably formed a rather continuous cover overlying the High Atlas and the surrounding areas (Meseta and Anti Atlas domains) are now exposed mainly in the areas bordering northward and southward the High Atlas (Haouz, Souss and Ouarzazate basins), while they have been largely eroded into the High Atlas being pre-served only in few limited outcrops.

The Alpine (Atlasic) tectonic evolution of the High Atlas fold and thrust belt has been gen-erally considered to be characterized by at least two main deformation steps spanning in age from Late Eocene to Oligocene-Miocene and from Pliocene to Early Quaternary, respectively (Gorler, 1988; Jacobshagen et al., 1988; Giese and Jacobshagen, 1992; Frizon de Lamotte et al., 2000; El Harfi et al., 2001, 2006; Missenard et al., 2007). Teson and Teixell (2006) documented a rather continous thrusting at the southern border of the High Atlas (Boumalne area) active from the Oligocene to the Pliocene. Nevertheless Quaternary tectonics, deforming alluvial terraces, has been also documented (e.g. Morel et al., 2000; Sebrier et al., 2006; Cerrina Feroni et al., 2007; Delcaillau et al., 2010).

The present-day seismicity along the marginal zones of the High Atlas testifies that the orogenic movements are still active.

The tectonic style characterizing the High Atlas is mainly thick-skinned, as the basement was involved in the compressional deformation (e.g. Frizon de Lamotte et al., 2000; Teixell et al., 2003; El Harfi et al., 2006). However, the structures of the southern border have been interpreted as evolved within a thin-skinned style of deformation (e.g. Beauchamp et al., 1999; Bennami et al., 2001; Teixell et al., 2003).

The limits that bound, to the north and to the south, the High Atlas orogenic system are represented by two tectonic lineaments of regional importance.

In particular, the northern boundary of the Western High Atlas is represented by a complex system of thrusts and high angle faults that divides the main range from the Haouz Plain.

The northern border of the Central High Atlas belt is marked by a N-verging thrust, associated with strike-slip faults, that finds its westernward extension with the thrust that subtends and bounds northward the Jebilet range (Figure 1). On the contrary, the southern boundary of the Central High Atlas matches the South Atlas Fault Zone (Russo and Russo, 1934), a major tectonic feature extending from Morocco, where the High Atlas belt is juxtaposed to the Anti Atlas, to Tunisia (Bracene et al., 1998). In the Western High Atlas the South Atlas Fault Zone corresponds to the western termination of the Tizi n'Test Fault system.

3. The studied sectors

We have focused our study in two selected sectors corresponding to the southern boundary of the Central High Atlas in the Tinerhir-Boumalne area and to the Western High Atlas transect between Imi n'Tanoute and Taroudant region (Figure 1).

In these sectors, the occurrence of Mesozoic-Cenozoic deposits allowed to describe the geometry and kinematics of the main Alpine structures and define the ages of the tectonic events that characterize the polyphase deformation of the High Atlas borders. The availability of industry seismic data, discussed in earlier papers (e.g. Mustaphi et al., 1997; Hafid et al., 2006; Sebrier et al., 2006), provided further useful information.

In particular, the structural study of the Tinerhir-Boumalne area is important in the understanding of the kinematics induced by the South Atlas Fault Zone. The Western High Atlas is a privileged sector where it is possible the study of the entire mountain belt along a relatively short transect (about 50 km) between its northern and southern boundaries, that is to say from the Imi n'Tanoute Fault to the Tizi n'Test Fault Zone.

Actually, whereas the Imi n'Tanoute Fault is located at the northern boundary of the Western High Atlas at the margin of the Haouz Plain, from a structural point of view it does not represent the northern belt front as the Jebilet range, north of Marrakech (Figure 1), must be considered as part of the High Atlas itself (Hafid et al., 2000, 2006; Michard et al., 2008). Nevertheless, along the Western High Atlas transect it is possible to acquire new data about the characterization of fault and thrust-fold systems evolving from northern to southern vergences crossing the belt from north to south.

3.1. Central High Atlas (Tinerhir-Boumalne area)

3.1.1. Geological setting

The study area corresponds to the NE-SW trending zone, from Boumalne and Tinerhir, between the Central High Atlas to the north, and the Eastern Anti Atlas (Saharan Craton) to the south (Figures 1 and 2). The geological setting is characterized by the eastern termination of the Ouarzazate basin, a topographic low where most of the Miocene-Pliocene and Quaternary deposits sedimented. It represents the physiographic separation between Central High Atlas and Eastern Anti Atlas belts. In the Tinerhir area, where the Ouarzazate basin ends, the two belts face each other, being separated only by the Cretaceous-Eocene succession.

The structural pattern of the southern boundary of the Central High Atlas between Boumalne and Tinerhir is dominated by the high angle faults belonging to the South Atlas Fault system showing an overall direction of about N70E. In particular the South Atlas Fault main zone corresponds to a strike-slip fault that juxtaposes a northern block represented by the Central High Atlas belt s.s. and a southern block characterized by the Mesozoic-Cenozoic succession unconformably overlapping the Paleozoic basement outcropping north the Proterozoic rocks of the Jebel Saghro. The Mesozoic-Cenozoic succession of the southern block is deformed by a south-verging fold-thrust system that affects also the underlying Paleozoic basement. The latter is formed by terrigenous clastics with subordinate carbonate sediments ranging in age from Lower Cambrian to Carboniferous (Visean-Namurian). The Paleozoic succession is characterized by a polyphase deformation consistent with the Late Carboniferous Variscan evolution (Michard et al., 1982; Hoepffner et al., 2006; Cerrina Feroni et al., 2010). Nevertheless, a post-Variscan tectonics, connected with the Cenozoic Atlasic orogeny, has been documented in the Eastern Anti Atlas (Malusà et al., 2007).

In the Tinerhir area, the Mesozoic-Cenozoic succession which unconformably overlies the Paleozoic basement is characterized by a basal unit formed by Upper Cretaceous continental conglomerates and sandstones ("Infracenomanian") which directly overlaps the Paleozoic and Proterozoic basement (Figure 2). The pebbles of the conglomerates are formed by Precambrian and Paleozoic rocks of the Anti Atlas domain; the basal unconformity sealed the structures generated by the Variscan deformation.

Above the Infracenomanian deposits the succession evolves upward to Upper Cretaceous-Eocene mainly marine deposits which can be subdivided in many lithostratigraphic units (Figure 2B). The overlying deposits consist of red pelites and fine sandstones with gypsum lenses and intercalations of sandstones and conglomerates (e6-e7); this lagoonal-continental succession corresponds to the Hadida-Ait Ouglif formation which has been referred to the Upper Eocene-Early Oligocene (El Harfi et al., 2001).

The uppermost part of the Mesozoic-Cenozoic succession is characterized by a very thick (up to more than 700 m) Miocene-Pliocene formation constituted by polygenic conglomerates with pebbles of Precambrian to Paleogene rocks. It is deposited in an alluvial fan environment evolving to shales and lacustrine limestones and to alluvial fan conglomerates mainly composed by Jurassic limestone clasts. The Miocene-Pliocene deposits lies unconformably over all the previous formations comprising the

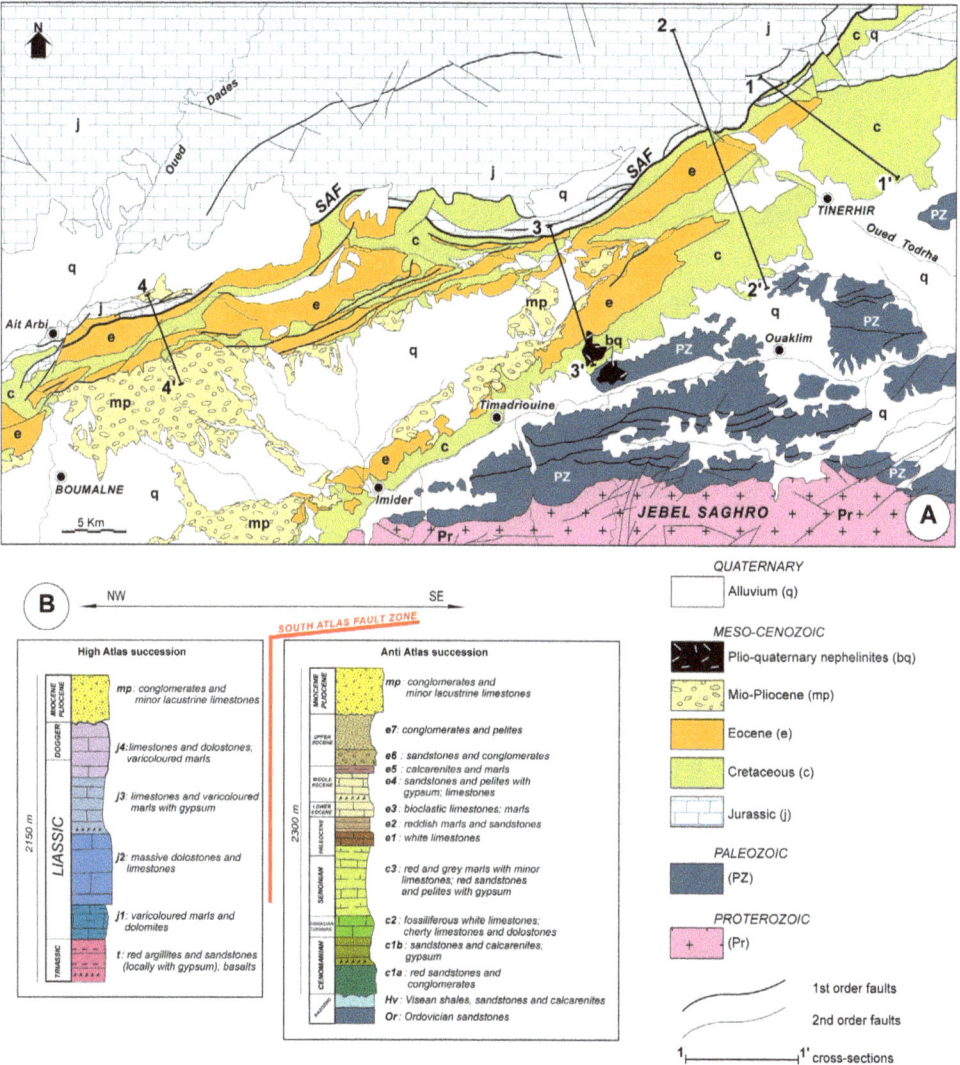

Figure 2. A: Geological map of the Tinerhir-Boumalne area (simplified after Hindermayer et al., 1977, and Milhi, 1997.) indicating the geological cross-sections of Figure 3. SAF: South Atlas Fault. B: Lithostratigraphy of geological cross-sections in Figure 3.

Jebel Saghro Precambrian rocks (outside the area represented in the geological map of Figure 2A). The Miocene-Pliocene basal contact represents a regional unconformity as the Miocene-Pliocene deposits directly overlap also the Middle Jurassic rocks belonging to the stratigraphic succession of the Central High Atlas, north of the South Atlas Fault zone (Figure 2B).

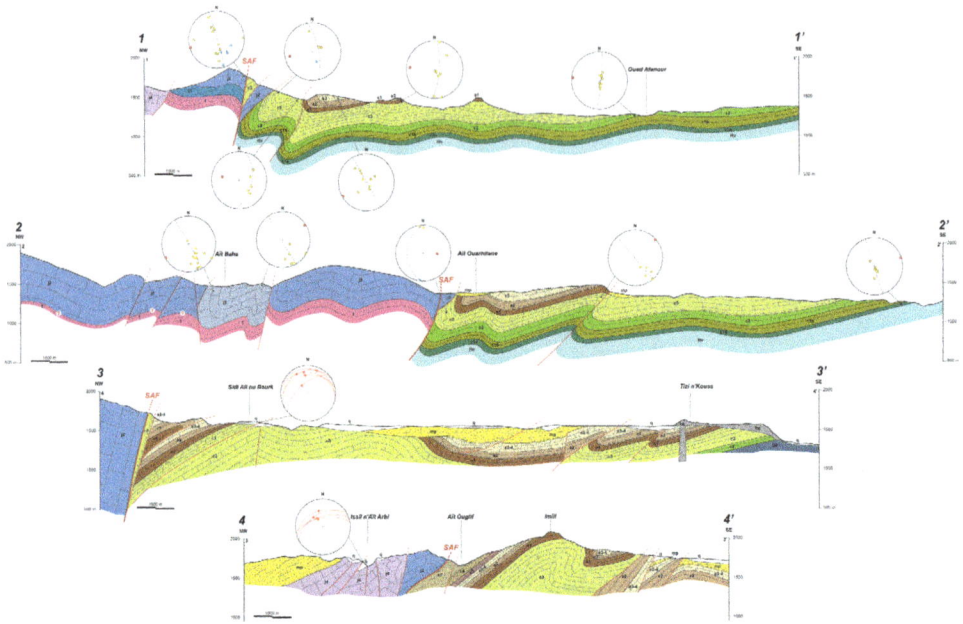

Figure 3. Geological cross-sections through the South Atlas Fault zone between Tinerhir and Boumalne (location in Figure 2). In sections 1-1' and 2-2' stereonet diagrams (Equal area projection, lower hemisphere) represent poles to bedding (yellow circles: normal beds; blue circles: overturned beds), and measured fold axes (yellow triangles); the β axes are indicated (red squares). In sections 3-3' and 4-4' stereonet diagrams represent fault planes and striae.

The Central High Atlas succession exposed in the study area consists of Triassic red siltstones, sandstones and basalts evolving upward to Liassic-Dogger mainly massive limestones and dolostones and varicolored marls, locally with gypsum (Figure 2B).

3.1.2. Structure geometry and kinematics

The geological-structural data collected in the field allowed to produce four cross-sections that describe the geometry and kinematic of the main structures occurring at the southern margin of the Central High Atlas between Boumalne and Tinerhir villages (Figure 3).

The outstanding structure is the South Atlas Fault zone along which the mainly Lower Jurassic rocks of the Central High Atlas succession are juxtaposed to the Cretaceous-Eocene succession of the Anti Atlas block. The main fault zone is characterized by NE-SW trending high angle fault planes displaying horizontal to oblique slickensides that indicate an overall dextral displacement (Figure 4A). The kinematic analysis allowed to obtain two palaeostress tensors by inversion of the fault data collected along the South Atlas Fault zone northwest of Tinehrir (Figure 5). The Imarirene site of measurement evidences an homogeneous population of NE-SW dextral strike-slip faults compatible with a strike-slip palaeostress tensor with a sub-horizontal maximum compression σ1 axis directed roughly E-W. The dextral strike-slip faults collected at Ait Snane are conjugated with NNW-SSE sinistral strike slip faults. This fault system is consistent with a strike slip palaeostress regime (sub-vertical intermediate compression axis σ2) and a sub-horizontal σ1 axis showing a NNW-SSE direction (Figure 5). This strike slip tensor is kinematically compatible with the ENE-WSW trending inverse faults occurring in the same site of measurement.

Figure 4. Geometrical and kinematic features along the South Atlas Fault zone in the Tinerhir-Boumalne area. A: Main South Atlas Fault plane corresponding to the Lower Jurassic rock wall. The direction of the red arrow corresponds to the movement of the missing block, indicating a dextral sense of movement. B: Thrust surface affecting the Quaternary deposits. The thrust plane is characterized by the same direction of the South Atlas Fault zone with a top to the South sense of movement. C: Hectometre-scale south-vergent fold developed in Lower Jurassic rocks of the High Atlas domain. D: Hectometre-scale south-vergent fold developed in the Upper Cretaceous-Eocene succession of the Anti Atlas domain.

Thrust folds are structures characterizing also the deformation pattern of the study area, affecting the Jurassic rocks and the Upper Cretaceous-Eocene succession, as well as the Miocene-Pliocene deposits and the Quaternary terrace gravels (Figures 3 and 4B). The thrusts are directed about ENE-WSW (~ N70E), broadly parallel to the South Atlas Fault zone, dipping toward NNW of 30°-40°. The fault surfaces display down-dip slickenside striations indicating a top to the S sense of movement. The resulting palaeostress tensors are consistent with pure compressive tectonic regimes with sub-horizontal σ1 axes directed NNW-SSE as shown in the Ait Arbi and Sidi Ali Ou Bourk stations (Figure 5).

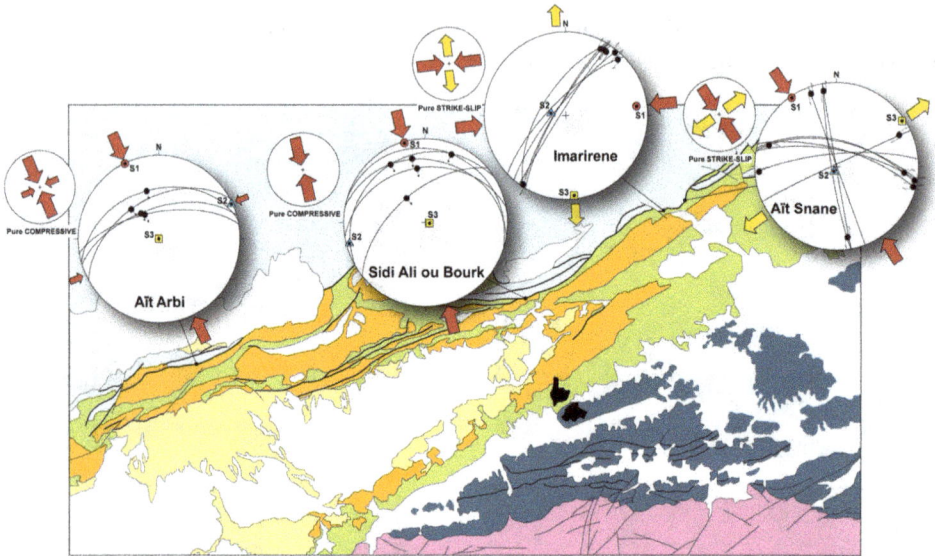

Figure 5. Geological sketch map with stress invertion results (TENSOR program; Delvaux, 1993). Stereograms (Schmidt net, lower hemisphere) with traces of fault planes, observed slip lines and slip senses; the principal stress axes (S1, S2, S3) and type of stress tensor are reported.

Northwestward this thrust system links to the high angle faults (South Atlas Fault system) displaying an overall asymmetrical positive flower structure geometry (Figure 3).

The thrust system is associated with a fold system characterized by anticlines and synclines showing steeply dipping axial planes; the fold axes, generally showing sub-horizontal plunging, trend from N70E to about E-W, again sub-parallel to the South Atlas Fault trend (Figure 4C and 4D). The fold asymmetry indicates a southward vergence. The sub-vertical limbs of the folds are often affected by thrust faults that cut off the anticline-syncline hinge zones (Figure 3).

The analysis of the geological cross-sections indicates that the deformation is not homogeneously distributed in the study area: deformation zones constituted by folds and thrusts are more developed close to the South Atlas Fault zone, while moving to the SE the deformation decreases generating more spaced open syncline and anticline folds.

The geometric-kinematic analysis suggests that thrusting and folding can be linked to the development of contemporaneous strike-slip faulting in a complex polyphase tectonic evolution. In fact, the relationships of the Miocene-Pliocene deposits that unconformably sealed the fold structures of the Upper Cretaceous-Eocene succession, and the deformation affecting the Quaternary deposits indicate two distinct tectonic phases characterizing the Alpine evolution of the southern boundary of the Central High Atlas belt.

3.2. Western High Atlas

The studied sector of the Western High Atlas develops between Imi n'Tanoute village and Menizla village (SE of Argana), at the northern limit of the Souss Plain (Figure 6).

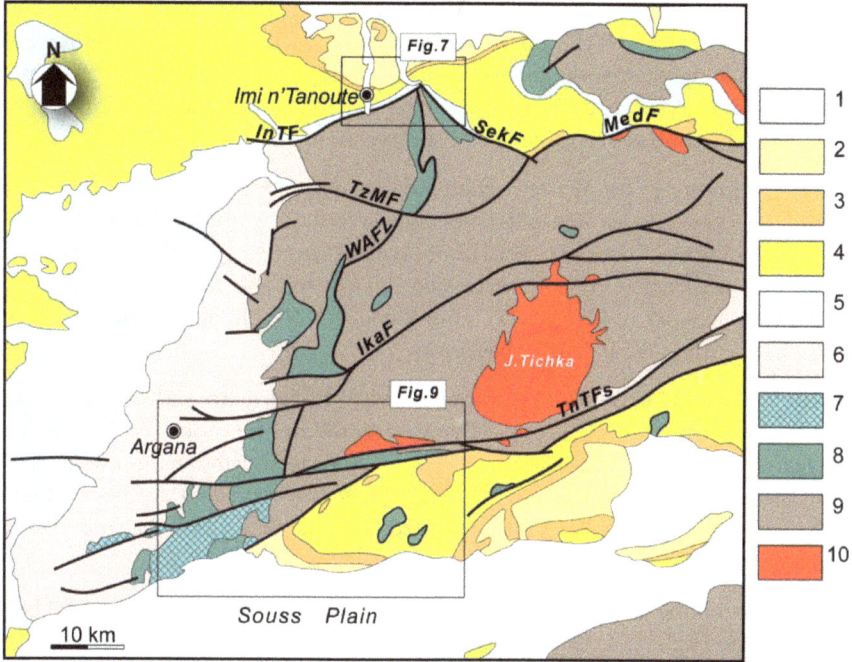

Figure 6. Schematic structural map of the Western High Atlas with location of Figures 7 and 9 (modified after Hollard et al, 1985). 1: Quaternary; 2: Neogene; 3: Eocene; 4: Cretaceous; 5: Jurassic; 6: Permian-Triassic, 7: Carboniferous; 8: Ordovician, Silurian and Devonian; 9: Cambrian; 10: Variscan Granites. InTF: Imi n'Tanoute Fault; SekF: Seksaoua Fault; MedF: Medinat Fault; TzMF: Tizi Maachou Fault; IkaF: Ikakern Fault; TnTFs: Tizi n'Test Fault system; WAFZ: Western Atlasic Fault Zone.

The central part of the belt is formed by a 10 km thick Paleozoic succession, mostly represented by the Cambrian metasediments (sandstones, schists and greywackes). East of the Western Atlasic Fault Zone (Cornée and Destombes, 1991), a major N-S trending Cambrian tectonic lineament (Figure 6), the Lower Cambrian succession is characterized by volcano-detritic schists with conglomeratic lenses overlying a complex of schists with arkoses, volcano-detritic and calcareous intercalations. The Paleozoic succession evolves up to Ordovician sandstones and to Silurian-Devonian units respectively composed of black/reddish schists and sandstones and conglomerates with lens of limestones.

The Carboniferous formations are restricted to the Ida Ou Zal basin in the southwestern sector of the Western High Atlas (Figure 6). It is composed by 1800 m thick succession of conglomerates followed by sandstones, pelites, coal seams and argillaceous sandstones alternating with dolomitic calcareous layers (De Koning, 1957). The succession was accumulated during Stephanian-Autunian time span in this basin which has been interpreted as a Late Variscan basin created along a strike-slip fault system in a transextensional regime (Saber et al., 2001). The Stephanian-Autunian deposits of the Ida Ou Zal basin were deformed by a folding phase with E-W to ESE-WNW axial direction developed in a transpressional tectonic regime (Saber et al., 2001). A more complex tectonic evolution has been proposed by Qarbous et al. (2003) consisting in superimposed folding and faulting deformations in alternating compressional and extensional regimes. The last deformation stage, connected to the Alpine tectonics, produced the reactivation of the ENE-WSW faults as reverse faults in the context of a roughly N-S compression.

The western boundary of the Western High Atlas is characterized by the Upper Permian-Triassic deposits (sandstones and siltstones) whose outcrops are limited to a NNE-SSW trending basin (Argana Corridor). This basin evolves westward to the Agadir-Essaouira basin where the Mesozoic-Cenozoic succession developed. The Argana Corridor deposits are affected by a network of ENE-WSW, NE-SW and WNW-ESE faults (Tixeront, 1974) that extend eastward cutting the Paleozoic basement.

The Mesozoic-Cenozoic succession characterizes also the northern and southern marginal sectors of the Western High Atlas, respectively north the Imi n'Tanoute Fault at the margin with the Haouz Plain and south the Tizi n'Test Fault system at the margin with the Souss Plain. The Souss Plain, which acted as High Atlas foreland basin during the Cenozoic, constitutes an E-W oriented depression separated from the Ouarzazate basin by the Siroua high plateau (Figure 1).

The Haouz Plain, on the contrary, is interpreted as an intra-mountain basin located between the Jebilet and the High Atlas (Michard et al., 2010) and characterized by Miocene-Pliocene molasse deposits.

In the following paragraphs we will discuss separately the structural geology of two sectors of the Western High Atlas, respectively the Imi n'Tanoute area (northern sector) and the Menizla area (southern sector).

3.2.1. Northern boundary (Imi n'Tanoute Fault)

The geological setting of the northern sector (Imi n'Tanoute area) is characterized by the contact between the Paleozoic basement and the Mesozoic-Cenozoic succession which develops northward in the Haouz Plain. This contact corresponds to the Imi n'Tanoute Fault, a major fault zone showing an about N70 direction (Figure 7). However, ESE of Imi n'Tanoute village the unconformable overlap of the Jurassic succession, starting with purple conglomerates and sandstones, above the Paleozoic rocks is preserved. This unconformity is well visible also at the map scale for the strong discordance between the sub-horizontal Jurassic beds and the Paleozoic succession. The latter displays a sub-vertical attitude, resulted from the Variscan tectonics which produced N-S directed folds associated with a pervasive sub-vertical cleavage. Further, this area is characterized by a N-S trending thrust, connected to the WAFZ, that duplicated the Paleozoic succession with an east vergence.

In the Paleozoic succession the Lower-Middle Cambrian schists and greywackes evolves to schists and schists with quartzite bars of Ordovician age. The Paleozoic succession ends with Silurian black/reddish schists and Devonian sandstones and conglomerates with lens of limestones.

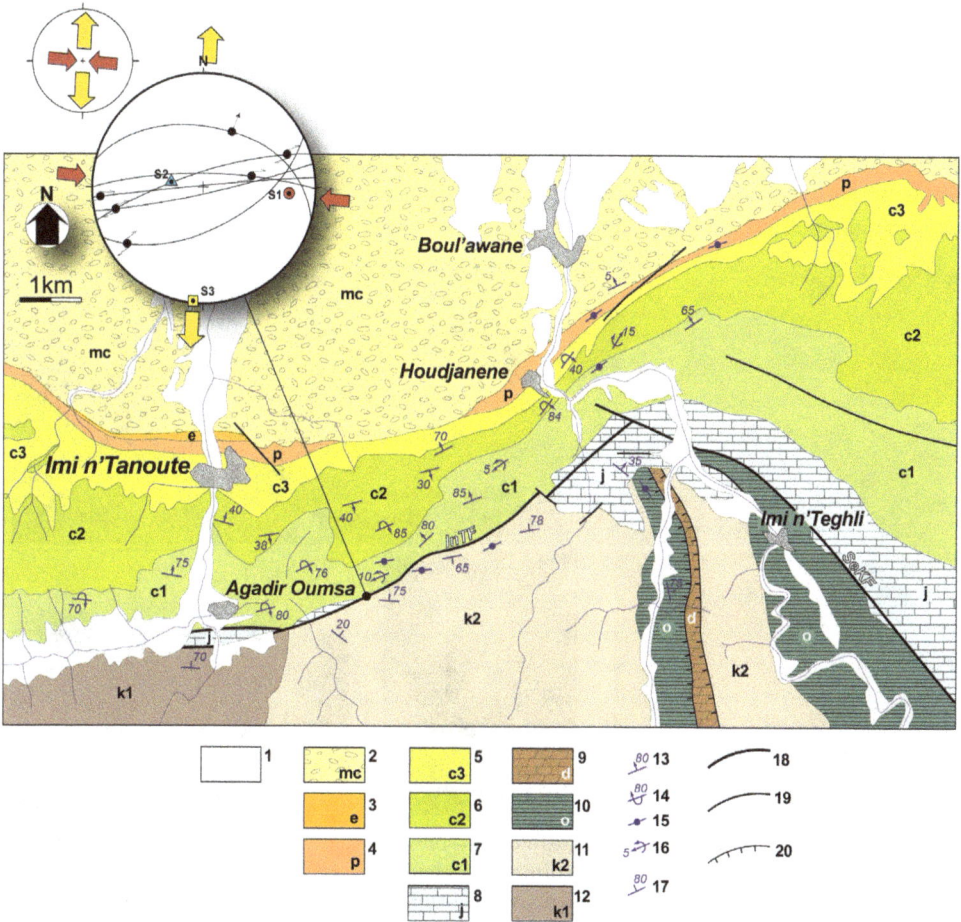

Figure 7. Geological map of the northern boundary of the High Atlas in the Imi n'Tanoute area (modified after Duffaud, 1981). 1: Quaternary; 2: Miocene; 3: Eocene; 4: Paleocene; 5: Upper Cretaceous; 6: Cenomanian-Turonian; 7: Lower Cretaceous; 8: Jurassic; 9: Devonian; 10: Ordovician; 11: Middle Cambrian; 12: Lower Cambrian; 13: normal beds, 14: overturned beds; 15: vertical beds; 16: fold axes, vergence is indicated; 17: main foliation in Paleozoic rocks; 18: main faults; 19: secondary faults; 20: tectonic boundaries. InTF: Imi n'Tanoute Fault; SekF: Seksaoua Fault. At the top left of the figure stereogram (Schmidt net, lower hemisphere) with traces of fault planes, observed slip lines and slip senses. The principal stress axes (S1, S2, S3) and type of stress tensor are indicated.

The post-Jurassic Mesozoic succession of the Imi n'Tanoute area begins with Lower Cretaceous marine deposits (yellow and reddish marls, limestones and sandstones with gypsum of Barremian FIGURE 8 age) followed by a thick Cenomanian-Turonian sequence formed by grey and red marls with anhydrite and by Senonian red and white sandstones with lumachellic limestones and white marlstones. The overlying Maastrichtian conglomerates and phosphatic sandy marls and limestones were deposited above an unconformity and evolved up to the Paleocene-Eocene reddish sandstones and brown marls. Slightly north of the Imi n'Tanoute village, up to 20 m thick white conglomerates with limestone and chert pebbles, considered to be Oligocene in age, are also documented unconformably overlapping the Eocene succession (Zuhlke et al., 2004). Nevertheless the main unconformity inside the Cenozoic succession corresponds to the basal contact of the Miocene conglomerates and sandstones (molassic deposits) that rest directly above different levels of the Upper Cretaceous-Eocene succession, sealing the thrusting and folding deformation of the first tectonic phase of the Western High Atlas belt.

Figure 8. Geometrical and kinematic features along the northern boundary of High Atlas in the Imi n'Tanoute area. A: Panoramic view of the north-vergent folding in the Cretaceous succession associated to the dextral sense of movement of the Imi n'Tanoute Fault. B: Detail of a fault plane (Middle Cambrian) bearing oblique slickenlines with a dextral strike-slip movement. The direction of the red arrow corresponds to the movement of the missing block. C) Shear zone developed in the Lower Cretaceous rocks along the Imi n'Tanoute Fault. D: Kilometer-scale north-vergent fold developed in Cretaceous-Paleogene successions, characterized by a secondary fold with rabbit-ear geometry. The deformed strata are unconformably overlain by the Miocene clastic deposits.

In the study area, the geological structures at the Western High Atlas northern border can be observed along a natural cross-section directed roughly N-S (Figure 8A). The Cambrian metamorphic rocks are juxtaposed to the Cretaceous sedimentary sequences along the Imi n'Tanoute Fault. A slice of Jurassic rocks is also isolated inside the fault zone. The foliation attitude of the Cambrian metasandstones and metapelites, that normally shows a sub-vertical N-S direction, close to the fault zone suffered a virgation produced by the fault activity becoming sub-parallel to the fault itself and showing a northward high angle dipping (Figure 7). North the Imi n'Tanoute Fault, the Lower Cretaceous red marlstones and sandstones belong to the southern vertical limb of a syncline-anticline-syncline system showing northward vergence. This folding system involved the entire Mesozoic-Cenozoic succession of the Imi n'Tanoute area up to the Paleocene-Eocene deposits, as can be observed northwest the Houdjanene village (Figure 8D) where the Turonian limestones evidence a spectacular secondary fold that can be interpreted as a rabbit-ear fold (Narr & Suppe 1994; Missenard et al., 2007). In the Houdjanene cross-section the sub-horizontal (slightly dipping) Miocene conglomerates are unconformable above the sub-vertical Paleocene-Eocene beds (limestones and sandstones).

The overall axial direction of the described fold system is directed sub-parallel to the Imi n'Tanoute Fault, i.e. N70E as evidenced also by the spatial arrangement of the bedding data collected in the Cretaceous-Eocene succession (Figure 7).

The kinematic data collected along the main fault zone of the Imi n'Tanoute Fault indicate a main dextral strike-slip movement (Figure 8B and 8C) consistent with a strike-slip tectonic regime displaying a sub-horizontal WNW-ESE directed maximum compression axis, where the associated roughly E-W directed normal faults are compatible too (Figure 7).

In the Cambrian metasandstones a population of NNE-SSW directed dextral strike-slip faults have been also collected, being compatible with a different palaeostress tensor where the sub-horizontal axis is directed ENE-WSW.

3.2.2. Southern boundary (Tizi n'Test Fault system)

The Paleozoic succession of the Menizla area is characterized by the occurrence of the Upper Carboniferous deposits of the Ida ou Zal basin. Unlike the northern boundary of the Western High Atlas, the Mesozoic-Cenozoic succession starts with the Lower Cretaceous red sandstones directly overlying the Ordovician rocks outcropping in two small inliers north of Addouz (Figure 9). The upper part of the succession is characterized by the Maastrichtian-Ypresian phosphate series and by the Miocene-Pliocene continental deposits.

The Mesozoic-Cenozoic succession at the southern boundary of the Western High Atlas was deformed by southward verging fold systems comprehensively formed by two wide anticlines separated by a wider syncline. The sub-vertical, locally reversed, southernmost limb of the southern anticline represents the margin with the Quaternary deposits of the Souss Plain.

For the main object of our study, we analyzed the relationships between the folding structures and the Tizi n'Test Fault system in the area of Tafrawtane (Figures 9 and 10). The main fault that juxtaposes the Cambrian rocks with sub-vertical principal foliation to the Upper Cretaceous deposits is a sub-vertical dextral strike-slip fault directed about E-W. The relative kinematic data allow defining a strike-slip palaeostress tensor with a sub-horizontal NW-SE trending σ1 axis (Figure 9).

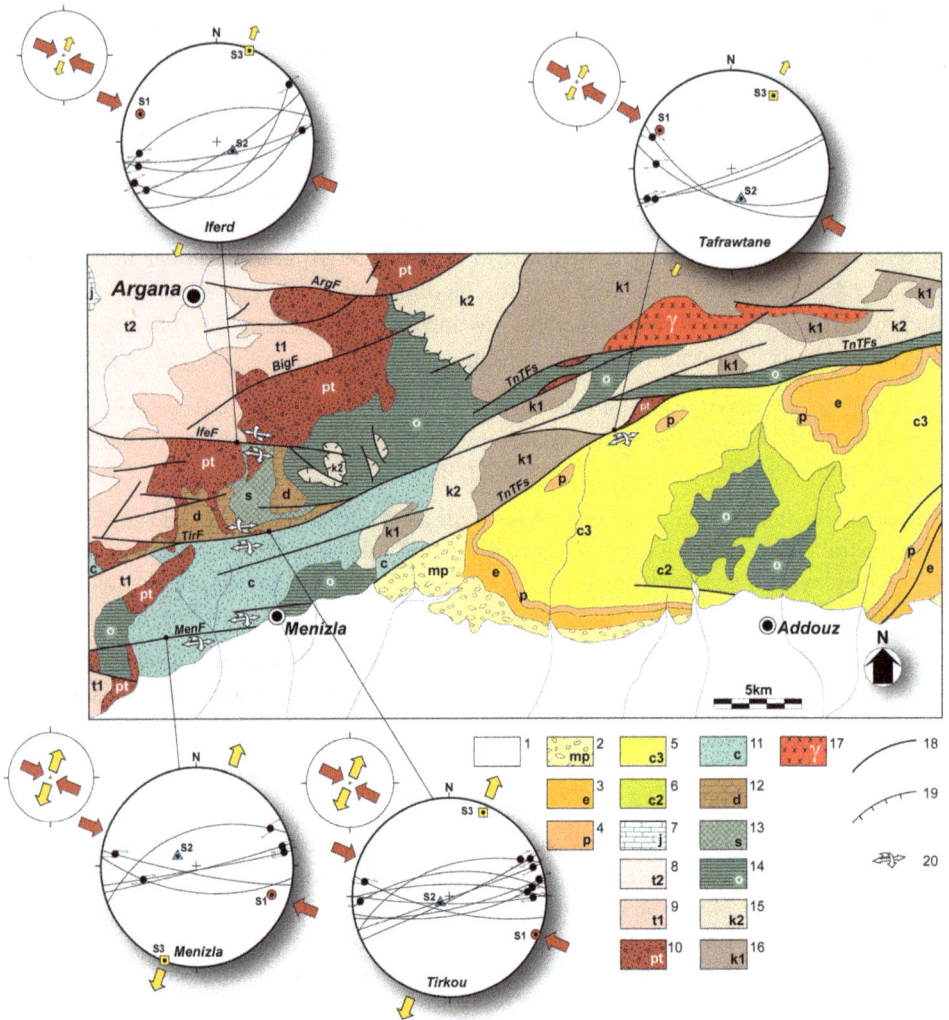

Figure 9. Geological map of the southern boundary of the High Atlas in the Menizla area (modified after Choubert, 1957 and Tixeront, 1974). 1: Quaternary; 2: Miocene-Pliocene; 3: Eocene; 4: Paleocene; 5: Upper Cretaceous; 6: Cenomanian-Turonian; 7: Jurassic; 8: Upper Triassic; 9: Middle Triassic; 10: Permian-Triassic; 11: Carboniferous; 12: Devonian; 13: Silurian; 14: Ordovician; 15: Middle Cambrian; 16: Lower Cambrian; 17: Variscan Granites; 18: Main faults; 19: Tectonic Boundaries; 20: Direction of tectonic transport. ArgF: Argana Fault; BigF: Bigoudine Fault; IfeF: Iferd Fault; TirF: Tirkou Fault; TnTFs: Tizi n'Test Fault system. Stereograms (Schmidt net, lower hemisphere) with traces of fault planes, observed slip lines and slip senses are reported for the Iferd, Menizla and Tirkou Faults and for Tizi n'Test Fault system in the Tafrawtane area. The principal stress axes (S1, S2, S3) and type of stress tensor are indicated.

Actually, the main fault zone is characterized by a slice of Permian-Triassic red sandstones interposed between the Cambrian and the Cretaceous rocks that are also affected by a south-verging thrust that roots into the main sub-vertical fault plane.

The Upper Cretaceous marls and limestones are affected by complex, disharmonic, fold structures showing a general southward vergence. As a result bedding is variable in orientation but it generally strikes ENE-WSW, sub-parallel to the main direction of the Tizi n'Test Fault system; bedding dips follow the fold structures, being progressively steeper approaching the fault zone. The axis of the fold system is directed about N70E, sub-parallel to the Tizi n'Test Fault system direction, showing a slight eastward plunging.

Figure 10. Panoramic view of the south verging folding in the Upper Cretaceous associated to the dextral sense of movement of the Tizi n'Test Fault system in the Tafrawtane zone.

However, from a structural point of view, the most important feature occurring in the Menizla area is the Tizi n'Test Fault system and in particular the western component of this major tectonic lineament consisting of different anastomizing branches, the main three of which (Menizla Fault, Tirkou Fault, Iferd Fault; Baudon et al., 2012) have been studied with more detail (Figure 9).

Starting from the margin with the Souss Plain, the southernmost fault is the Menizla Fault characterized by strike-slip movements associated with mostly south-verging thrusts sub-tending sub-vertical limbs of folds that deformed the Carboniferous rocks. In particular, the Menizla Fault developed dextral strike-slip faulting along high-angle planes directed about E-W. The relative palaeostress tensor indicates a maximum compression directed WNW-ESE (Figure 9); in this tectonic context oblique-normal faults can be also compatible and some of these have been detected.

Moving toward the north, the first main fault is the Tirkou Fault showing a direction varying from E-W to NE-SW. The structural analysis was performed where the Tirkou Fault trends ENE-WSW between the Carboniferous deposits and the Devonian dolostones and shows a very thick (300-400 m) fault zone cutting off a slice of Permian-Triassic red sandstones (Figure 9). The mesoscale observations evidenced that the Tirkou Fault is characterized by folds linked to the development of double-verging thrust systems that root in sub-vertical fault planes comprehensively describing a positive flower structure (Figure 11A, C). Fold styles vary from structures with rounded hinges to kink-like or chevron folds with steeply dipping axial planes. On average, the fold axes trend sub-parallel to the direction of the Tirkou Fault zone (~N100E) with shallow plunging. The fold limbs are cut by interlinked faults producing imbricate zones; the fault array associated with these folds provided striations indicating inverse-oblique displacement. As a consequence, the SSE-dipping and the NNW-dipping thrusts display roughly north and south vergences respectively. The occurring sub-vertical faults are characterized by nearly horizontal slickensides showing dextral displacements (Figure 11B). The fault data collected for the strike-slip faults of the Tirkou Fault zone indicate a dominant ENE-WSW fault direction and allow to obtain a palaeostress tensor characterized by a sub-vertical σ2 axis and sub-horizontal σ1 and σ3 axes oriented N110E and N10E respectively (Figure 9). Within the obtained palaeostress tensor, ESE-WNW directed normal faults, such as that observable in Figure 11A, can be also compatible.

Figure 11. Geometrical and kinematic features along the southern boundary of High Atlas in the Tirkou Fault zone.

A: Mesoscale positive flower structure developed in the Carboniferous succession. This structure is characterized by double-verging thrust systems that root in sub-vertical fault planes, associated to the development of double-verging folds that trend subparallel to the faults.

B. Detail of a fault plane bearing oblique slickenlines with a dextral strike-slip movement (Carboniferous). The direction of the red arrow corresponds to the movement of the missing block.

C. Shear zone developed in the Carboniferous rocks with top to the south sense of movement.

Likewise, the Iferd Fault is outlined by double-verging structures within an overall dextral strike-slip fault zone, locally juxtaposing Silurian rocks and Permian-Triassic deposits (Figure 9). The flower structures that characterize the Iferd Fault zone are well developed in the Silurian schists consisting of a series of anastomosing convex-upward reverse faults which steepen progressively at depth into sub-vertical strike-slip faults. The reverse faults are directed about E-W steeply dipping toward N and S and display opposite senses of shear, that is southward and northward respectively (Figure 12A). The inversion of strike-slip fault data has resulted in a palaeostress tensor similar to that obtained for the Tirkou Fault (Figure 9). In the Ifern Fault zone the possible development of normal faults, due to the permutation of σ1 and σ2 axes, has been documented by the occurrence of curved striations on a single fault plane suggesting a dextral-oblique to oblique-normal movement (Figure 12B).

Figure 12. Geometrical and kinematic features along the southern boundary of High Atlas in the Iferd Fault zone. A. Anastomosing reverse faults root progressively at depth into sub-vertical strike-slip faults in the Silurian schists (positive flower structure). The reverse faults are characterized by opposite senses of shear. B. Fault plane bearing curved slickenlines suggesting the transition from dextral-oblique to oblique-normal movement.

3.2.3. The major faults of the inner belt sectors

In the Western High Atlas, between the Tizi n'Test Fault system and the Imi n'Tanoute Fault, other two major faults, cutting both the Paleozoic basement and the overlying Permian-Triassic succession, occur: the Ikakern Fault and the Tizi Machou Fault (Figure 13A).

In the studied outcrops the Ikakern Fault zone is formed by nearly E-W dextral strike-slip/oblique faults (Figure 13C) consistent with a palaeostress tensor with a sub-vertical σ2 axis and sub-horizontal σ1 and σ3 axes directed WNW-ESE and NNE-SSW respectively (Figure 13A). Several thrust faults associated to the main fault zone have been observed, showing north and south vergences.

The Tizi Machou Fault shows a dextral map-scale offset evidenced by the displacement of the Western Atlasic Fault Zone. However the sense of shear along individual faults can be rarely deduced at the outcrop scale since kinematic indicators are only sporadically preserved. The kinematically defined structures suggest a predominance of high-angle dextral-oblique faults clustered along two trends: NNE-SSW and NE-SW (Figure 13A and 13B). In addition, few about E-W normal faults occur, showing moderate south-dipping.

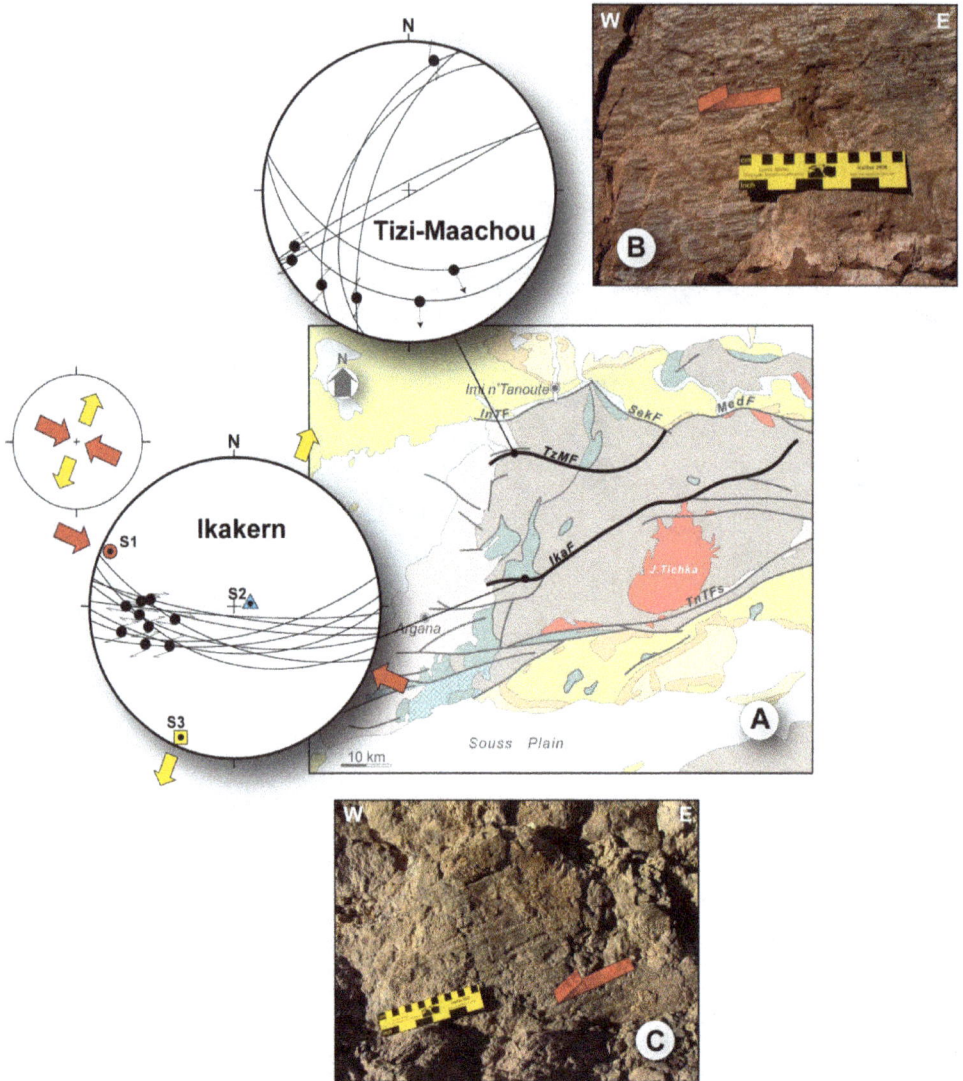

Figure 13. Kinematic data from the inner belt sector. A. Schematic structural map of Figure 6. The two principal fault zones, the Tizi Maachou and Ikakern Fault, are evidenced, for which stereograms (Schmidt net, lower hemisphere) with traces of fault planes, observed slip lines and slip senses are reported. The principal stress axes (S1, S2, S3) and type of stress tensor are indicated for the Ikakern Fault only. B. Fault plane bearing oblique slickenlines with a dextral strike-slip movement developed in the Triassic rocks along Tizi Maachou Fault. The direction of the red arrow corresponds to the movement of the missing block. C. Fault plane bearing oblique slickenlines with a dextral strike-slip movement developed in the Cambrian rocks along Ikakern Fault. The direction of the red arrow corresponds to the movement of the missing block.

4. Discussion

The results of our field study highlight widespread Cenozoic deformation on both the southern boundary of the Central High Atlas and on the northern and southern boundaries of the Western High Atlas. The main deformation structures are represented by NE-SW trending high angle dextral strike-slip faults and sub-parallel thrust faults linked together forming asymmetric positive flower structures. The overall deformation of these structures is completed by fold systems associated with the fault and thrust systems and involving mainly the Mesozoic-Cenozoic successions. The flower structures are typical structures developed under a transpressional tectonic regime where the deformation is partitioned between high-angle strike-slip faults and lower angle reverse faults (Wilcox et al, 1973; Sanderson and Marchini, 1984; Tikoff and Teyssier, 1994). Large scale geometries and mesoscale data indicate that reverse faults merge into the main NE-SW oriented strike-slip faults. Strike-slip faults are not offset by the thrusts and vice versa, this supporting that thrusts are genetically related to the sub-parallel strike-slip faults. The observed structure relationships support therefore that the analysed sectors of the High Atlas belt were affected by a transpressional evolution during the Alpine tectonics.

Along the southern boundary of West and Central High Atlas the fault-thrust-fold systems, belonging respectively to the Tizi n'Test Fault system and the South Atlas Fault, show clear southward vergences. On the contrary, the structures of the northern margin of the Western High Atlas are connected with the Imi n'Tanoute Fault activity, being characterized by northward vergences.

The kinematic inversion of the collected fault-slip data in the Western High Atlas indicates that deformation is controlled by sub-horizontal maximum and minimum stress axes, within a strike-slip tectonic setting with a WNW-ESE directed $\sigma1$ axes (Figure 7, 9 and 13). The lack of pure compressive tensors is probably due to the general scarce preservation of kinematic indicators on the thrust planes of the analysed outcrops. Pure compressional stresses of NW-SE direction have been sometime documented in the northern part of the Western High Atlas (Amrhar, 2002).

The transpressional character of deformation is confirmed by the relevant occurrence of oblique striations on the fault planes, thus that pure strike-slip and/or pure inverse faults are relatively few.

The WNW-ESE trending normal faults, which have been collected within some transpressional fault zones, should be also inserted in the documented tectonic pattern of the Western High Atlas attesting a NNE-SSW extension in the late stages of the Cenozoic Alpine evolution.

On the contrary, palaeostress determinations from the Boumalne-Tinehrir area provide both strike-slip and compressional tensors, with a quite steady sub-horizontal $\sigma1$ axis trending NNW-SSE. The maximum compression $\sigma1$ axes obtained from the palaeostress tensors from the Central High Atlas are consistent with the palaeostress fields reconstructed in the same area for the Pliocene-Quaternary stage of the High Atlas tectonic evolution by Ait Brahim et al. (2002).

Our observations documented a WNW-ESE compression that was never detected before as individual phase for the Cenozoic deformation of the Western High Atlas and comes to enriche palaeostress evolution of the High Atlas. In fact, analogous WNW-ESE directions of compression have been also evidenced by Qarbous et al. (2003) but referred to the Carboniferous (Namurian-Westphalian) phase and therefore to the Variscan orogeny, and to the Middle Permian tectonics of the Tizi n'Test Fault system. As the WNW-ESE compression derived from the fault-slip data collected along fault zones clearly affecting also the Mesozoic-Cenozoic successions at the High Atlas belt boundaries (Tizi n'Test Fault system at Tafrawtane; Imi n'Tanoute Fault) we can consider this compressional direction referable to the Alpine orogeny.

In the inner belt sectors, the fault zones that do not cut the Cretaceous-Eocene deposits but only the Paleozoic basement and the Permian-Triassic rocks show geometric-kinematic features similar to those of the bordering fault zones. Therefore they have been interpreted as Cenozoic Alpine faults, admitting the possible reactivation of older high-angle shear zones.

Regional data and mesostructural analyses suggest the superposition of younger Cenozoic deformation on older structural trends producing reactivation of previous major fault zones. As a consequence, it is a generally shared opinion that Tizi n'Test Fault system and Imi n'Tanoute Fault were active since the Early Paleozoic and in turn reactivated more times during the successive tectonic events up to the Alpine Cenozoic phases.

Following Baudon et al. (2012), also the other about E-W oriented faults of the Western High Atlas (from south to north Tirkou, Iferd and Argana faults; Figure 9) can be interpreted as faults reactivated during the Alpine transpressional phase, after deposition of the Late Triassic deposits.

About the timing of deformation, the results of our study define two main episodes of deformation separated by the basal unconformity of the Miocene-Pliocene molassic deposits. The first episode occurred post-Eocene time, as the Upper Cretaceous-Eocene successions were deformed by fold-thrust systems that were sealed by the unconformable Miocene-Pliocene deposits.

The second tectonic event deformed the Miocene-Pliocene deposits as well as the Quaternary deposits and therefore can be assigned to a Quaternary age.

We documented the complete polyphase evolution in the Boumalne-Tinehrir area (Central High Atlas) whereas we have not new data about the Quaternary deformation in the Western High Atlas that has been already documented by Sebrier et al. (2006) in the Souss Plain. Likewise, Quaternary reactivations consisting of thrusting associated with strike-slip faulting characterize also the northern boundaries of Western and Central High Atlas and the Houaz Plain (Morel et al., 2000).

Seismic data suggest that several brittle structures along northern and southern margins of the High Atlas belt are still active. The few available focal solutions show that the southern boundary of the High Atlas is characterized mostly by strike slip faulting and subordinate transpressional mechanisms, with the direction of the maximum compression P axes ranging from NW-SE to N-S (Serpelloni et al., 2007). In particular the focal mechanism solutions obtained for the earthquakes of magnitude Mb=5.2 occurred on October 23 and 30, 1992, in the Rissani region (Eastern Anti Atlas) indicate for both events a pure strike slip faulting (Hanou et al., 2003; Bensaid et al., 2009). The seismogenic zone has been interpreted to be an E-W trending structure that could corresponds to the South Atlas Fault and/or associated structures activated as a dextral strike slip fault. However, the largest earthquake ever recorded within the Atlas system corresponds to the event of M=5.7 occurred on February 29, 1960, which destroyed the Agadir city causing 12.000 victims.

Also the GPS data indicate that the deformation along the High Atlas is still active and accommodates about 1.5 mm/year of NW-SE compression related to the convergence between Africa and Iberia plates (Serpelloni et al., 2007).

In addition, as more general result, we propose two regional geological cross-sections representing at the belt scale the structural patterns of Western and Central High Atlas, performed on the basis of our field studies integrated with literature data (Figure 14) (Hollard et al, 1985; Froitzheim et al, 1988; Teixell et al., 2003).

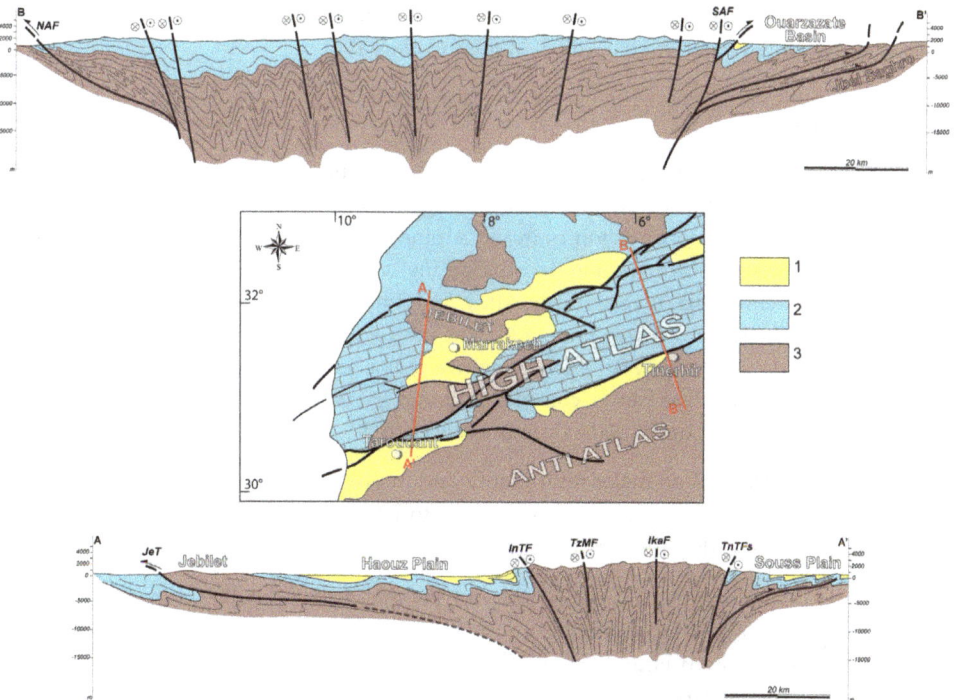

Figure 14. Simplified geological cross-sections through Western (A-A') and Central (B-B') High Atlas. 1: Neogene and Quaternary; 2: Mesozoic and Paleogene, 3: Precambrian and Paleozoic. SAF: South Atlas Fault; NAF: North Atlas Front; TnTFs: Tizi n'Test Fault system; JeT: Jebilet Thrust; InTF: Imi n'Tanoute Fault; TzMF: Tizi Maachou Fault; IkaF: Ikakern Fault.

Comprehensively, these schematic cross-sections represent a possible model of Alpine transpressional evolution for the whole High Atlas belt. The main structures are characterized by high-angle geometries and dextral strike-slip kinematics along an ENE-WSW direction. These fault zones generated also thrust planes and folds that in the inner belt sectors show double vergences at the mesoscale whereas along the bounding areas they are more developed characterizing the entire High Atlas belt by a double vergence, as at the northern margin the vergence is toward north (North Atlas Fault and Imi n'Tanoute Fault zones in A-A' and B-B' cross-sections, respectively) while at the southern margin it is toward south (South Atlas Fault zone and Tizi n'Test Fault system in A-A' and B-B' cross-sections, respectively).

The proposed geological cross-sections evidence also that the Alpine deformation was/is not limited to the High Atlas mountain range but involved/involves wider sectors. In the B-B' cross-section the Jebilet Thrust has been interpreted as a low-angle surface that roots into the high-angle transpressional zone of the Imi n'Tanoute Fault (Figure 14) and therefore Paleozoic basement and Mesozoic-Cenozoic deposits of the Haouz Plain are faulted and folded by the Cenozoic tectonic events. Along the southern boundary of the Western High Atlas we considered the Alpine tectonics deforming the Paleozoic rocks of the Anti Atlas belt as well as the Souss and Ouarzazate deposits of which they represent the basement, as already documented (Sebrier et al., 2006; Malusà et al., 2007).

5. Conclusion

The field study, mainly consisting in detailed mesostructural analyses, from the Western High Atlas transect and the Boumalne-Tinerhir region in the Central High Atlas, indicates a major role for transpressional tectonics in the Alpine structural evolution of the High Atlas belt. In particular, the deformation in the studied regions is controlled by two regional right-lateral fault systems (Tizi n'Test-South Atlas and Imi n'Tanoute-North Atlas) and their associated structures that involved, at south, the Anti Atlas belt and, at north, the Western Meseta domain (Jebilet range). Between these two major tectonic lineaments, a 50 to 100 km wide region is characterized by a complex tectonic framework, dominated by strike-slip faulting, in which strong uplift and exhumation occurred.

Kinematic measurements on major fault planes and mesoscale structural analysis reveal that the prevailing structural associations correspond to ENE-WSW trending dextral strike-slip faults and sub-parallel thrust faults, describing typical positive flower structures.

In the proposed transpressional model (Figure 14) the High Atlas belt appears to have a flower structure cross-sectional geometry with greater amount of thrust displacement along its northern and southern boundaries. This pattern showing double-verging structures re-quires a downward extrapolation of surface thrusts rooting into high-angle fault zones.

Fault analyses and palaeostress reconstructions suggest that flower structures and fold systems evolved into a right-lateral transpression which is related to a direction of maximum compression varying from about E-W to about N-S.

The High Atlas is a significant example of transpressional belt dominated by strain partitioning between distinct strike-slip and thrust faults which result from reactivation of pre-existing structures inherited from the pre-Alpine complex evolution.

The Cenozoic reactivation occurred during two main deformation events: Late Eocene-Oligocene and Pliocene-Pleistocene, the latter being still active.

The High Atlas can be therefore considered an example of active transpressional belt as defined by Cunningham (2005).

This study, although incomplete, furnished interesting results that indicate as structural and kinematic analyses are important methodologies and that their future development can provide new data for the better understanding crustal architecture, history of structural reactivation, partitioning of strain and distribution of present-day tectonic activity within the orogens. Particularly, in this case, the definition of the orogenic mechanisms represents the main step for interpreting the origin of topographic elevation, the principal still debated topic of the High Atlas geology.

Author details

Alessandro Ellero and Giuseppe Ottria
CNR Institute of Geosciences and Earth Resources, Pisa, Italy

Marco G. Malusà
Department of Geological Sciences and Geotechnology, University of Milano Bicocca, Milano, Italy

Hassan Ouanaimi
Université Cadi Ayyad Ecole Normale Supérieure Département de Géologie, Marrakech, Morocco

Acknowledgement

This paper is dedicated to Piero Elter, outstanding geologist and our professor who died when we were preparing the final draft of the manuscript.

This work was partly carried out within the CNR Short Term Mobility Project 2009 "Study of the structural-geological evolution of the High Atlas (Morocco)".

6. References

Aït Brahim L, Chotin P, Hinaj S, Abdelouafi A, El Adraoui A, Nakcha C, Dhont D, Charroud M, Sossey Alaoui F., Amrhar M, Bouaza A, Tabyaoui H, Chaouni A (2002) Paleostress evolution in the Moroccan African margin from Triassic to Present. Tectonophysics 357: 87-205.

Amrhar M (2002) Paléocontraintes et déformations syn- et post-collision Afrique–Europe identifiées dans la couverture mésozoïque et cénozoïque du Haut Atlas occidental (Maroc). C. R. Géoscience 334: 279–285.

Arboleya ML, Teixell A, Charroud M, Julivert MA (2004) Structural transect through the High and Middle Atlas of Morocco. Journal of African Earth Sciences 39: 319-327.

Ayarza P, Alvarez-Lobato F, Teixell A, Arboleya ML, Teson E, Julivert M, Charroud M (2005) Crustal structure under the central High Atlas Mountains (Morocco) from geological and gravity data. Tectonophysics 400: 67-84.

Barbero L, Teixell A, Arboleya ML, Rio PD, Reiners PW, Bougadir B (2006) Jurassic-to-present thermal history of the central High Atlas (Morocco) assessed by low-temperature thermochronology. Terra Nova 19: 58-64.

Baudon C, Redfern J, Van Den Driessche J (2012) Permo-Triassic structural evolution of the Argana Valley, impact of the Atlantic rifting in the High Atlas, Morocco. Journal of African Earth Sciences 65: 91-104.

BeauchampW, Barazangi M, Demnati A, El Alji M (1996) Intracontinental rifting and inversion: Missour Basin and Atlas Mountains, Morocco. AAPG Bull. 80: 1459-1482.

Beauchamp W, Allmendinger RW, Barazangi M, Demnati A, El Alji M, Dahmani M (1999) Inversion tectonics and the evolution of the High Atlas mountains, Morocco, based on a geological geophysical transect. Tectonics 18: 163-184.

Benammi M, Toto EA, Chakiri S (2001) Les chevauchements frontaux du Haut Atlas central marocain: styles structuraux et taux de raccourcissement diff'erentiel entre les versants nord et sud C R Acad Sci Paris 333: 241-247.

Bensaid I, Medina F, Cherkaoui TE, Buforn E, Hahou Y (2009) New P-wave first motion solutions for the focal mechanisms of the Rissani (Morocco) earthquakes of October 23d and 30th, 1992. Bulletin de l'Institut Scientifique, Rabat, section Sciences de la Terre.31: 57-61.

Bracène R, Bellahcene A, Bekkouche D, Mercier E, Frizon de Lamotte D (1998) The thin-skinned style of the South Atlas Front in central Algeria. In: Macgregor DS, Moody RTJ, Clark-Lowes DD, editors. Petroleum Geology of North Africa. Geol. Soc. Spec. Publ. 133: 395-404.

Delcaillau B, Laville E, Amhrar M, Namous M, Dugué O, Pedoja K (2010) Quaternary evolution of the Marrakech High Atlas and morphotectonic evidence of activity along the Tizi N'Test Fault, Morocco. Geomorphology 118: 262-279.

Cerrina Feroni A, Ellero A, Ottria G, Malusà M, Polino R, Musumeci G, Pertusati PC (2007) Kinematic analysis of the High Atlas in the Tinerhir Area (Southern Morocco): Evidence of a transpressional fold-thrust belt. The First MAPG International Convention Conference & Exhibition, Marrakech, October 28-31, 2007. Abstract Book, pp. 165.

Cerrina Feroni A., Ellero A., Malusà M.G., Musumeci G., Ottria G., Polino R., Leoni L. (2010) - Transpressional tectonics and nappe stacking along the Southern Variscan Front of Morocco. International Journal of Earth Sciences, (Geol Rundsch) 99: 1111-1122.

Choubert G (1957), Carte Géologique du Maroc au 1/500.000, Feuille Marrakech. Notes et Mém. Serv. géol. Maroc 70.

Cornée JJ, Destombes J (1991) L'ordovicien De La Partie W Du Massif Ancien Du Haut-Atlas Occidental (maroc Hercynien). Geobios 24: 403-415.

Cunningham D (2005) Active intracontinental transpressional mountain building in the Mongolian Altai: Defining a new class of orogen. Earth and Planetary Science Letters 240: 436–444.

De Koning G (1957) Geologic des Ida ou Zal (Maroc). Stratigraphie, pétrographie et tectonique de la partie Sud-Ouest du bloc occidental du massif hercynien du Haut-Atlas (Maroc). Leidese Geology Meded 23: 209p.

Delvaux D (1993) The TENSOR program for reconstruction: examples from the East Africa and the Baikal Rift System. Terra Abstract 5: 216.

Dewey JF, Helman MNL, Turco E, Hutton DHW, Knott SD (1989) Kinematics of the western Meditettanean. In: Coward M, editor. Alpine Tectonics. Geological Society London Special Publication 45. pp. 265-283.

Dias R, Hadani M, Leal Machado I, Adnane N, Hendaq Y, Madih K, Matos C (2011) Variscan structural evolution of the western High Atlas and the Haouz plain (Morocco). Journal of African Earth Sciences 61: 331-342.

Duffaud F (1981) Carte géologique du Maroc au 1/100000, feuille Imi n'Tanout. Notes et Mém. Serv. géol. Maroc 203.

El Harfi A, Lang J, Salomon J, Chellai EH (2001) Cenozoic sedimentary dynamics of the Ouarzazate foreland basin (Central High Atlas Mountains, Morocco). Int. J. Earth Sci. 90: 393-411.

El Harfi A, Guiraud M, Lang J (2006) Deep-rooted "thick skinned" model for the High Atlas Mountains (Morocco). Implications for the structural inheritance of the southern Tethys passive margin. Journal of Structural Geology 28: 1958-1976.

Ellouz N, Patriat M, Gaulier JM, Bouatmani R, Saboundji S (2003) From rifting to Alpine inversion: Mesozoic and Cenozoic subsidence history of some Moroccan basins. Sedim. Geol. 156: 185–212.

Ettachfini E, Andreu B (2004) Le Cénomanien et le Turonien de la Plate-forme Préafricaine du Maroc. Cretaceous Research 25: 277-302.

Fabuel-Perez I, Redfern J, Hodgetts D (2009) Sedimentology of an intra-montane rift-controlled fluvial dominated succession:The Upper Triassic Oukaimeden Sandstone Formation, Central High Atlas, Morocco. Sedimentary Geol. 218: 103-140.

Fiechtner L, Friedrichsen H, Hammerschmidt K (1992) Geochemistry and geochronology of early Mesozoic tholeiites from Central Morocco. Geol. Rund. 81: 45-62.

Fraissinet C, Zouine EM, Morel JL, Poisson A, Andrieux J, Faure-Muret A (1988) Structural evolution of the southern and northern Central High Atlas in Paleogene and Mio-Pliocene times. In: Jacobshagen V, editor. The Atlas system of Morocco. Lect. Notes Earth Sci. pp. 273-291.

Frizon de Lamotte D, Saint Bézar B, Bracène R, Mercier E (2000) The two main steps of the Atlas building and geodynamics of the western Mediterranean. Tectonics 19: 740-761.

Froitzheim N, Stets J, Wurster P (1988) Aspects of Western High Atlas tectonics. In: Jacobshagen V, editor. The Atlas system of Morocco. Lect. Notes Earth Sci. pp. 219-244.

Gauthier H (1960) Contribution à l'étude géologique des formations post-liasiques des basins du Dadès et du Haut Todra (Maroc méridional). Notes et Mém. Serv. Géol. Maroc 119.

Giese P, Jacobshagen V (1992) Inversion tectonics of intracontinental ranges: High and Middle Atlas, Morocco. Geol. Rundsch. 81: 249-259.

Gomez F, Allmendinger R, Barazangi M, Beauchamp W (2000) Role of the Atlas Mountains (northwest Africa) within the African-Eurasian plate-boundary zone. Geology 28: 769-864.

Görler K, Helmdach FF, Gaemers P, Heissig K, Hinsch W, Mädler K, Schwarzhans W, Zucht M (1988) The uplift of the Central High Atlas as deduced from Neogene continental sediments of the Ouarzazate province, Morocco. In: Jacobshagen V, editor. The Atlas system of Morocco. Lect. Notes Earth Sci. pp. 361-404.

Hafid M (2000) Triassic-early Liassic extensional systems and their Tertiary inversion, Essaouira Basin (Morocco). Marine Petro. Geol. 17: 409-429.

Hafid M, Ait Salem A, Bally AW (2000) The western termination of the Jbilet -High Atlas system (Offshore Essaouira Basin, Morocco). Marine Petrol. Geol. 17: 431-443.

Hafid M, Zizi M, Bally AW, Ait Salem A (2006) Structural styles of the western onshore and offshore termination of the High Atlas, Morocco. C. R. Géoscience 338: 50-64.

Hahou Y, Jabour N, Oukemeni D, El Wartiti M (2003) The October 23; 30, 1992 Rissani earthquakes in Morocco: Seismological, macroseismic data. Bull. Int Inst. Seismol. Earthq. Eng. Special edition: 85-94.

Hindermeyer J, Gauthier H, Destombes J, Choubert G., Faure-Muret A. (1977) Carte géologique du Maroc, Jbel Saghro-Dadès (Haut Atlas central, sillon sud-atlasique et Anti-Atlas oriental) – Echelle 1/200000. Notes et Mém. Serv. géol. Maroc 161.

Hoepffner C, Soulaimani A, Piqué A (2005) Moroccan Hercynides. Journal of African Earth Science 43: 144–165.

Hoepffner C, Houari MR, Bouabdelli M (2006) Tectonics of the North African Variscides (Morocco, Western Algeria), an outline. In: Frizon de Lamotte D, Saddiqi O, Michard A, editors. Recent Developments on the Maghreb Geodynamics. C. R. Géoscience 338: pp. 25-40.

Hollard H, Choubert G, Bronner G, Marchand J, Sougy J (1985) Carte géologique du Maroc – Echelle 1/1000000. Notes et Mém. Serv. géol. Maroc 260.

Houari MR, Hoepffner C (2003) Late Carboniferous dextral wrench-dominated transpression along the North African craton margin (Eastern High Atlas, Morocco). Journal of African Earth Sciences 37: 11-24.

Jacobshagen V, Brede R, Hauptmann M, Heinitz W, Zylka R (1988) Structure and post-Paleozoic evolution of the Central High Atlas. In: Jacobshagen V, editor. The Atlas system of Morocco. Lect. Notes Earth Sci. pp. 245-271.

Jossen JA, Filali-Moutei J (1992). A new look at the structural geology of the southern side of the central and eastern High Atlas Mountains. Geol. Rundsch. 81: 143-156.

Knight B, Nomade S, Renne PR, Marzoli A, Bertrand H, Youbi N (2004) The Central Atlantic Magmatic Province at the Triassic–Jurassic boundary : paleomagnetic and 40Ar/39Ar evidence from Morocco for brief, episodic volcanism. Earth Planet. Sci. Let. 228: 143-160.

Laville E, Lesage JL, Séguret M (1977) Géométrie, cinématique, dynamique de la tectonique atlasique sur le versant sud du Haut Atlas marocain : aperçu sur les tectoniques hercyniennes et tardi-hercyniennes. Bull. Soc. geol. Fr. 19: 527-539.

Laville E, Petit JP (1984) Role of synsedimentary strike-slip faults in the formation of the Moroccan Triassic basins. Geology 12: 424-427.

Laville E, Piqué A (1991) La distension crustale atlantique et atlasique au Maroc au début du Mésozoique: le rejeu des structures hercyniennes. Bull. Soc. geol. France 162: 1161–1171.

Laville E, Piqué A (1992) Jurassic penetrative deformation and Cenozoic uplift in the central High Atlas (Morocco). A tectonic model Structural and orogenic inversions. Geol. Rundsch. 81: 157-170.

Laville E, Piqué A, Amrhar M, Charroud M (2004) A restatement of the Mesozoic Atlasic rifting (Morocco). Journal of African Earth Sciences 38: 145-153.

Malusà M, Polino R, Cerrina Feroni A, Ellero A, Ottria G, Baidder L, Musumeci G (2007) Post-Variscan tectonics in eastern Anti-Atlas (Morocco). Terra Nova 19: 481–489.

Marzoli A, Bertrand H, Knight KB, Cirilli S, Buratti N, Verati C, Nomade S, Renne PR, Youbi N, Martini R, Allenbakh K, Neuwerth R, Rapaille C, Zaninetti L, Bellieni G (2004) Synchronism of the Central Atlantic magmatic province and the Triassic-Jurassic boundary climatic and biotic crisis. Geology 32: 973-976.

Marzoqi M, Pascal A (2000) Séquences de dépots et tectono-eustatisme à la limite Crétacé/Tertiaire sur la marge sud-téthysienne (Atlas de Marrakech et bassin de Ouarzazate, Maroc). Newslett. Stratigr., 38: 57-80.

Mattauer M, Tapponier P, Proust F (1977) Sur les mécanismes de formation des chaines intracontinentales L'exemple des chaines atlasiques du Maroc. Bull. Soc. geol. France 19: 521-526.

Michard A, Yazidi A, Benziane F, Hollard H, Willefert S (1982) Foreland thrusts and olistostromes on the presaharian margin of the variscan orogen, Morocco. Geology 10:253-256.

Michard A, Saddiqi O, Chalouan A, Frizon de Lamotte D (2008) Continental Evolution: The Geology of Morocco. Structure, Stratigraphy, and Tectonics of the Africa-Atlantic-Mediterranean Triple Junction. Springer-Verlag, Berlin Heidelberg 116: 404 p.

Milhi A. (1997) Carte géologique du Maroc, Tinerhir – Echelle 1/100000. Notes et Mém. Serv. géol. Maroc 377.

Missenard Y, Zeyen H, Frizon de Lamotte D, Leturmy P, Petit C, Sébrier M, Saddiqi O (2006) Crustal versus asthenospheric origin of the relief of the Atlas mountains of Morocco. J. Geophys. Res. 111 (B03401) doi:101029/2005JB003708.

Missenard Y, Taki Z, Frizon de Lamotte D, Benammi M, Hafid M, Leturmy P, Sebrier M (2007) Tectonic styles in the Marrakesh High Atlas (Morocco): the role of heritage and mechanical stratigraphy. Journal of African Earth Sciences 48: 247-266.

Morel JL, Zouine EM, Andrieux J, Faure-Muret A (2000) Déformations néogènes et quaternaires de la bordure nord-haut- atlasique (Maroc); rôle du socle et conséquences structurales. Journal of African Earth Sciences 30: 119-131.

Mustaphi H, Medina F, Jabour H, Hoepffner C (1997) Le bassin du Souss (Zone de faille du Tizi n'Test, Haut Atlas occidental, Maroc): résultat d'une inversion tectonique contrôlée par une faille de d'etachement profonde. Journal of African Earth Sciences 24: 153-168.

Narr W, Suppe J (1994) Kinematics of basement-involved compressive structures. American Journal of Science 294: 802-860.

Ouanaimi H, Petit JP (1992) The southern limit of the Hercynian belt in the High Atlas (Morocco): reconstitution of an undeformed projecting block. Bull. Soc. Geol. France 163: 63-72.

Piqué A, Tricart P, Guiraud R, Laville E, Bouaziz S, Amrhar M, Aït Ouali R (2002) The Mesozoic-Cenozoic Atlas belt (North Africa): An overview. Geodinamica Acta 15: 185-208.

Proust F (1973) Etude stratigraphique, petrographique et structurale du bloc oriental du Massif Ancien du Haut Atlas (Maroc). Notes et Mém. Serv. géol. Maroc 34, 254: 15-54.

Proust F, Petit JP, Tapponnier P (1977) L'accident de Tizi n'Test et le rôle des décrochements dans la tectonique du Haut Atlas occidental (Maroc). Bulletin de la Societé Géologique de France 7: 541-551.

Qarbous A, Medina F, Hoepffner C, Ahmamou, M, Errami A, Bensahal A (2003) Apport de l'étude des bassins stéphano-autuniens et permo-triasiques du Haut Atlas occidental (Maroc) à la chronologie du fonctionnement de la zone de failles de Tizi n'Test. Bulletin de l'Institut Scientifique (Rabat), Sciences de la Terre. 25: 43-53.

Russo P, Russo L (1934) Le grand accident sud - atlasien . Bull. Soc. géol. France 5: 375-384.

Saber H, El Wartiti M, Broutin J (2001) Dynamique sédimentaire comparative dans les bassins stéphano-permiens des Ida Ou Zal et Ida ou Ziki (Haut Atlas occidental, Maroc). Journal of African Earth Sciences 32:573–594.

Sanderson DJ, Marchini WRD (1984) Transpression. Journal of Structural Geology 6: 449–458.

Saint-Bézar B, Frizon de Lamotte D, Morel JL, Mercier E (1998) Kinematics of large scale tip line folds from the High Atlas thrust belt, Morocco. J. Struct. Geol. 20: 999-1011.

Sébrier M, Siame L, El Mostafa Z, Winter T, Missenard Y, Leturmy P (2006) Active teconics in the Moroccan High Atlas. C. R. Géoscience 338: 65-79.

Serpelloni E, Vannucci G, Pondrelli S, Argnani A, Casula G, Anzidei M, Baldi P, Gasperini P (2007) Kinematics of theWestern Africa-Eurasia plate boundary from focal mechanisms and GPS data. Geophys. J. Int. 169: 1180-1200.

Teixell A, Arboleya ML, Julivert M, Charroud M (2003) Tectonic shortening and topography of the central High Atlas (Morocco). Tectonics 22: 1051, doi:10.1029/2002TC001460.

Teixell A, Ayarza P, Zeyen H, Fernandez M, Arboleya ML (2005) Effects of mantle upwelling in a compressional setting: the Atlas Mountains of Morocco. Terra Nova 17: 456-461.

Teson E, Teixell A (2008) Sequence of thrusting and syntectonic sedimentation in the eastern thrust belt (Dadès and Mgoun Valleys, Morocco). Int. J. Earth Sci. 97: 103-113.

Tikoff B, Teyssier C (1994) Strain modeling of displacement-field partitioning in transpressional orogens. Journal of Structural Geology 16: 1575–1588.

Tixeront M (1974) Carte géologique et minéralisations de couloir d'Argana (Haut Atlas occidental) au 1/100 000. Notes et Mém. Serv. géol. Maroc 205.

Wilcox RE, Harding TP, Seely DR (1973) Basic wrench tectonics. American Association of Petroleum Geologists Bulletin 57: 74–96.

Zühlke R, Bouaouda MS, Ouajhain B, Bechstädt T, Reinfelder R (2004) QuantitativeMeso-Cenozoic development of the eastern Central Atlantic Continental shelf, western High Atlas, Morocco. Marine and Petroleum Geology 21: 225-276.

Plate Tectonics in Paleozoic and Paleoproterozoic

Chapter 10

Plate Tectonic Evolution of the Southern Margin of Laurussia in the Paleozoic

Jan Golonka and Aleksandra Gawęda

Additional information is available at the end of the chapter

1. Introduction

The role of an active margin of Eurasia during Mesozoic and Cenozoic times was well defined (Golonka, 2004). The trench-pulling effect of the north dipping subduction, which developed along the new continental margin caused rifting, creating the back-arc basin as well as transfer of plates from Gondwana to Laurasia. The present authors applied this model to the southern margin of Laurussia during Paleozoic times. The preliminary results of their work were presented during the Central European Tectonic Group (CETEG) in 2011 in Czech Republic. The supercontinent of Laurussia, defined by Ziegler (1989), included large parts of Europe and North America. The southern margin of this supercontinent stretched out between Mexico and the Caspian Sea area. The present authors attempted to characterize the entire margin, paying the special attention to Central and Eastern Europe.

2. Methods

The present authors were using a plate tectonic model, which describes the relative motions plates and terranes during Paleozoic times. This model is based on PLATES, GPLATES and PALEOMAP software (see Golonka *et al.* 1994, 2003, 2006a,b, Golonka 2000, 2002, 2007a,b,c, 2009a,b). The plate tectonic reconstruction programs generated palaeocontinental base maps. It takes tectonic features in the form of digitised data files, assembles those features in accordance with user specified rotation criteria (Golonka *et al.* 2006a).

The rigid, outer part of the earth divided into many pieces known as lithospheric plates, comprising both the continental landmasses and oceanic basins These plates are in motion relative to each other and to the earth itself. Assuming the earth is a sphere, the motion of a plate across the earth's surface can be described as motion about the axis of a pole of rotation that goes through the centre of the earth. The intersection of the pole's axis with the earth's surface is referred to by its latitude/longitude coordinates. The distance the plate travels about the pole is an angular distance and is recorded in degrees. A stage pole of rotation describes the distance a plate moved from one time to the next time (i.e. from 20 Ma to 10 Ma). A finite pole of rotation describes the total distance a plate moved from some time in the past to the present day (i.e. from 20 Ma to 0 Ma). A rotation file contains a list of poles of rotation for various plates. The rotation files used by applied software contain finite poles of rotation. Thus for each plate, there are several finite poles of rotation for different times in the past. Plate models that use rotation files that describe the motion of plates relative to other plates are called relative framework models. Among, the data that show the relative motions between plates are fracture zones. Fracture zones are essentially flowlines between plates. For example, in the South Atlantic, the fracture zones show the motion of South America relative to (or away from) Africa. The finite pole of rotation describing this motion is a relative pole. South America is referred to as the moving plate and Africa as the fixed plate (Golonka *et al.* 2006a). The rotation file contains a list of finite rotations between pairs of tectonic elements, at different episodes of time, with brief bibliographic notes or general comments for each individual rotation.

3. Paleozoic major continental plates and ocean related to the Laurussia supercontinent

The break-of the supercontinent Pannotia (Dalziel at al., 1994) during the latest Precambrian times (Golonka, 2000, 2002) lead to the formation of the new continents. Fig 1. depicts the position of these continents at the beginning of the Paleozoic.

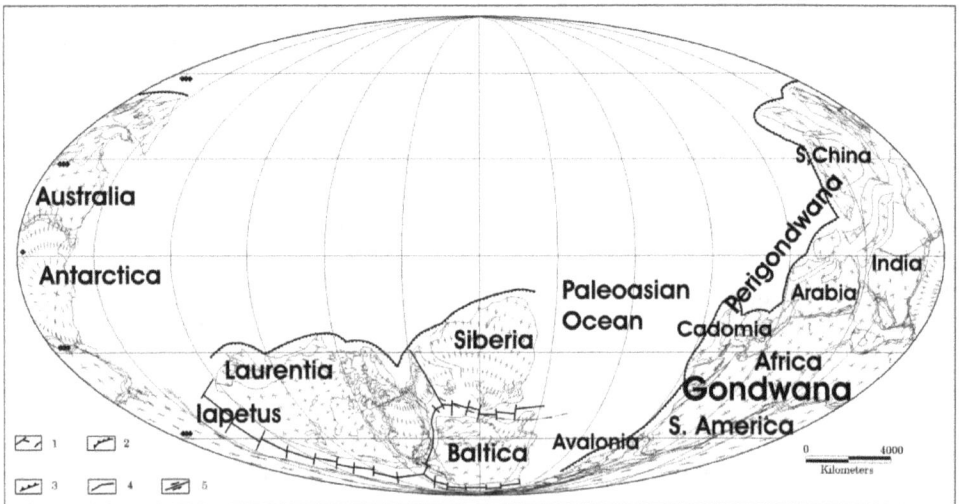

Figure 1. Plate tectonic global map of Early Cambrian (plates position as of 544 Ma). Mollweide projection. Modified from Golonka (2012). 1 - oceanic spreading center and transform faults, 2 - subduction zone, 3 - thrust fault, 4 - normal fault, 5 - transform fault..

3.1. Baltica

The continent Baltica was named after the Baltic Sea. It consisted of a major parts of northern and eastern Europe. It was bounded by the north by the border of the shelf of Norway, southern Barents Sea and Novaya Zemlya, on the east by the Ural suture, and on the southwest by a suture located close, but not quite along the Teisseyre-Tornquist line (Scotese & McKerrow, 1990; Golonka *et al.*, 1994). The southern boundary is more controversial. The Ukrainian shield certainly belonged to Baltica (Zonenshain et al, 1990). Perhaps also fragments of the North European platform like the Malopolska block, Bruno-Vistulicum, Moesia and other small blocks located now around the Baltic Sea belonged to Baltica (Kalvoda & Bábek 2010, Żelaźniewicz et al., 2009, Besutiu, 2001). In Central and southwestern Europe the possible boundary of Baltica is marked by the extent of North European plate below the Carpathian and Balkan nappes (Figs. 1-3)

3.2. Gondwana

The supercontinent Gondwana, also known as Gondwanaland (Vevers, 2004) was named after the ancient Indian tribes Gonds; in Sanskrit Gondwana means "the forest of Gonds". The continents forming the core of Gondwana include South America, Africa, Madagascar, India, Antarctica and Australia. The location of numerous smaller continental blocks that bordered Gondwana is less certain. The following were adjacent to Gondwana at some time during the Paleozoic: Yucatan, Florida, Avalonia, central European (Cadomian) terranes between the Armorica and Bohemian Massif, Moesia, Iberia, Apulia and the smaller, southern European terranes, central Asian terranes (Karakum and others), China (several separate blocks), and the Cimmerian terranes of Turkey, Iran, Afghanistan, Tibet and Southeast Asia (Figs, 1, 2).

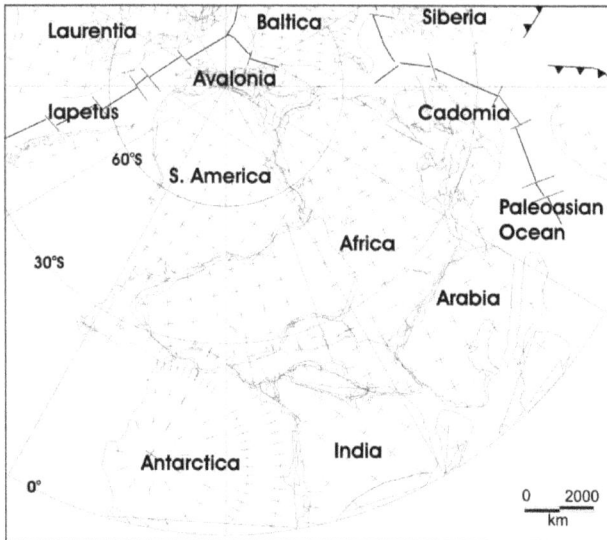

Figure 2. Plate tectonic map of Gondwana. Late Vendian - Early Cambrian (plates position as of 544 Ma). Stereographic polar projection. Modified from Golonka (2012). Legend as in Fig. 1

3.3. Laurentia

The continent Laurentia was named after the Laurentia Shield, in turn after St. Lawrence (Laurentius in Latin) River. North America was a major component of Laurentian plate. This plate also included Greenland, Chukotka peninsula, Svalbard and large part of the Barents Sea (Barentsia) fragment of Alaska (North Slope), northwest Ireland, and Scotland, (Golonka, 2000, 2002, Ford & Golonka, 2003, Golonka *et al.*, 2003). (Figs 1-3). Its southern (present day eastern boundary is located within Appalachians, its northern (present day western) boundary is located within Rocky Mountains. The relationship between Laurentia and Siberia remains speculative (Figs 1-5).

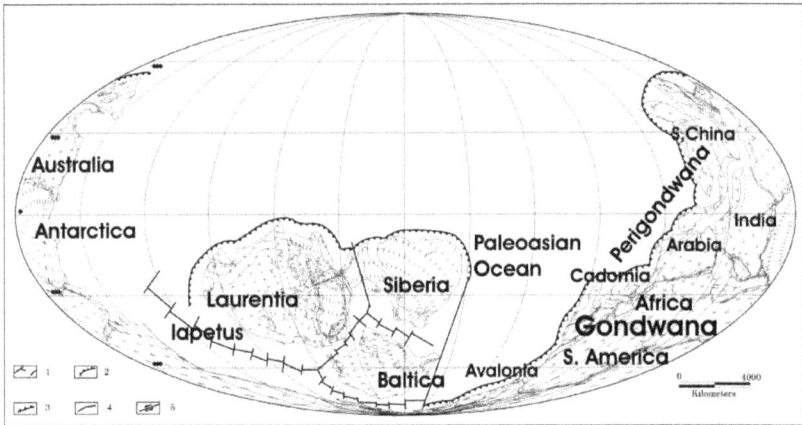

Figure 3. Plate tectonic map of Middle Cambrian (plates position as of 510 Ma). Mollweide projection. Modified from Golonka (2012). 1 - oceanic spreading center and transform faults, 2 - subduction zone, 3 - thrust fault, 4 - normal fault, 5 - transform fault.

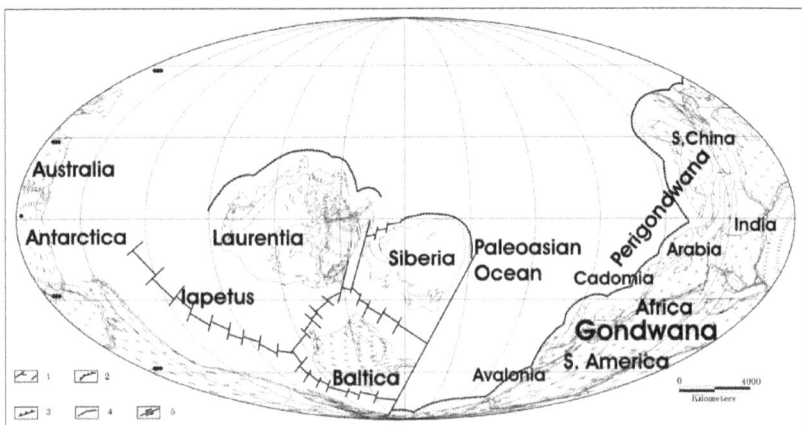

Figure 4. Plate tectonic map of Late Cambrian (plates position as of 498 Ma). Mollweide projection. Modified from Golonka (2012). 1 - oceanic spreading center and transform faults, 2 - subduction zone, 3 - thrust fault, 4 - normal fault, 5 - transform fault.

3.4. Avalonia

The name Avalonia is derived from the Avalon peninsula, Newfoundland, eastern Canada. The names 'Avalon Composite Terrane' (Keppie, 1985), or Superterrane were also used. Western Avalonia included terranes in northern Germany, the Ardennes in Belgium and northern France, England, Wales, southeastern Ireland, eastern Newfoundland, much of Nova Scotia, southern New Brunswick and some coastal parts of New England (Golonka 2009, McKerrow et al. 1991). The inclusion of terranes in the eastern part of Avalonia is more speculative (Fig. 5).

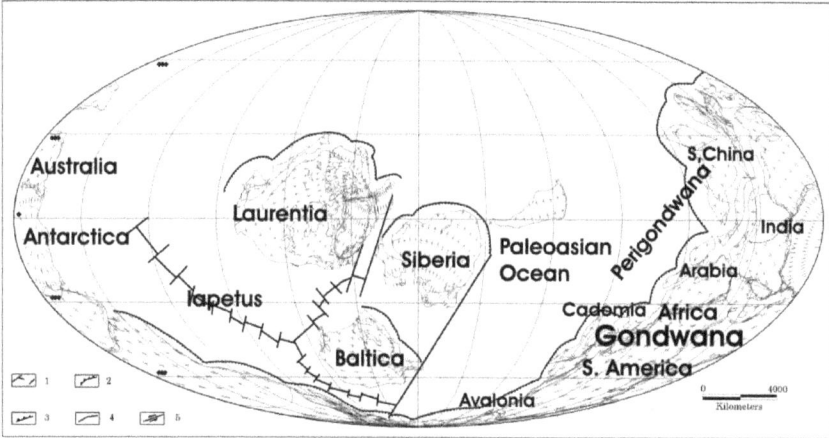

Figure 5. Plate tectonic map of Early Ordovician (plates position as of 485 Ma). Mollweide projection. Modified from Golonka (2012). 1 - oceanic spreading center and transform faults, 2 - subduction zone, 3 - thrust fault, 4 - normal fault, 5 - transform fault.

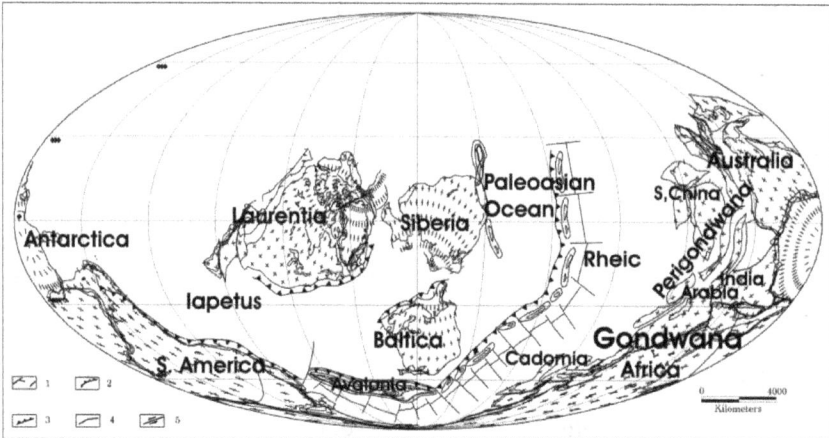

Figure 6. Plate tectonic map of Middle Ordovician (plates position as of 472 Ma). Mollweide projection. Modified from Golonka (2012). 1 - oceanic spreading center and transform faults, 2 - subduction zone, 3 - thrust fault, 4 - normal fault, 5 - transform fault.

Perhaps eastern Avalonia consisted of northwestern Poland fragments now included into Sudety Mountains and Bohemian Massif, terranes accreted in Carpathian-Balkan and Pannonian regions, containing fragments consolidated during Cadomian orogeny, and also Serbo-Macedonian massif, Rhodopes and Pontides in eastern Europe and adjacent part of Asia (Golonka, 2012). Avalonia originated after Gondwana break-up during Early-Middle Ordovician times (Fig.6).

3.5. Cadomia

Cadomia is named after city of Caen, Roman/Latin Cadomia in northern France also known as Armorica or Armorica Group Terrane (e.g. Lewandowski 2003 and references therein) or Gothic terranes (Stampfli, 2001). It consisted of the fragments of western and Central Europe between northwestern France (Brittany) and Czech Republic consolidated during Cadomian orogeny (Figs.1-6). It includes Saxoturingian zone/terrane in Germany and northwestern Czech Republic. Cadomia belonged to Perigondwana during Early Paleozoic times. Perhaps it was detached from Gondwana during Silurian times (Golonka et al. 2006a, Lewandowski 2003), the existence and nature of this detachment as well as extension of Cadomia eastward and westward remain quite speculative.

3.6. Iapetus Ocean

The Iapetus Ocean is named after Titan, son of Uranus, the sky, and Gaia, the Earth, father of Atlas. It preceded Atlantic off the American coast, therefore Atlantic was derived from Atlas and proto-Atlantic was named after Atlas' father. The Iapetus Ocean was located between Gondwana and Laurentia (Figs 1-6), later also between Avalonia and Laurentia (Fig. 6). It started to open as a rift between Laurentia and Rodinia in late Neoproterozoic, about 760-700 Ma, and the maximum obtained at about 600-520 Ma (Kamo et al. 1989; Cawood et al. 2001). Its closure is connected with the rotation and collision of Baltica, Laurentia and the fragment of Gondwana – Avalonia (Hartz & Torsvik 2002). The Iapetus suture was formed Caledonian Orogeny was formed in the time interval 480-440 Ma in the west while the eastern part (in Europe) closed at 440-420 Ma. The eastern extension of Iapetus is not so certain (Golonka, 2002, 2006a, b). Perhaps it was located between Baltica and Gondwana. Part of the Iapetus Ocean is known as Tornquist Sea, the oceanic basin located between the southwest and southern margin of Baltica and Gondwana, and later also Avalonia plate. The name is derived from the Tornquist zone in central Europe, already mentioned border of Baltica. The position of Tornquist Sea is speculative, because Baltica rotated during Early Paleozoic times.

3.7. Rheic Ocean

The Rheic Ocean was named after Rhea, Titaness-Goddess, daughter of Uranos, the sky, and Gaia, the Earth, wife of Cronos and mother of Zeus. Rheic Ocean originated between Gondwana and Avalonia (Fig. 6), later between Gondwana and Laurussia, during Early Paleozoic times (Nance et al. 2010 and references therein). It was opened during Late

Cambrian – Early Ordovician times, around 500 Ma. System of back-arc basins developed along the northern branch of Rheic Ocean after formation of Laurussia during Devonian times (von Raumer & Stampfli, 2008). These basins are considered either as a part of the Rheic Ocean (e.g. Nance et al. 2010, Golonka, 2007b, McKerrow et al., 1991) or as a separate entity known as Rheno-Hercynian or Moldanubian basin in Central Europe (e.g. Golonka et al. 2006b, Golonka 2002, Schulmann et al., 2009). Rheic Ocean narrowed during Devonian and was closed during Carboniferous times as a result of collision of Laurussia with Gondwana.

4. Global Early Paleozoic plate tectonic evolution leading to the assembly of Laurussia

Gondwana supercontinent was located around the South Pole at the beginning of Paleozoic (Figs 1-2). Baltica and Laurentia were located at the high latitude in the southern hemisphere, their southern margins close to the South Pole. They drifted apart from Gondwana during the Late Vendian times (Torsvik et al., 1996, Golonka et al, 2002, 2009b, 2012). Their breakup led to the formation of new oceans, including Iapetus (Figs. 1-2). Continued seafloor spreading occurred in this ocean during Cambrian times (Figs. 2-4). The fragmentation of northern margin of Gondwana was also marked by magmatic activity in the 550-500 Ma (Dörr et al., 1998, Turniak et al., 2000, Tichomirowa, 2002, Burda and Klötzli, 2011). Laurentia drifted rapidly northward and rotated counter-clockwise, reaching low latitudes (Golonka, 2000, 2002). Seafloor spreading also occurred within the Pleionic Ocean between East Siberia and Baltica. The relationship between Laurentia and Siberia remains quite speculative. Latest Cambrian – earliest Ordovician was the time of maximum dispersion of continents during the Paleozoic. Baltica, Laurentia and Siberia drifted further northward (Fig. 5). The subduction along the central margin of Gondwana caused the onset of rifting of the Avalonian terranes (Golonka et al., 1994, Golonka, 2000, 2002). The subduction along the northern margin of Baltica was perhaps related to the Ordovician rotation of this plate (see Golonka et al., 1994, 2006b, Torsvik et al., 1996, McKerrow et al., 1991, Golonka, 2000, 2002, cocks & Torsvik, 2011). The distance between Gondwana and Laurentia, which was situated on equator reached 5000 km (Golonka, 2002, Golonka et al. 2006).

Early-Middle Ordovician were the times of a major plate reorganization (Golonka, 2000, 2002, 2009b, 2012, Golonka et al. 2006b). Avalonia probably started to drift from Gondwana and move northward toward Baltica and Laurentia (Golonka, 2000, 2002, 2009b, 2012). This movement was related to the origin of Rheic Ocean. (Fig. 6). The Iapetus Ocean had begun to narrow.

During Late Ordovician times (Fig. 7 the Rheic Ocean between Gondwana and Avalonia widened significantly (Golonka, 2000, 2002, 2009b, 2012). The position of Cadomia remains uncertain. On the presented reconstruction, the Cadomian blocks are positioned relatively close to Gondwana. This is mainly based on the paleobiogeographical data (Scotese and McKerrow, 1990, Robardet et al., 1993, Golonka, 2002, 2009b, 2012). The alternative

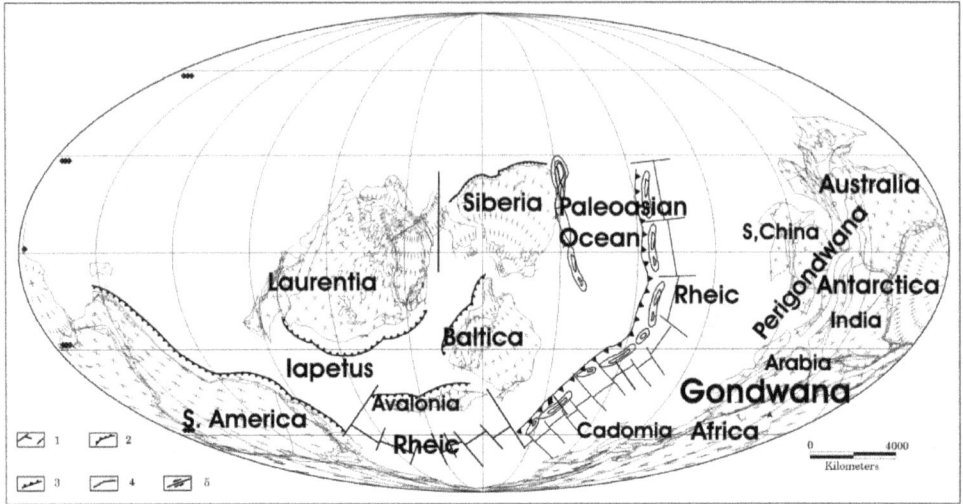

Figure 7. Plate tectonic map of Late Ordovician (plates position as of 452 Ma). Mollweide projection. Modified from Golonka (2012). 1 - oceanic spreading center and transform faults, 2 - subduction zone, 3 - thrust fault, 4 - normal fault, 5 - transform fault.

reconstructions assumed that these terranes were rifted away and formed separate Cado-mia plate floating within the Rheic Ocean (Lewandowski, 1993, see also Golonka et al., 2006a). Latest Ordovician- Early Silurian were the times of collision between Avalonia and Baltica (Fig. 8).

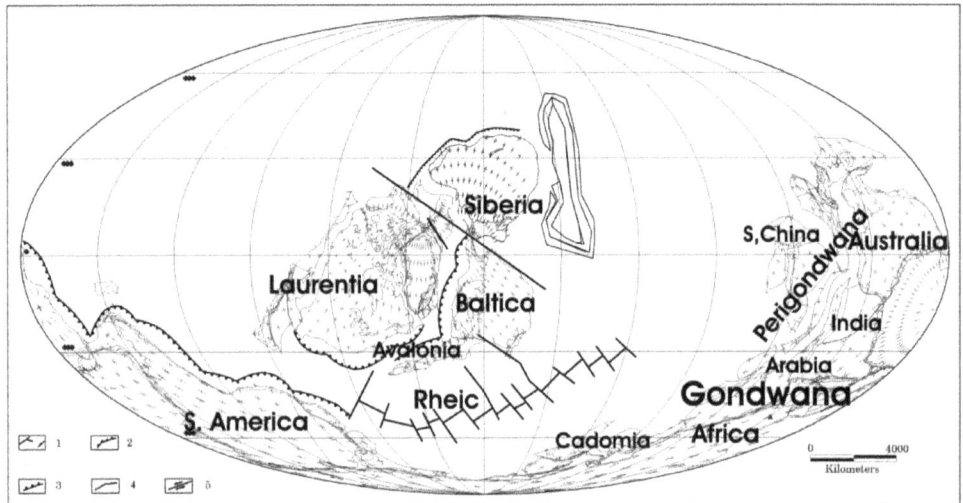

Figure 8. Plate tectonic map of Early Silurian (plates position as of 435 Ma). Mollweide projection. Modified from Golonka (2012). 1 - oceanic spreading center and transform faults, 2 - subduction zone, 3 - thrust fault, 4 - normal fault, 5 - transform fault.

This convergence was dominated by a strike-slip suturing of the two continents, rather than by full-scale continent-continent collision (Golonka, 2000, 2002). Northwestern Poland and adjacent part of Germany was joined with Baltica, along a strike-slip fault zone known as the Tornquist-Teisseyre line forming new continent - Balonia. Perhaps the Brunovistulicum and Malopolska terranes of southern Poland also belonged to Avalonia and joined Baltica (Moczydłowska, 1997, Bełka et al., 2000, Golonka, 2002, 2009b). In the Central Western Carpathians Late Ordovician – Early Silurian tonalitic gneisses of calc-alkaline character, associated with meta-gabbros, revealed the presence of magmatic episodes at 470-435 Ma (Kohut et al. 2008, Janák et al. 2002, Gaab et al. 2003, Gawęda & Golonka, 2011). These rocks and granites present results of docking of Avalonia to Baltica. Also in the East Carpathians in Romania the intrusion of 459-470 Ma granitoids (Munteanu & Tatu, 2003, Pana et al., 2002, Ballintoni et al., 2010) document the collision-related tectono-magmatic effects of docking of eastern prolongation of Avalonia to Baltica. The Scandian Orogeny was the result of the collision between Balonia and Laurentia. the onset of the orogeny occurred during the Early Silurian times and by late Silurian the orogeny was concluded (Golonka et al., 1994, Golonka, 2000, 2002, 2009b). The main phase of the Scandian orogeny is marked by nappes in Norway and Greenland as well as large crustal thickening (Dewey & Burke 1973, Torsvik et al., 1996, Golonka, 2000, 2002, 2009b). During the Mid-Silurian Avalonia collided with Laurentia (Fig. 9). This collision is known as the Caledonian orogeny, named after Roman name of Scotland – Caledonia. This name is also extending into other related orogenic events in Scanidinavia, Greenland and Ventral Europe. After the complete closure of the Iapetus Ocean, the continents of Baltic, Avalonia, and Laurentia formed the continent of Laurussia (Ziegler, 1989).

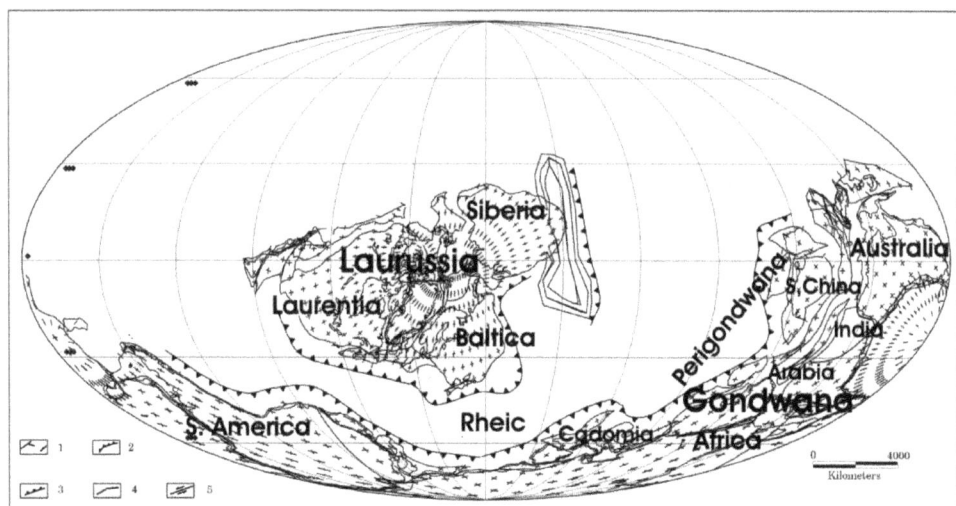

Figure 9. Plate tectonic map of Late Silurian (plates position as of 425 Ma). Mollweide projection. Modified from Golonka (2012). 1 - oceanic spreading center and transform faults, 2 - subduction zone, 3 - thrust fault, 4 - normal fault, 5 - transform fault.

5. Laurussia during Devonian times

Supercontinent Laurussia existed during Late Silurian and Devonian times. Late Silurian Global paleogeography depicted on Fig. 9 id showing separation of Gondwana and Laurussia as well as subduction along the southern margin of Laurussia. Fig 10 depicts the global paleogeography during Early Devonian times. It is showing a possibility of Early Devonian collision between South and North America according to Golonka (2002, 2007b, 2012 see also Keppie 1989, McKerrow *et al.* 1991, Dalziel *et al.* 1994, Keppie *et al.* 1996). This collision was marked   by orogenic events in Venezuela, Columbia, Peru, and northern Argentina (Gallagher & Tauvers 1992, Williams 1995). Paleomagnetic data (Kent & Van der Voo 1990, Van der Voo 1993, Lewandowski 1998, 2003) and paleobiogeography (Young 1990) also support the hypothesis about the proximity of South and North America (Golonka, 2007, 20012 b). This proximity is also shown by recent maps of Cocks & Torsvik (2010), however without the collision. The Carolina region (Rast & Skehan 1993) also contains an element of collision and transpression between South and North America, acting as an indenter, with a dextral strike-slip component (Golonka, 2012).

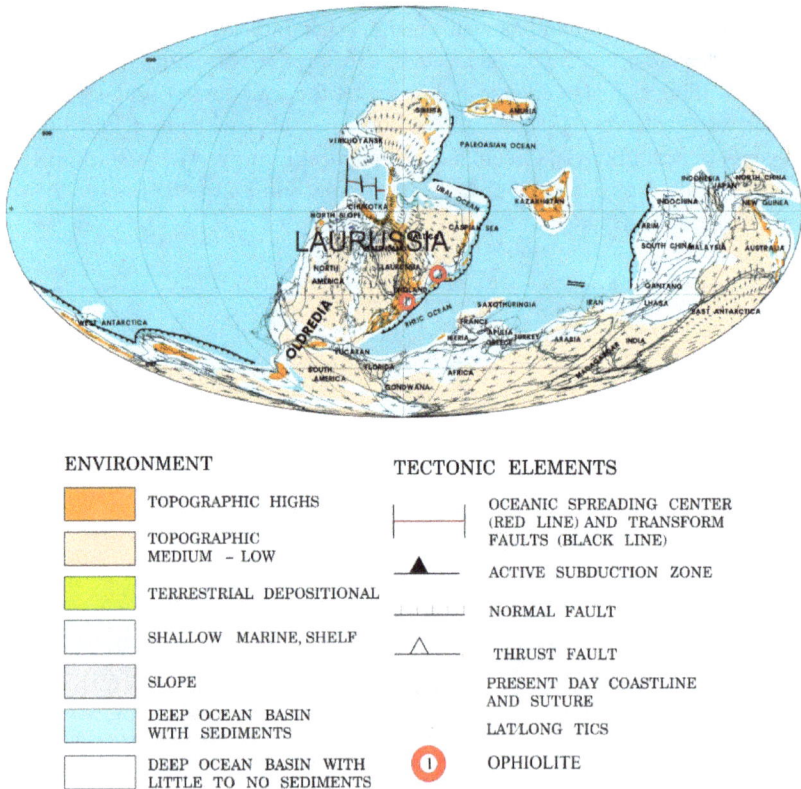

Figure 10. Paleogeographic c map of Early Devonian (plates position as of 401 Ma). Mollweide projection. Modified from Golonka 2007b, 2012) Ophiolites – 1 Lizard ophiolite, 2 – central and eastern European ophiolites (depicted in details on Fig. 11).

The accretion of Avalonia was followed by the development of north dipping subduction, along the new continental margin of Laurussia. That subduction caused rifting and creating the new back-arc basin. That basin was first recognized in central Europe as Rheno-Hercynian zone (Ziegler, 1989, Franke, 1992, Franke et al., 1995, Golonka, 2007b and references therein). Its paleogeography resembled present-day marginal Seas of East Asia. Fragment of Rheno-Hercynian zone, with oceanic crust, was also named Lizard-Giessen Ocean (Zeh & Gerdes, 2010). The name Lizard is derived from the well known Lizard Ophiolitic Complex in Cornwall, United Kingdom (Bromley, 1975, 1976, Kirby, 1976), marked as number 1 on Fig.10. This Devonian complex contain peridotites, serpentinites dolerite dikes as well as amphibolites (Bromley, 1975, 1976, Kirby, 1976, Cook et al., 2002, Clark et al., 2003). Probably El Castillo volcanic rocks from Spain are related to the Lizard ophiolites (Gutiérez- Alonso et al., 2008). They yield Devonian, 394,7±1,4 Ma age of magmatism.

Figure 11. Paleogeography of Laurussia margin in central-eastern Europe during Early Devonian). Mollweide projection. Legend as on Fig. 10. Abbreviations: T – L - Tepla Barrandien – Lugia terranes, Carp. – Central Carpathian and Balkan terranes. Ophiolites: 1- Central-Sudetic ophiolite, 2 - Western Carpathian ophiolite, 3 - Balkan-South Carpathian ophiolite.

The eastern extension of the Rheno-Hercynian Basin is marked by ophiolites in Sudety area in Poland and in the Carpathian-Balkan area (Figs. 10, 11). During Devonian times the Central-Sudetic ophiolites were located between Avalonian terrenes, sutured to Laurussia, and several microplates/terranes now included in the mosaic structure of Bohemian Massif and Sudety Mountains. These terranes are marked on the Fig. 11 as T – L – Tepla-Barrandien – Lugia. They include several small blocks like Tepla-Barrandien, Góry Sowie and others, with uncertain, speculative position between Cadomia- Saxoturungia and Laurussia (see. e.g. Aleksandrowski et. al., 2000, Franke & Żelaźniewicz, 2000, Winchester et al., 2002, Mazur, et al., 2006, Schulmann et al., 2009, Kryza & Pin, 2010, Nanece et al., 2010). According to Kryza & Pin (2000) the SHRIMP zircon data supplied evidence for Devonian age of the Central-Sudetic ophiolites, around 400 Ma.

The age of the ophiolitic remnants within the Carpathians from Tatric and Gemeric units Unit yield various ages: 371 Ma - 385-383 Ma (Putiš et al. 2009), 391 Ma (Gawęda, 2008), and 394 Ma (Kohut et al., 2006). The recorded events are related to the development of a back-arc basin with ocean crust as a result of the slab roll-back and derivation of ribbon-like Proto-Carpathian Terrane from Laurussia (see Gawęda & Golonka, 2011). The associated granitoid magmatism post-dated that event (387 Ma; Burda & Klötzli 2011, Putiš et al., 2008). The Balkan-South Carpathian ophiolite belt yield the isotopic age of 406-399 Ma (Zakariadze et al, 2007), comparable with the Central-Sudetic ophiolites age being around 400 Ma (Kryza, Pin, 2010).

Gondwana drifted northward and rotated clockwise during Devonian times (Scotese and McKerrow, 1990, Scotese and Barret, 1990, Golonka, 2000, 2002, 2007b). At the same time, Laurussia was rotating clockwise (Torsvik et al., 1996) at a somewhat faster rate. Figure 12 depict global paleogeography during the Late Devonian times.

Figure 12. Paleogeographic c map of Late Devonian (plates position as of 370 Ma). Mollweide projection. Legend as on Fig. 10. Modified from Golonka 2007b, 2012) .Ophiolites – 1 Lizard ophiolite, 2 – central and eastern European ophiolites (depicted in details on Fig.13).

The first contact between Laurussia and the Cadomian promontory of Gondwana occurred in Central Europe This contact marks the onset of Hercynian orogeny. The Saxoturingian part of the Cadomian plate collided with small terranes in Germany, Czech Republic and Poland (Fig. 13). Collisional events were marked also in Carpathians and Balkans. The Rheno-Hercynian Basin changed its character from extensional into compressional one. By latest Devonian – Early Carboniferous it displayed synorogenic features, filling with turbiditic flysch and culm facies (Golonka, 2007b).

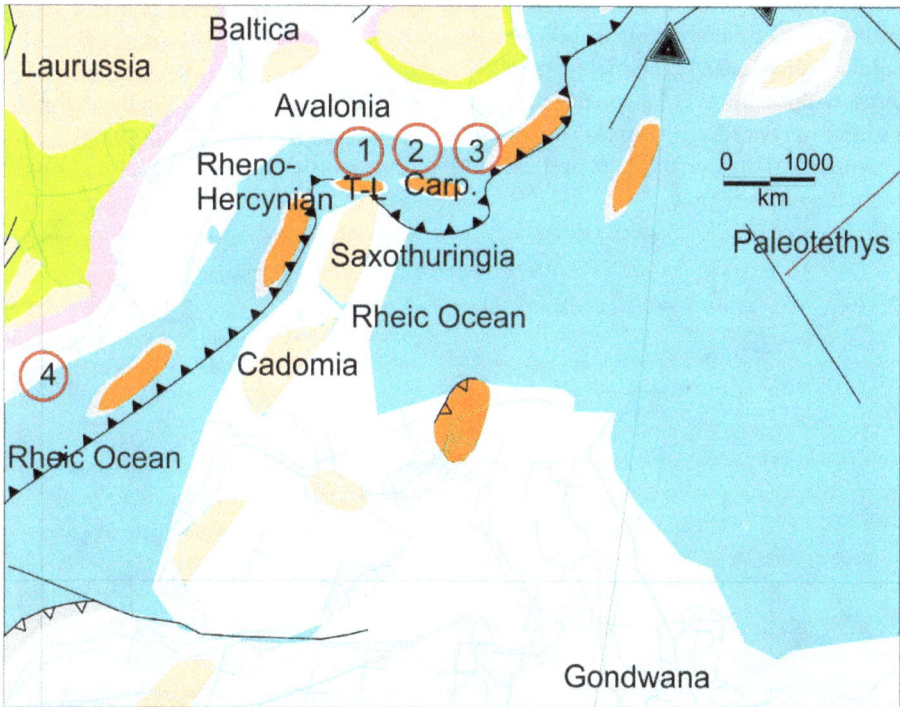

Figure 13. Paleogeography of Laurussia margin in central-eastern Europe during Late Devonian). Mollweide projection. Legend as on Fig. 10. Abbreviations: T – L - Tepla Barrandien – Lugia terranes , Carp. – Central Carpathian and Balkan terranes. Ophiolites: 1- Central-Sudetic ophiolite, 2 - Western Carpathian ophiolite, 3 - Balkan-South Carpathian ophiolite, 4 – Lizard ophiolite.

The new subduction zone on the southern margin Laurussia and the formation of a back-arc resulted in the intensive partial melting and magmatic activity from 370 to 340 Ma (Burda & Gawęda 2009, Schulmann et al. 2009). The continuous subduction along both old and new subduction zones resulted in docking of the Cadomia- Saxoturingia Terrane to Laurussia and significant consumption of the Rheic Ocean (Fig. 10). Finally the subduction caused formation of the Variscan Orogenic Suture, extending from Turkey to Mexico. That event resulted in voluminous granitoid magmatism and amphibolite-granulite facies metamor- phism – the most prominent features of all crystalline cores, present in all the Variscan com- plexes (Stipska et al., 1998; Schulmann et al. 2009). These Variscan granitoid magmas were formed and intruded in the interval of 370-340 Ma, contemporaneously with Variscan nappes formation (Dallmayer et al., 1996; Burda et al., 2011) and synchronous with uplift of the continental blocks during prolonged collision (Janak et al., 1999, Gawęda et al., 2000). The resulting collision was possibly associated with basaltic underplating or slab break-off (Broska & Uher, 2001; Finger et al., 2009). Most of the granitoid bodies show hybrid character, with both mantle and crustal components involved (e.g. Słaby & Martin, 2005, Burda et al. 2011). Their VAG and CAG affinities suggest that melted metasediments, representing crustal component of the magma, were originally deposited both in the volcanic arc and intracontinental basins during subduction at the active, Andean-type continental margin (Schulmann et al., 2009). The melted material represented mainly recycled Proterozoic components, with addition of Paleozoic ones and influence of DM (depleted mantle) component as a source of heat (Poller et al., 2000). The subsequent uplift caused the exhumation of eclogitic remnants (Janak et al., 1996) and caused their retrograde metamorphism in granulite – amphibolite facies regime.

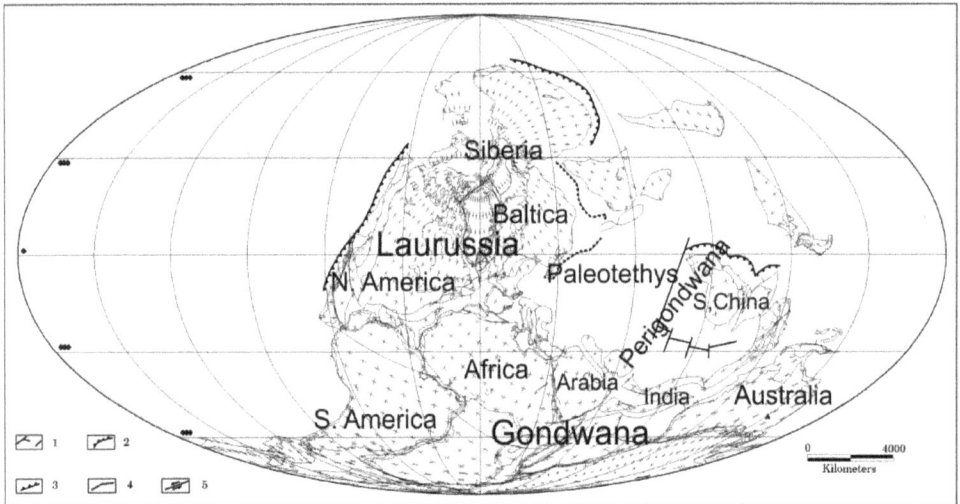

Figure 14. Plate tectonic map of Early Carboniferous (late Visean – Serpukhovian), plates position as of 328 Ma). Mollweide projection. Modified from Golonka (2012). 1 – oceanic spreading center and transform faults, 2 – subduction zone, 3 – thrust fault, 4 – normal fault, 5 – transform fault.

The ongoing Hercynian convergence in Europe led to large scale dextral shortening, overthrusting and emplacement of parts of the accretionary complexes (Edel & Weber, 1995). The amount of convergence was modified by large, dextral and sinistral transfer faults. The thrusting took place in the Tatra Mts. in the Carpathians (Gawęda et al., 2000, Golonka, 2000, 2002, 2007b). The collision between Gondwana and Laurussia continued to develop during Carboniferous times (Figs. 14, 15). The intercontinental collision began to affect the northwestern part of Africa, developing Mauretinides, Bassarides, Rokelides orogens is. The Alleghenian orogeny in North America continued (Hatcher et al., 1989, Rast & Skehan, 1993), prograding westwards to the Ouachita fold belt in Arkansas, Oklahoma,

Texas and adjacent part of Mexico (Arbenz, 1990, Golonka, 2000, 2002). The clockwise rotation of Gondwana resulted in the involvement of the deformation. This Gondwanian influence resulted in the convoluted shape of the Hercynian orogen, strike-slip zones (Franke et al., 1995) and Hercynian deformation at the eastern end in Poland. The European foreland basin was elevated or changed its sedimentation regime from flysch to molasse. The central Pangean mountain range was formed, which extended from Mexico to Poland (Golonka, 2000, 2002, 2007b). The Laurussia continent ceased its independent existence becoming a part of the supercontinent Pangea.

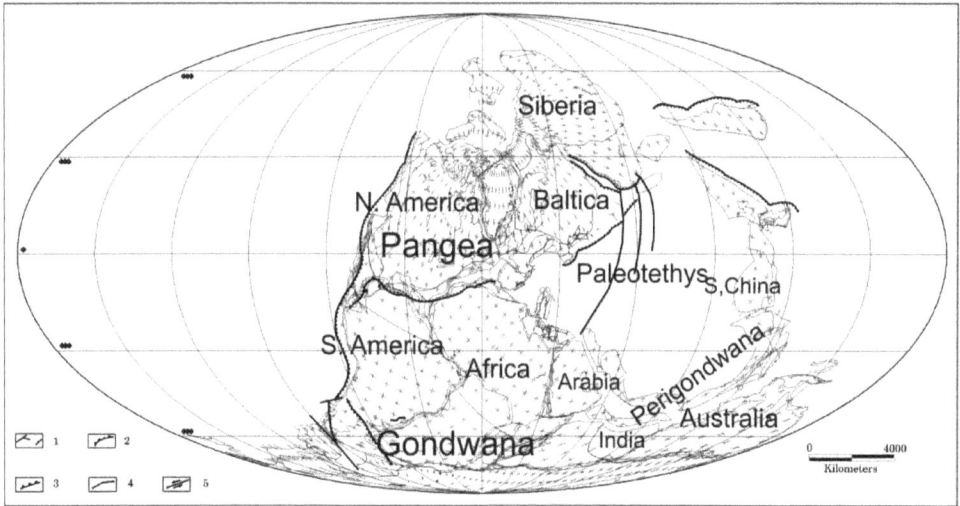

Figure 15. Plate tectonic map of Late Carboniferous (plates position as of 302 Ma). Mollweide projection. Modified from Golonka (2012). 1 – oceanic spreading center and transform faults, 2 – subduction zone, 3 – thrust fault, 4 – normal fault, 5 – transform fault.

6. Conclusions

1. The Late Precambrian (Vendian) to present plate tectonic processes contributed to the complex structure of Western and Central Europe.
2. The supercontinent of Laurussia, originated as a result of a closure of Iapetus Ocean and collision of Baltica, Avalonia and Laurentia.
3. Laurussia originated during Late Silurian times
4. The accretion of Avalonia was followed by the development of north dipping subduction, forming the new Rheno-Hercynian Basin back-arc basin during Devonian times
5. The oceanic crust of the Rheno-Hercynian Basin is recognised by ophiolites in Western, Central and Eastern Europe. The Lizard ophiolite in U.K. and Central-Sudetic ophiolite in Poland represent the best developed Devonian oceanic complexes.
6. The new geological research (including dating) in the Carpathian-Balkan area supports the idea about the prolongation of the Rheic Suture to the east.

7. The new geological research (including dating) in the Carpathian-Balkan area supports the idea about the prolongation of the Rheic Suture to the east.

8. The tectonic events and associated magmatism point out the development of the back-arc at the southern margin of Laurussia in the time interval 406-371 Ma.

Laurussia cease to exist during Hercynian orogeny in Carboniferous times and was included into supercontinent Pangea.

Author details

Jan Golonka
AGH University of Science and Technology, Kraków, Poland

Aleksandra Gawęda
Faculty of Earth Sciences, University of Silesia, Sosnowiec, Poland

Acknowledgement

This research has been partially financially supported by Ministry of Science and Higher Education grants (Statutory Activities) through the AGH University of Science and Technology in Krakow grants no. 11.11.140.447 (J. Golonka) and Ministry of Education and Science grant No N307 027837 (A. Gawęda).

7. References

Aleksandrowski, P., Kryza, R., Mazu,r S., Pin, C. & Zalasiewicz, J.A,. (2000). The Polish Sudetes: Caledonian orVariscan? *Transactions of the Royal Society, Edinburgh, Earth Sciences (1999)*, Vol. 90, 127–146.

Arbenz, J. K., 1990, The Ouachita system, *The Geology of North America*, Bally, A. W. & Palmer, A. R., (Eds.,) Vol. . A., 371-396, Boulder, Geological Society of America,

Ballintoni, I., Balica, C., Seghedi, A. & Ducea, M.N. (2010). Avalonian and Cadomian terranes in North Dobrogea, Romania. Precambrian Research, Vol. 182, No. 3, 217-229.

Bełka, Z., Ahrendt, H., Franke, W. & Wemmer, K. (2000). The Baltica-Gondwana suture in central Europe: Evidence from K-Ar ages of detrital muscovites and biogeographical data, *Geological Society, London, Special Publications*, Vol. 179, 87-102.

Besutiu L. (2001). Moesia – a Baltica derived terrane? *Pancardi 2001 II. Abstracts*, pp. CP29. Ádám, A., Szarka, L., Szendröi, J. (Eds.) Hungarian Academy of Sciences, Sopron, Humgary.

Bromley, A.V. (1975). Is the Lizard Complex, South Cornwall, a fragment of Hercynian oceanic crust?. *The Lizard: A Magazine of Field Studies*, Vol., No. 3, 2-11.

Bromley, A.V. (1976). A new interpretation of the Lizard Complex, S. Cornwall, in the light of the ocean crust model. *Proceedings. Geological. Society London*, Vol. 132, 114.

Broska, I, & Uher, P. (2001). Whole-rock chemistry and genetic typology of the West-Carpathian Variscan granites. *Geologica Carpathica*, Vol. 52 no. 2, 79-90.

Burda, J. & Gawęda, A. (2009). Shear-influenced partial melting in the Western Tatra metamorphic complex: geochemistry and geochronology. *Lithos*, Vol. 110, . 373-385.

Clark, A. H., Sandeman, H. A. I., Nutman, A.P., Green, D.H & Cook, A. C.. (2003). Discussion on SHRIMP U–Pb zircon dating of the exhumation of the Lizard Peridotite and its emplacement over crustal rocks: constraints for tectonic models. *Journal of the Geological Society, London*, Vol. 160, 331–335.

Cocks, L.R.M. & Torsvik T.H. (2011). The Palaeozoic geography of Laurentia and western Laurussia: a stable craton with mobile margins. *Earth Science Reviews*, Vol. 106, 1-51.

Cook, C. A., Holdsworth, R.E. & Styles, M.T. (2002). The emplacement of peridotites and associated oceanic rocks from the Lizard Complex, southwest England, *Geological Magazine*, Vol. 139, 27-45.

Dallmeyer, D., Neubauer, F., Handler, R., Fritz, H., Muller, W., Pana, D., & Putis, M. (1996). Tectonothermal evolution of the internal Alps and Carpathians: evidence from ^{40}Ar-^{39}Ar mineral and whole-rock data. In: Schmid S.M., Frey M., Froitzheim N., Heilbronner R., Stuenitz, H. (eds.): Alpine geology, proceedings of the second workshop: 2nd workshop on Alpine geology, *Eclogae Geologica Helvetica*, Vol. 89, pp. 203-227.

Dalziel, I.W.D., Dalla Salda, L.H & Gahagan, L.M. (1994). Paleozoic Laurentia-Gondwana interaction and the origin of the Appalachian-Andean mountain system, *Geological Society of America Bulletin*, Vol. 106, 243-252.

Dewey, J. F. & Burke, V. B. S., (1973). Tibetan, Variscan, and Precambrian basement reactivation, products of continental collision, Journal of Geology, Vol. 81, 683-692.

Dörr, W., Fiala, J., Vejnar, Z., & Zulauf, D. (1998). U-Pb zircon ages and structural development of metagranitoids of the Tepla crystalline complex: evidence for pervasive Cambrian plutonism within the Bohemian Massif (Czech Republiv). *International Journal of Earth Sciences*, Vol. 87, 135-149.

Edel, J. B., Weber, K., 1995. Cadomian terranes, wrench faulting and thrusting in the central Europe Variscides: geophysical and geological evidence, *Geologische Rundschau*, Vol. 84, 412-432.

Finger, F., Gerdes, A., Rene, M., & Riegler G. (2009). The Saxo-Danubian Granite Belt: magmatic response topost-collisional delamination of mantle lithosphere below the south-west sector of the Bohemian Massif (Variscan Orogen). *Geologica Carpathica*, Vol. 60, No. 3, 205-212.

Ford, D. & Golonka J. (2003). Phanerozoic paleogeography, paleoenvironment and lithofacies maps of the circum-Atlantic margins, *Thematic set on paleogeographic reconstruction and hydrocarbon basins: Atlantic, Caribbean, South America, Middle East, Russian Far East, Arctic*. Golonka J. (ed.), *Marine and Petroleum Geology*, Vol. 20, 249-285.

Franke, W. (1992), Phanerozoic structures and events in central Europe, *The European Geotraverse, A continent revealed*, Blundell, D., Freeman, R. & Mueller, S. (Eds.), 164-180, University of Cambridge, Cambridge.

Franke, W. & Żelaźniewicz, A.,. (2000).The eastern termination of the Variscides: terrane correlation and kinematic evolution, *Orogenic processes:quantification and modelling in the Variscan belt, Geological Society Special. Publication*, Vol. 179, 63–86.

Franke, W., Dallmeyer, R. D. & Weber, K., 1995, Geodynamic Evolution, *Pre-Permian geology of Central and Eastern Europe, IGCP 233 international conference: Gottingen, Federal Republic of Germany: Berlin*, Dallmeyer, R. D.., Franke, W. & Weber, K., (Eds.), 579-593, Springer-Verlag.

Gaab, A., Poller, U., Todt, W. & Janák M. (2003). Geochemical and isotopic characteristics of the Murán Gneiss Complex, Veporic Unit (Slovakia), *Journal of the Czech Geological Society*, Vol., 48, No. 1-252.

Gallagher, J. J. & Tauver,s P. R. (1992). Tectonic evolution of northwestern South America, *Basement tectonics*. Mason, R. (Ed), 123-137.Kluver Academic Publishers..

Gawęda, A. (2008). An apatite-rich enclave in the High Tatra granite (Western Carpathians): petrological and geochronological study. Geologica Carpathica, Vol. 59, No. 4, 295-306.

Gawęda, A., Golonka J., (2011). Variscan plate dynamics in the Circum-Carpathian area. Travaux Géophysiques XL (2011), Abstracts of the 9th Central European Tectonic Groups meeting, Hotel Skalský Dvůr, Czech Republic, 13 - 17. April 2011, 19. Prague. Institute of Geophysics Academy of Sciences of the Czech Republic,.

Gawęda, A., Kozłowski, K., & Piotrowska, K. (2000). Early-Variscan collision and generation of leucogranite melts in the Western Tatra Mountains (S-Poland, W-Carpathians). *Journal of the Czech Geological Society, Vol.* 45, 230.

Golonka, J. (2000). *Cambrian-Neogene Plate Tectonic Maps*. 1-125, Wydawnictwa Uniwersytetu Jagiellońskiego, Kraków.

Golonka, J. (2002). Plate-tectonic maps of the Phanerozoic. *Phanerozoic reef pattern*, Kiessling W., Flügel E. & Golonka J. (eds.), *SEPM (Society for Sedimentary Geology) Special Publication*, Vol. 72, 21-75.

Golonka, J. (2004). Plate tectonic evolution of the southern margin of Eurasia in the Mesozoic and Cenozoic, *Tectonophysics*, Vol. 381, 235-273.

Golonka, J. (2007a). Late Triassic and Early Jurassic paleogeography of the world. *Palaeogeography, Palaeoclimatology, Palaeoecology*, Vol. 244, 297-307.

Golonka, J. (2007b). Phanerozoic Paleoenvironment and Paleolithofacies Maps. Late Paleozoic. Mapy paleośrodowiska i paleolitofacje fanerozoiku. Późny paleozoik. *Kwartalnik AGH. Geologia*, Vol .33, No. 2,: 145-209.

Golonka, J. (2007c). Phanerozoic Paleoenvironment and Paleolithofacies Maps. Mesozoic. Mapy paleośrodowiska i paleolitofacje fanerozoiku. Późny mezozoik. *Kwartalnik AGH. Geologia*, Vol. 33, No. 2, 211-264.

Golonka, J. (2009a). Phanerozoic Paleoenvironment and Paleolithofacies Maps. Cenozoic. Mapy paleośrodowiska i paleolitofacje fanerozoiku. Kenozoik. *Kwartalnik AGH. Geologia*, , Vol. 35 No. 4, 507-587.

Golonka, J. (2009b). Phanerozoic Paleoenvironment and Paleolithofacies Maps. Early Paleozoic. Mapy paleośrodowiska i paleolitofacje fanerozoiku. Wczesny paleozoik. *Kwartalnik AGH. Geologia*, , Vol. 35, No. 4, 589-654.

Golonka, J. (2012). *Paleozoic paleoenvironment and paleolithofacies maps of Gondwana*, Wydawnictwa AGH Publishing House. Kraków. In press

Golonka, J., Ross M. I. & Scotese C. R., 1994. Phanerozoic paleogeographic and paleoclimatic modeling maps. *Pangea: Global environment and resources*, Embry A. F., Beauchamp, B. & Glass, D. J., (Eds.),. *Canadian Society of Petroleum Geologists Memoir*, 17, 1-47.

Golonka, J., Krobicki, M.,Oszczypko, N., Ślączka, A. & Słomka, T. (2003). Geodynamic evolution and palaeogeography of the Polish Carpathians and adjacent areas during Neo-Cimmerian and preceding events (latest Triassic - earliest Cretaceous), *Tracing tectonic deformation using the sedimentary record*, McCan, T. & Saintot, A. (Eds.),*Geological Society Special Publications*. 208, 138-158.

Golonka J. Gahagan L., Krobicki M., Marko F., Oszczypko N. & Slaczka A. 2006a. Plate Tectonic Evolution and Paleogeography of the Circum-Carpathian Region. In: Golonka, J. & Picha, F. (Eds.) The Carpathians and their foreland: Geology and hydrocarbon resources. *American Association of Petroleum Geologists Memoir*, 84, 11-46.

Golonka, J., Krobicki, M., Pająk, J, Nguyen Van Giang & Zuchiewicz, W. (2006b). *Global plate tectonics and paleogeography of Southeast Asia*. Faculty of Geology, Geophysics and Environmental Protection, AGH University of Science and Technology, Arkadia, Kraków.

Gutiérrez-Alonso G., Murphy J.B., Fernández-Suárez J. & Hamilton M.A. (2008). Rifting along the northern Gondwana margin and the evolution of the Rheic Ocean: A Devonian age for the El Castillo volcanic rocks (Salamanca, Central Iberian Zone). *Tectonophysics*, Vol. 461, 157-165.

Hartz E.H., & Torsvik T.H. (2002). Baltica upside down: a new plate tectonic model for Rodinia and Iapetus Ocean. Geology, Vol. 30, No. 3, 255-258.

Hatcher R. D., Jr, Thomas. W. A., Geiser, P. A., Snoke, A. W., Mosher S. & Wiltschko D. V., (1989). Alleghenian orogen, *The Appalachian-Ouachita Orogen in the United States*, Hatcher R. D., Jr, Thomas, W. A. & Viele, G. W., (Eds.),V. F, . 233-318, Boulder, Geological Society of America, The Geology of North America,

Janák, M., Hurai, V., Ludhova, L., O`Brien, P.J., & Horn, E.E. (1999). Dehydration melting and devolatilization during exhumation of high-grade metapelites: the Tatra Mountains, Western Carpathians. *Journal of Metamorphic Geology*, Vol. 17, 379-395.

Janák, M., Finger, F., Plašienka, D., Petrik, I., Humer, B., Meres, S. & Luptak, B., (2002). Variscan high P-T recrystallization of Ordovician granitoids in Veporic Unit (Nizke Tatry Mountains, Western Carpathians): new petrological and geochronological data. *Geolines*, Vol. 14, 38-39.

Kalvoda, J. & Bábek O. (2010) .The Margins of Laurussia in Central and Southeast Europe and Southwest Asia. *Gondwana Research*, Vol. 17, 526-545.

Kamo S.L., Gowar C.F., & Krogh T.E. (1989). Birthdate for the Iapetus Ocean ? A precise U-Pb zircon and baddeleyite age for Long Range dikes, southeast Labrador. *Geology*, Vol. 17, No. 7, 602-605.

Kent, D. V. & Van der Voo, R., (1990). Palaeozoic palaeogeography from palaeomagnetism of the Atlantic-bordering continents, *Palaeozoic palaeogeography and biogeography*, McKerrow, W. S. & Scotese, C. R., (Eds.), *Geological Society Memoir*, Vol. 12, 49-56.

Keppie, J.D. (1985). The Appalachian collage, The Caledonide orogen, Scandinavia, and related areas, 1217-1226, Gee, D.G., & Sturt, B. (Eds)., New York, J. Wiley and Sons.

Keppie, J. D. (1989). Northern Appalachian terranes and their accretionary history, *Terranes in the Circum-Atlantic Paleozoic orogens*. Dallmeyer, R. D. (Ed), *Geological Society of America, Special Paper*, Vol. 230, 159-192.

Keppie, J. D., Dostal J., Murphy, J. B., & Nance, R. D. (1996). Terrane transfer between eastern Laurentia and western Gondwana in the Early Paleozoic: Constraints on global reconstructions, *Avalonia and Related Peri-Gondwanan Terranes of the Circum-North Atlantic*, Nance R. D. & Thompson M. D. (Eds), *Geological Society of America Special Paper*, Vol. 304, 369-380.

Kirby, G. A. (1979). The Lizard Complex as an ophiolite. *Nature*, Vol. 282, 58-61.

Kohut M., Konecny P. & Siman P. (2006) The first finding of the iron Lahn-Dill mineralization in the Tatric Unit of the Western Carpathians. *Mineralogia Polonica – Special Papers*, Vol.28, 112-114.

Kohut, M., Poller, U., Gurk, Ch. & Todt W. (2008). Geochemistry and U-Pb detrital zircon ages of metasedimentary rocks of the Lower Unit, Western Tatra Mountains (Slovakia), *Acta Geologica Polonica*, Vol. 58, 371-384.

Kryza, R. & Pin C. (2010). The Central Sudetic ophiolites (SW Poland): petrogenetic issues, geochronology and paleotectonic implications. *Gondwana Research*, Vol. 17, 292-305.

Lewandowski, M. (1998). Assembly of Pangea: Combined Paleomagnetic and Paleoclimatic Approach, *Circum-Arctic Palaeozoic Faunas and Facies*, Ginter, M. & Wilson, M. H. (Eds), *Ichtyolith Issues Special Publication*, Vol. 4, 29-32.

Lewandowski, M. (2003). Assembly of Pangea: Combined Paleomagnetic and Paleoclimatic Approach. *Advances in Geophysics*, Vol. 46, 199-236.

Mazur, S., Aleksandrowski, P., Kryza ,R. & Oberc-Dziedzic, T,.(2006). The Variscan Orogen in Poland, *Geological Quarterly*, Vol. 50 89–118.

McKerrow, W. S. , Dewey, J. F. & Scotese, C. R. (1991). The Ordovician and Silurian development of the Iapetus Ocean *The Murchison symposium; proceedings of an international conference on the Silurian System in* Bassett, M. G., Lane, P. D. & Edwards, D. (Eds.), *Special Papers Palaeontology*, Vol. 44, 165-178.

Moczydłowska, M. (1997). Proterozoic and Cambrian successions in Upper Silesia: an Avalonian terrane in southern Poland, *Geological Magazine*, Vol. 134, 679-689.

Munteanu M. & Tatu M. (2003.) The East-Carpathian Crystalline-Mesozoic Zone (Romania): Paleozoic amalgamation of Gondwana- and East European Craton- derived terranes. Gondwana Research, Vol. 6, No. 2, 185-196.

Nance, R.D., Gutiérrez-Alonso, G., Keppie, J.D., Linnemann, U., Murphy, J.B., Quesada, C., Strachan, R.A. & Woodcock, N.H., (2010). Evolution of the Rheic Ocean. *Gondwana Research*, Vol. 17, 194-222.

Pana, D., Ballintoni, I., Heaman, L. & Creaser R. (2002). The U-Pb and Sm-Nd dating of the main lithotectonic assemblages of the East Carpathians, Romania. *Geologica Carpathica*, Vol. 53, 177-180.

Putiš, M, Ivan, P, Kohút, M, Spišiak, J, Siman, P, Radvanec, M, Uher, P, Sergeev, S, Larionov, A, Méres, Š, Demko, R. & Ondrejka M. (2009). Meta-igneous rocks of the West-Carpathian basement, Slovakia: indicators of Early Paleozoic extension and shortening events. *Bulletin Societe Géolologique France*, Vol. 180, No.6, 461-471

Rast, N. & Skehan, J. W., 1993. Mid-Paleozoic orogenesis in the North Atlantic, the Acadian orogeny. *The Acadian Orogeny, recent studies in New England, Maritime Canada, and the autochtohonous foreland.* , Roy C. & Skehan J. W. (Eds), *Geological Society of America Special Paper*, Vol. 275, 1-25.

Robardet, M., Blaise, J., Bouyx, E., Gourvennec, R., Lardeux, H., Le Hérissé, A., Le Menn, J., Melou, M., Paris, F., Plusquellec, Y., Poncet J., Régnault, S., Rioult, M. & Weyant M. (1993). Paléogeographie de l'Europe occidentale de l'Ordovicien au Dévonien; Paleogeography of Western Europe from the Ordovician to the Devonian, *Bulletin Societe géologique de France*, Vol. 164, 683-695.

Schulmann, K., Konopásek, J., Janoušek, V., Lex, O., Lardeaux, J.-M., Ede,l J.-B., Štípská & P., Ulrich S. (2009). An Andean type Palaeozoic convergence in the Bohemian Massif. *Comptes Rendus – Geoscience*, Vol. 341, 266-286.

Scotese, C. R. & McKerrow, W. S. (1990). Revised world maps and introduction, *Palaeozoic palaeogeography and biogeography*, McKerrow W. S. & Scotese C. R. (Eds), *Geological Society of London Memoir*, 12, 1-21.

Scotese, C.R. & Barret, S.F. (1990). Gondwana's movement over the South Pole during the Paleozoic: evidence from lithologic indicators of climate. In: W.S. McKerrow and C.R. Scotese (Eds.) *Paleozoic Paleogeography and Biogeography*, , *Geological Society of London, Memoir* Vol. 12, pp. 75-85.

Sears, J.W. (2012). Transforming Siberia along the Laurussian margin. *Geology*, doi: 10.1130/G32952.1

Słaby, E., & Martin, H. (2005). Mafic and felsic magma interaction in granites: the karkonosze Hercynian pluton (Sudetes, Bohemian Massif). *Journal of Petrology*, Vol. 49, 353-391.

Štipská, P., Schulmann, K., Kroener, A. (1998). From Cambro-Ordovician rifting to Variscan collision at the NE margin of the Bohemian Massif: petrological, geochronological and structural constraints. *Paleozoic orogenesis and crustal evolution of the European lithosphere – post-conference excursion*, pp. 24-31.

Tichomirova, M., (2002). Zircon inheritance in diatexite granodiorites and its consequence on geochronology – a case study in Lusatia and Erzgebirge (Saxo-Thuringia, eastern Germany). *Chemical Geology*, Vol. 191, 209-224.

Torsvik, T. H. , Smethurst, M. A., Meert, J. G., Van der Voo, R., McKerrow, W. S., Brasier, M. D. & Sturt, B. A., Walderhaug, H. J. (1996). Continental break-up and collision in the Neoproterozoic and Palaeozoic; a tale of Baltica and Laurentia, *Earth-Science Reviews*, Vol. 40, No.3-4, 229-258.

Turniak, K., Mazur, S., & Wysoczański, R. (2000). SHRIMP zircon geochronology and geochemistry of tyhe Orlica snieznik gneisses (Variscan Belt of Central Europe) and their tectonic implications. *Geodinamica Acta*, Vol. 13, 293-312.

Veevers, J.J. (2004). Gondwanaland from 650-500 Ma assembly through 320 Ma merger in Pangea to 185-100 Ma breakup: Supercontinental tectonics via stratigraphy and radiometric dating. *Earth-Science Reviews*, Vol. 68, 1-132.

von Raumer, J.F., & Stampfli, G.M. (2008). The birth of the Rheic Ocean – Early Paleozoic subsidence patterns and subsequent tectonic plate scenario, *Tectonophysics*, Vol. 461, 9-20.

Williams, K. E., (1995). Tectonic Subsidence Analysis and Paleozoic Paleogeography of Gondwana, *Petroleum basins of South America*, Tankard, A.J., Suarez, S. & Welsink, H. J. (Eds),*American Association of Petroleum Geologists Memoir*, Vol. 62, 79-100.

Winchester J.A., Floyd P.A., Crowley Q.G., Piasecki M.A.J., Lee M.K., Pharaoh T.C., Williamson P., Banka D., Verniers J., Samuelsson J., Bayer U., Marotta A.-M., Lamarche J., Franke W., Dörr W., Valverde-Vaquero P., Giese U., Vecoli M., Thybo H., Laigle M., Scheck M., Maluski H., Marheine D., Noble S.R., Paarish R.R., Evans J., Timmerman H., Gerdes A., Guterch A., Grad M., Cwojdzinski S., Cymerman Z., Kozdroj W., Kryza R., Alexandrowsk, P., Mazur S., Stedrá V., Kotková J., Belka Z., Patoćka F. & Kachlik V.,(2002). Palaeozoic amalgamation of Central Europe: New results from recent geological and geophysical investigations. *Tectonophysics*, Vol. 360, 5-21.

Young, G. C., (1990). Devonian vertebrate distribution patterns and cladistic analysis of paleogeographic hypothesis, *Palaeozoic palaeogeography and biogeography*. McKerrow W. S. & Scotese C. R. (Eds), *Geological Society of London Memoir*, Vol. 12, 243-255.

Zakariadze, G. S.; Karamata, S. O. & Dilek, Y. (2007). Significance of E. Paleozoic Paleo-Tethyan Ophiolites in the Balkan Terrane and the Greater Caucasus for the Cadomian-Hercynian Continental Growth of Southern Europe . *American Geophysical Union, Fall Meeting 2007*, Abstract #V43B, 1367, American Geophysical Union, Washigton D.C.

Zeh, A. & Gerdes, A. (2010). Baltica- and Gondwana-derived sediments in the Mid-German Crystalline Rise (Central Europe): Implications for the closure of the Rheic ocean, *Gondwana Research*, Vol. 17, 254-263).

Ziegler, P.A., 1989. *Evolution of Laurussia*. Kluwer Academic Publishers, Dordrecht.

Zonenshain, L. P., Kuzmin, M. L. & Natapov, L. N., 1990. Geology of the USSR: A Plate-Tectonic Synthesis. Page, B. M. (Ed), *Geodynamics Series, American Geophysical Union*, Vol. 21, 1-242.

Żelaźniewicz, A., Seghedi, A., Żaba, J., Fanning, M. &, Buła Z. (2009) More evidence on Neoproterozoic terranes in Southern Poland and southeastern Romania, *Geological Quarterly*, Vol. 51, 93-124.

Chapter 11

Was the Precambrian Basement of Western Troms and Lofoten-Vesterålen in Northern Norway Linked to the Lewisian of Scotland? A Comparison of Crustal Components, Tectonic Evolution and Amalgamation History

Steffen G. Bergh, Fernando Corfu, Per Inge Myhre,
Kåre Kullerud, Paul E.B. Armitage, Klaas B. Zwaan, Erling K. Ravna,
Robert E. Holdsworth and Anupam Chattopadhya

Additional information is available at the end of the chapter

1. Introduction

Temporal and spatial linkage of Archaean and Palaeoproterozoic crustal provinces in the North Atlantic realm requires a well-established geological and geodynamic framework. Such a framework is well established for the Fennoscandian Shield of Finland, Sweden and northwestern Russia [1-4], for Greenland/Laurentia [5, 6] and for the Lewisian of NW Scotland [7, 8], but not yet for the Precambrian crystalline rocks within and west of the Scandinavian Caledonides in North Norway (Figure 1).

In western Troms (Figure 2) and in the Lofoten-Vesterålen areas of North Norway (Figure 3) Neoarchaean and Palaeoproterozoic continental crust (2.9-1.67 Ga) is preserved as an emerged basement horst bounded to the east by thrust nappes of the Scandinavian Caledonides (Figure 1b) [9-11] and to the west by offshore Mesozoic basins [12]. These basement outliers are believed to be part of the Archean-Palaeoproterozoic Fennoscandian Shield [1, 11, 13] that stretches from NW Russia, through Finland and Sweden (Figure 1a). Similarly, a pronounced magmatic suite in the Lofoten area [9, 10] corresponds in age (1.80-1.78 Ga) and structural position with the NNW-trending Transscandinavian igneous belt of Sweden [14, 15]. In spite of the internal position relative to the Caledonides, and in great contrast to the basement inliers in southern Norway where Caledonian high-grade metamorphic reworking is widespread, the geotransect in western Troms and Lofoten displays only modest Caledonian reworking and, thus, provides a reliable framework for regional correlation of Neoarchaean and Palaeoproterozoic crust [9,10, 16,17].

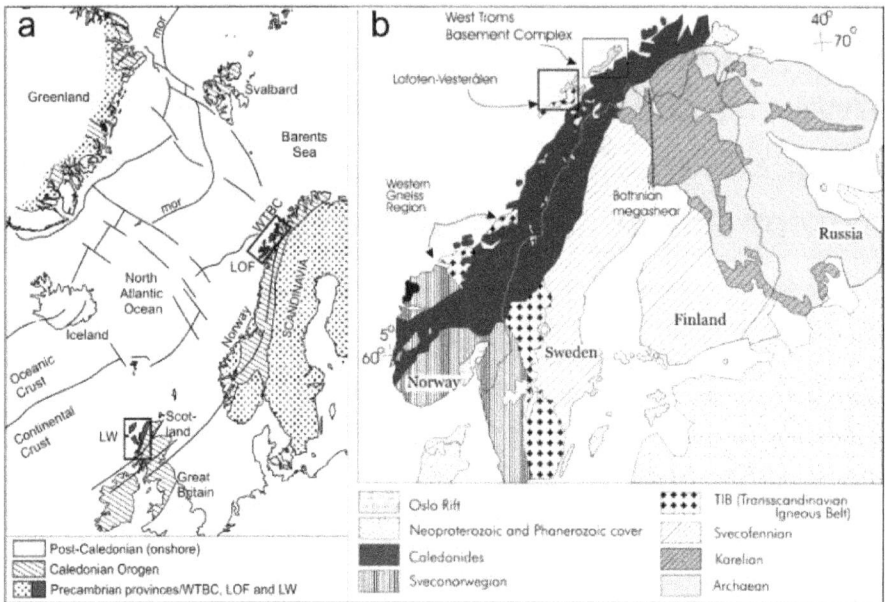

Figure 1. (a) Location of Precambrian basement outliers in western Troms (WTBC), Lofoten-Vesterålen (LOF) and NW Scotland (LW) and related basement provinces within today's plate setting of the North Atlantic Ocean. (b) Geologic map of Fennoscandia with location of the West Troms Basement Complex and the Lofoten-Vesterålen province west of the Scandinavian Caledonides [16].

In NW Scotland, the Lewisian Complex is situated to the west of the Caledonian Moine thrust and covers the Outer Hebrides, the NW Scottish mainland and part of the Inner Hebrides (Figures 1a and 4). The Lewisian rocks also form inliers in the Caledonian orogenic belt, possibly continuing under the Moine Group up to the Great Glenn fault in the southeast, while Mesozoic to early Cenozoic extensional basins offshore Scotland largely disrupted the continuity of the Lewisian outcrops [18, 19, 20].

A possible linkage of the western Troms and Lofoten-Vesterålen basement rocks with the Lewisian basement inliers of the Caledonides in NW Scotland (Figure 1a) and with Laurentia-Greenland has been raised [3, 4, 8, 21, 22], but excact correlation of these cratonic-marginal provinces in the North Atlantic realm, their role during assembly of Fennoscandia and Laurentia in the Neoarchaean [23, 24], and the situation prior to Palaeoproterozoic orogenies [3, 25, 26], still remain enigmatic.

This paper reviews the current knowledge of the crustal components, tectono-magmatic evolution and amalgamation history of the basement rocks in western Troms and Lofoten-Vesterålen, North Norway, and compares them with the Lewisian of Scotland (Figures 2-4, Table 1). New and focused structural and geochronological work in the West Troms

Basement Complex [16, 17, 27-32] has sparked off new interest in these provinces. Questions specifically raised for these basement suites concern the age and nature of supracrustal units, and the character of crustal-scale ductile shear zones, either as potential terrane boundaries between assembled older crustal blocks or just reflecting episodes of basin formation and later reworking. Such boundaries can, in general, help to restore the outline and correlation of each craton and the cratonic margin characteristics and to unravel cycles of tectono-magmatic events [33].

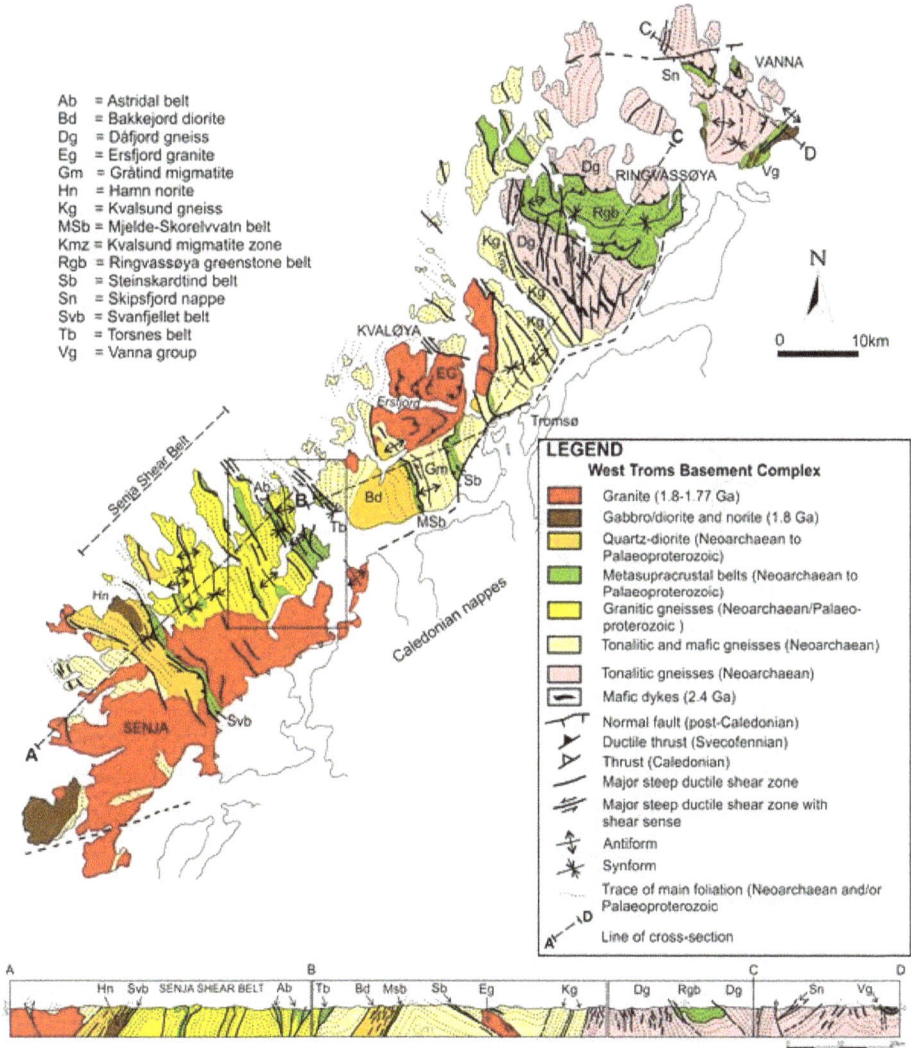

Ab = Astridal belt
Bd = Bakkejord diorite
Dg = Dåfjord gneiss
Eg = Ersfjord granite
Gm = Gråtind migmatite
Hn = Hamn norite
Kg = Kvalsund gneiss
MSb = Mjelde-Skorelvvatn belt
Kmz = Kvalsund migmatite zone
Rgb = Ringvassøya greenstone belt
Sb = Steinskardtind belt
Sn = Skipsfjord nappe
Svb = Svanfjellet belt
Tb = Torsnes belt
Vg = Vanna group

LEGEND

West Troms Basement Complex

Granite (1.8-1.77 Ga)
Gabbro/diorite and norite (1.8 Ga)
Quartz-diorite (Neoarchaean to Palaeoproterozoic)
Metasupracrustal belts (Neoarchaean to Palaeoproterozoic)
Granitic gneisses (Neoarchaean/Palaeoproterozoic)
Tonalitic and mafic gneisses (Neoarchaean)
Tonalitic gneisses (Neoarchaean)
Mafic dykes (2.4 Ga)
Normal fault (post-Caledonian)
Ductile thrust (Svecofennian)
Thrust (Caledonian)
Major steep ductile shear zone
Major steep ductile shear zone with shear sense
Antiform
Synform
Trace of main foliation (Neoarchaean and/or Palaeoproterozoic
Line of cross-section

Figure 2. Regional map of the West Troms Basement Complex, North Norway, that shows the main crustal components and tectonic features, with a generalized cross-section. Frame shows location of Figure 10a. The map is revised after [17, 34].

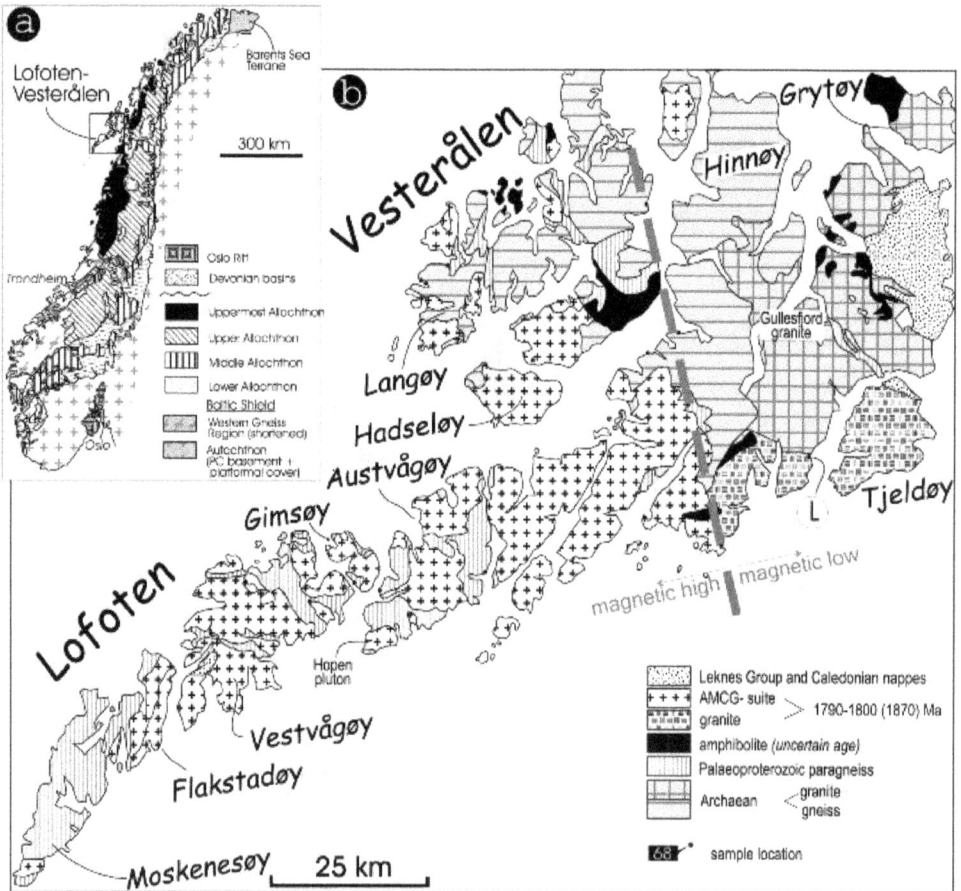

Figure 3. Geological map of the Lofoten-Vesterålen province, showing the Neoarchaean-Palaeoproterozoic basement rocks and the anorthosite-mangerite-charnockite-granite igneous suite [10, 35, 36]

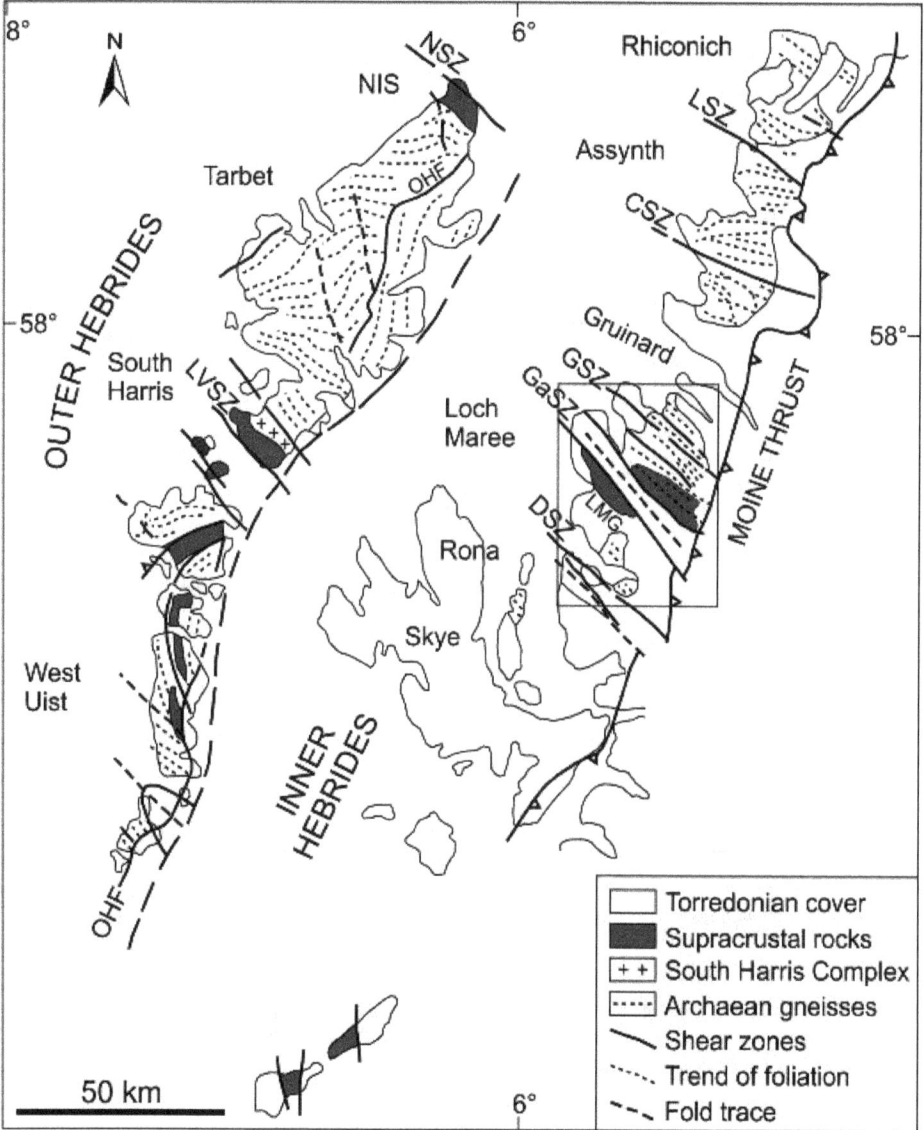

Figure 4. Simplified geologic and structural map of the Lewisian Complex in northwest Scotland showing the main rock units and the overall subdivision of the mainland into multiple regions (or terranes)
separated by main Palaeoproterozoic ductile shear zones. The map is modified from from [20, 37].
Abbreviations: LSZ=Laxford shear zone, CSZ=Canisp shear zone, GaSz=Gairloch shear zone,
DSZ=Diabaig shear zone. Frame shows outline of figure 18.

West Troms Basement Complex		Lofoten-Vesterålen Province		Lewisian Complex	
Age (Ga)	Components and events	Age (Ga)	Components and events	Age (Ga)	Components and events
2.92-2.8 Ga	**Neoarchaean cratonization:** -Tonalite crystallization (*Dåfjord & Kvalsund gneisses*) Volcanism and sedimentation: - *Ringvassøya greenstone belt*	2.85-2.7 Ga	**Neoarchaean cratonization:** Accretion, convergence and crustal thickening	3.145-2.75 Ga	**Neoarchaean cratonization:**
2.85-2.83 Ga	**Continued Neoarchaean cratonization** -Mafic plutonism (*Bakkejord*		-Tonalite magmatism	2.9-2.7 Ga	-Various types of *TTG*-gneisses, protoliths of tonalite (*Scourian Gneiss*), granites (*Laxfordian Gneiss*),
2.75-2.6 Ga	*diorite*) in the southwest	2.75-2.68 Ga			Regional high-grade metamorphism, crustal
2.75-2.7 Ga	Neoarchaean deformation and metamorphism: -Magmatism, migmatization (*Gråtind migmatite*) and ductile shearing (in *Dåfjord*	2.72-2.66 Ga	**Neoarchaean deformation and metamorphism:** - Various orthogneisses (e.g. *Bremnes gneiss*) formed by crustal	2.8 Ga?	accretion (?) and thickening **Volcanism and sedimentation:** - Mafic/ultramafic
2.7-2.67 Ga	and *Kvalsund* gneisses) - Main gneiss foliation (initially flat-lying), ductile shear zones, tight folds and dip-slip stretching lineation. - Medium/high-grade metamorphism, ENE-WSW crustal contraction and thickening by accretion and/or underplating	2.64 Ga	shortening - Emplacement of tonalities, followed by high-grade metamorphism and localized migmatization and ductile crustal shearing (*Sigerfjord migmatite, Ryggedalen granulite*)	2.7-2.6 Ga	volcanics and supracrustal rocks (*Eanruig paragneiss, Claisfearn supracrustals*). Neoarchaean deformation and metamorphism: - Subhorizontal foliation & tight folding and thrusting, regional granulite facies metamorphism. NE-SW shortening
2.69-2.56 Ga	-High-grade metamorphism and resetting			2.49-2.4 Ga	-Open macro-folds and axial-planar dextral transpressive shear zones (*Canisp* and *Laxford*? shear zones). - Reworking and retrogression to amphibolite facies
2.40 Ga	**Crustal extension and intrusion** of the *Ringvassøya* mafic dyke swarm	?		2.4-2.0 Ga	**Crustal extension and intrusion** of the *Scourie mafic dyke swarm* and Na-rich pegmatites. Dextral transtensional setting.
2.4-2.2 Ga	**Deposition** of *Vanna group* clastic sediments in a marine subsiding basin	?	**Deposition** of supracrustal units, heterogeneous mafic gneisses, banded iron formations, quartzite, marbles, graphite schists	2.2-1.9 Ga	**Deposition** of the *Loch Maree Group* clastic and volcanic succession in marine extensional and/or arc-settings -Marine mudstones (*Flowerdale schists, Aundrary amphibolites*)
2.22 Ga	**Intrusion** of *Vanna diorite* sill	?		?	
2.2-1.9 Ga	**Deposition** of *Mjelde-Skorelvvatn, Torsnes* and possibly, the *Astridal* supracrustal belts	1.87-1.86 Ga	**Precursory stage** magmatism, AMCG-suite, *Lødingen granite*	1.9-1.8 Ga	**Intrusion** of early-stage *granites*

West Troms Basement Complex		Lofoten-Vesterålen Province		Lewisian Complex	
1.993 Ga	Intrusion/volcanism in the *Mjelde-Skorelvvatn belt*	1.8-1.79 Ga	Main stage intrusion of AMCG-suite plutonic rocks and the *Tysfjord Granite*	1.9-1.87 Ga	Kalk-alkaline volcanism: *Ard kalk-alkaline gneisses* Magmatism: *South Harris Igneous Complex* (Hebrides) Intrusion of late-stage granites
1.80 Ga	Magmatism/intrusion of granites and *norite* in Senja				
1.79 Ga	Magmatism/intrusion of *Ersfjord Granite* in Kvaløya	1.77 Ga	Intrusion of granite pegmatite dykes		
				1.7-1.68 Ga	
c. 1.9-1.7 Ga	Palaeoproterozoic deformation:		Palaeoproterozoic deformation	1.97-1.67 Ga	Palaeoproterozoic deformation:
1.9-1.8 Ga?	Early: -Mylonitic foliation (initially flat-lying), NW-SE trending gently-plunging isoclinal folds, NE-directed ductile thrusts with dip-slip stretching lineation. Prograde medium/high-grade metamorphism in the southwest. NE-SW orthogonal shortening, NE-directed thrusting/accretion	1.87-1.79 Ga	Strong ductile deformation and high-grade (granulite facies) metamorphism and reworking	1.8-1.85 Ga?	Early stage -Deep thrusting (*Gairloch* shear zone) and amphibolite facies metamorphism, accretion of the Loch Maree Group onto the continental crust, in subduction or arc-setting.
		1.83 Ga	Major ductile shear zone (suture)? Arc-related and/or collisional setting		
					Mid/main stage - Isoclinal folding and thrusting, upright macro-folding and transpressive shear zones (*Canisp, Laxford* and *Shieldaig* shear zones), Accretionary tectonic event
1.78-1.77Ga	Mid: -Regional open upright folding, NW-SE trend, flat-lying hinges and steep limbs. Medium/low grade metamorphism. Continued NE-SW orthogonal crustal shortening	1.78-1.76 Ga	Retrogressive metamorphism of AMCG-suite rocks	1.85 Ga	
c. 1.75 Ga?	Late: -Regional steep N-plunging folds, NW-SE striking, steep ductile shear zones (strike-slip). -Retrogressive low grade metamorphism.				Main/late stage - Partitioned deformation, thrust and steep dextral-oblique strike-slip shear zones on a flat detachment (on *Laxford* and *Canisp* shear zones), Amphibolite facies metamorphism, likely collisional event
				1.70-1.67 Ga	
1.7-1.67 Ga?	Latest: -NE-SW trending upright folds of the *Vanna group* and SE-directed thrusts, steep semi-ductile strike-slip shear zones. Retrogressive low grade metamorphism Partitioned NE-SW shortening and orogen-parallel (NW-SE) strike-slip shearing				Late-stage NW-SE steep transpressive shear zones, retrogression to greenschist facies, crustal rejuvenatiion.
	Intrusion of felsic pegmatites and retrogression				
1.57 Ga				1.6 Ga?	Crustal uplift, retrograde metamorphism, cooling

Table 1. Summary and comparison of tectono-magmatic components and events in the West Troms Basement Complex, the Lofoten- Vesterålen area and the Lewisian of Scotland. The data is based on references listed and discussed in the text.

2. Geological features of western Troms and Lofoten

Archaean and Palaeoproterozoic crust underlies much of the northeastern part of the Fen-noscandian Shield [1, 3], including also basement outliers west of the much younger, Palae-ozoic Scandinavian Caledonides (Figure 1). Here, the West Troms Basement Complex [17] and basement in Lofoten-Vesterålen [9, 10] emerge as a c. 300 km long horst, separated from the Caledonian nappes by Mesozoic rift-normal faults [11, 12]. The West Troms Basement Complex (Figure 2) is composed of various Mesoarchaean to Palaeoproterozoic plutonic rocks and orthogneisses (2.9-1.7 Ga), metasupracrustal rocks, mafic dyke swarms, and net-works of ductile shear zones [17, 38]. The basement of Lofoten and Vesterålen (Figure 3) consists of similar metamorphic Neoarchaean rocks intruded by a very extensive suite of 1.80-1.78 Ga plutonic rocks of the anorthosite-mangerite-charnockite-granite (AMCG) suite [9, 10] that appears to link up with the Transscandinavian Igneous Belt of southern Sweden (Figure 1a). Both areas display dominant NW-SE structural trends parallel with Archaean and Palaeoproterozoic orogenic belts of the Fennoscandian Shield that stretch from Russia, through Finland and northern Sweden into the Bothnian basin of central Scandinavia [1].

2.1. The West Troms Basement Complex

The West Troms Basement Complex (Figure 2) is underlain by Meso to Neoarchaean gneisses, various generations of metasupracrustal rocks and mafic dyke swarms that were later intruded by felsic and mafic plutonic suites and variably reworked, deformed and metamorphosed during the main Palaeoproterozoic (Svecofennian) orogeny [17].

2.1.1. Archaean crust

The Meso-Neoarchaean rocks of the West Troms Basement Complex (Figure 2) consist of tonalite-trondhjemite and anorthositic gneisses with mafic and ultramafic layers and banded intercalations (Figure 5a) and are overlain by the Neoarchaean Ringvassøya greenstone belt. These rocks were deformed and metamorphosed up to granulite/migmatite facies prior to deposition of Palaeoproterozoic cover units and intrusion of a 2.4 Ga mafic dyke swarm [17, 39]. A steep NW-SE trending transposed gneiss foliation with dip-slip stretching lineations (Figure 5b) and tight ENE-vergent intrafolial folds (Figure 5c) attests for WSW-ENE contraction and thrusting during the Meso/Neoarchaean [17]. Prominent high-grade migmatite zones interpreted as a ductile shear zone (Figure 5d) separate compositionally different gneisses [17], e.g. the Kvalsund migmatite zone separating the Dåfjord and Kvalsund gneisses on Ringvassøya [36] and similar zones within the Senja Shear Belt [38, 40]. Polyphase refolding and thrusting is common and suggests protracted Neoarchaean deformation [17]. In Ringvassøya, tonalitic orthogneisses and granitoids (Dåfjord gneiss) reveal U-Pb zircon crystallization ages of 2.92-2.77 Ga (Figure. 6) [36, 41], and these rocks are considered to be related to tonalites on the island of Vanna farther north, where a U-Pb crystallization age of 2885 ± 20 Ma has been obtained [33]. This Mesoarchaean basement was also intruded by the 2695 ±15 Ma Mikkelvik alkaline stock [42]. The overlying Ringvassøya Greenstone belt [43] comprises arc-related meta-volcanic rocks with MORB-transitional,

Was the Precambrian Basement of Western Troms and Lofoten-Vesterålen in Northern Norway Linked to
the Lewisian of Scotland? A Comparison of Crustal Components, Tectonic Evolution and Amalgamation History

291

tholeiitic to calk-alkaline affinity. Two meta-volcanic rocks of the Ringvassøy greenstone
belt yield ages of c. 2.85 Ga [44].

Figure 5. Outcrop features of Meso-Neoarchaean tonalitic gneisses in the West Troms Basement Complex: (a) Foliated tonalitic and mafic gneisses (Dåfjord gneiss) in central part of Ringvassøya. (b) Steep
SW-dipping foliation in tonalitic Dåfjord gneiss with dip-slip stretching lineation. (c) Banded tonalitic
gneiss with tightly folded mafic inclusions, cut by granitic pegmatite veins presumed to be related to
the 1.79 Ga Ersfjord granite. (d) Zone of major migmatized Kvalsund gneiss in southwestern part of
Ringvassøya. The zone is cut by a mafic dyke, which is part of the Ringvassøya dyke swarm dated at 2.4
Ga [39].

By contrast, all the dated metaplutonic rocks on the islands of Kvaløya and Senja farther
south (Figure 2) are Neoarchaean. The Bakkejord pluton, neosome in the Kattfjord gneiss,
and granodiorite units bordering the Torsnes belt on Kvaløya, as well as several major intrusive bodies on Senja all yield ages between 2.72 and 2.68 Ga [16, 32]. A somewhat younger element at 2.67 Ga is shown by mafic dykes that cut the Bakkejord pluton on Kvaløya
[32]. The only potentially younger Archaean event is the formation of migmatites in southern Senja, where zircon in a neosome suggests crystallization at c. 2.6 Ga [32]. The Kvalsund
migmatite zone in southwestern Ringvassøya (Figure 2) appears to represent a boundary
separating the Mesoarchean crust to the north from Neoarchean crust to the south. The time
of deformation is not yet dated, but dynamic melting structures in the migmatite are cut by
mafic dykes considered to belong to the 2.4 Ga swarm, hence indicating an Archean age of
shearing. One sample of neosome has a primary age of about 2.7 Ga indicated by zircons,
which, however, also records an event ≤ 2.55 Ga possibly reflecting the time of deformation
[32].

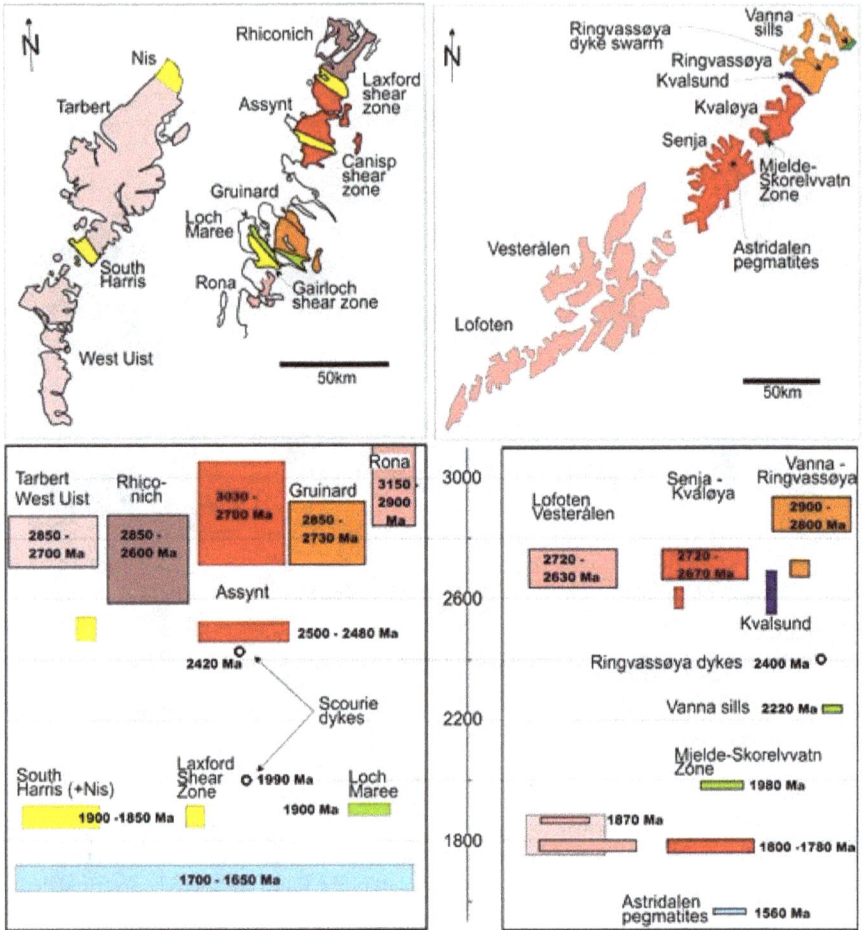

Figure 6. Summary of the main geochronological features in the Lewisian Complex (a) and the West-Troms Basement Complex and Loften province (b). The age compilation is based principally on U-Pb data. The references are listed and discussed in the text.

2.1.2. Ringvassøya mafic dyke swarm

The Neoarchaean gneisses and the oldest metasupracrustal belts of the West Troms Basement Complex have been intruded by a huge mafic, plagioclase phyric and gabbronoritic dyke swarm (Figure 7), the Ringvassøya dykes [39]. These dykes are widely distributed and display offset, shearing and reworking, thus providing a good time-marker for resolving the subsequent Svecofennian deformation [17, 27]. Zircon and baddeleyite from a dyke swarm on Ringvassøya provides a crystallization age of 2403 ± 3 Ma, and the dykes are classified as transitional between MORB and within-plate basalts with an affinity to continental tholeiites [39].

Figure 7. Outcrop features of the 2.4 Ga mafic dyke swarm that intruded Neoarchaean massive tonalites in Ringvassøya [39]. Note irregular and varied dyke orientations (a, b) and that the dykes truncate the main tonalitic gneiss fabric (c).

2.1.3. Palaeoproterozoic supracrustal rocks

The metasedimentary Vanna group represents unconformable continental deposits [29, 45]. Common rock types include layered meta-psammites locally exhibiting pronounced crossbedding (Figure 8a, b). The age of deposition is constrained between 2403 Ma, the age of the underlying Ringvassøya dykes, and 2221 ± 3 Ma, the age of a diorite sill in the supracrustal rocks [29, 39].

Figure 8. Outcrop features of meta-supracrustal rocks in the West Troms Basement Complex: (a) Layered meta-psammites with interbedded mudstones/mica-schists of the Vanna group [29, 45]. (b) Subvertical thinly bedded meta-sandstone of the Vanna group, with pronounced trough and planar cross-bedding in internal lenses. Up is to the right. (c) Basal meta-conglomerate of the Torsnes supracrustal belt. Note that the dominant clast-type is tonalite, tonalitic gneiss and granitoid gneisses (d) Rythmical laminated quartz-feldspatic meta-psammite with enrichment of iron-hydroxide staining from the Astridal belt, Senja. The beds are steeply dipping and subvertically folded. The fold hinge is located near person.

Arc-related meta-volcanic rocks with MORB-transitional, tholeiitic to calk-alkaline affinity occur in the central and southwestern parts, e.g. in the Astridal and Torsnes belts [17, 27, 31]. The Mjelde-Skorelvvatn belt (Figure 2) is dominated by metabasaltic rocks together with ultramafic rocks, meta-psammites, marble and calc-silicate gneisses. A differentiated gab-broic portion of the meta-basaltic pile yields an age of 1992 ± 2 Ma [31]. The Torsnes belt comprises a basal conglomerate (Figure 8c) overlain by meta-psammite and a thick sequence of mafic metavolcanic rocks. Detrital zircons indicate a maximum-age of deposition of 1970 ±14 Ma [31]. The Astridal belt is compositionally similar to the Mjelde-Skorelvvatn belt, but there are no direct radiometric dates yet. It contains abundant mica-schists, local graphite schists, mafic meta-volcanic rocks and widespread sulphide ore deposits (Figure 8d).

2.1.4. Late palaeoproterozoic igneous suites

The Neoarchaean crust in Kvaløya and Senja was intruded by an extensive suite of felsic and mafic plutonic rocks [17]. The most prominent are the Ersfjord granite (Figure 9a, b) on Kvaløya [46] and large granites and mafic plutons (Hamn norite) on Senja (Figure 2). The Ersfjord granite has a U-Pb zircon crystallization age of 1792 ± 5 Ma [16] and the Hamn norite 1802.3 ± 0.7 Ma [28], whereas the granitic masses farther south in Senja give Rb-Sr [47] and zircon-titanite ages of 1805 ± 2.5 Ma [16]. Metamorphic overprints of the Ersfjord granite are recorded by U-Pb titanite ages of 1769 ± 3 Ma and 1756 ± 3 Ma [16]. Pronounced and widespread granite pegmatite dykes (Figure 9b, c) formed syn-tectonically with shear zones in the metasupracrustal belts at c. 1768 ± 4 Ma, probably genetically related to the main intrusive activity [16]. All these ages are within the interval when most known Precambrian juvenile crust generated by arc-related magmatism [48].

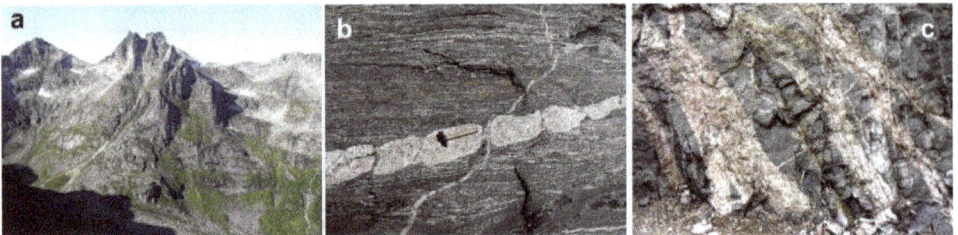

Figure 9. (a) Aerial view of the Ersfjord granite with its rugged mountains and presence of both steep and gently dipping planar fabrics. (b) Ersfjord granite pegmatite dykes and veins cutting Neoarchaean tonalitic gneiss foliation in southwestern ringvassøya, and later boudinaged during Palaeoproterozoic tectonism. (c) Ersfjord granite pegmatite dykes cutting dioritic gneisses near its western boundary against the Kattfjord gneiss.

2.1.5. Palaeoproterozoic deformation and metamorphism

Strong deformation and metamorphism at 1.8-1.76 Ga produced mega-blocks or segments delineated by NW-SE trending, variously mylonitized metasuprascrustal belts and lens-shaped ductile shear zones as outlined in Figure 10 [17, 38]. This shear belt deformation was characterized by high-strain, complex and multiphase deformation and up to amphibolite facies metamorphism and reworking. The most distinct one, the Senja Shear Belt is c. 30 km wide (Figures 2 and 10) and is delimited by the Svanfjellet belt to the south [38, 49] and the Torsnes belt in the north [17]. This linear crustal structure is thought to continue beneath the Caledonian nappes into parallelism with the Bothnian-Senja shear zone of the Swedish part of the Fennoscandian shield, but gravity and magnetic anomaly patterns do not uniquely confirm such a correlation [11, 13, 50].

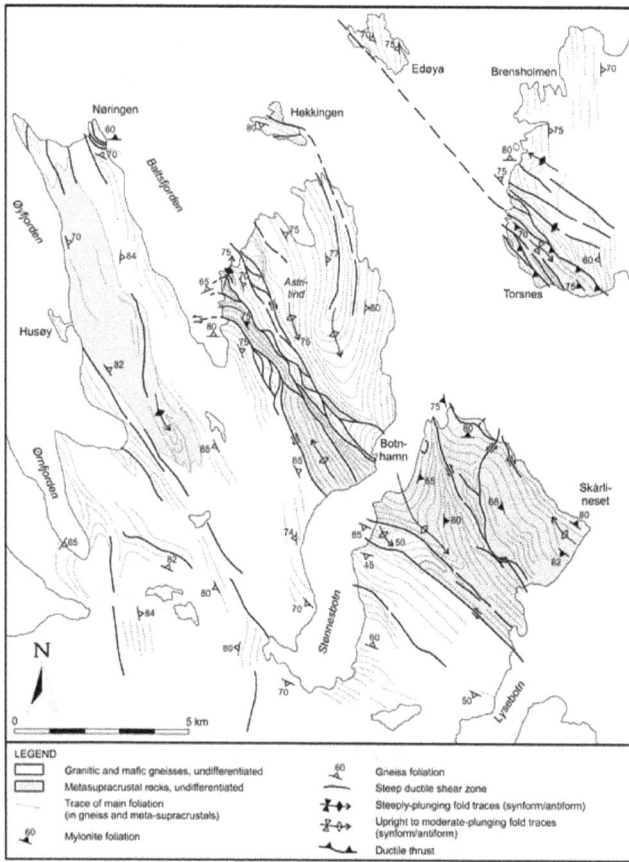

Figure 10. Tectonic map of the Senja Shear Belt in central part of the West Troms Basement Complex (for location see figure 2), illustrating the lens-shaped architecture of the metasupracrustal belts. Note macro-scale, steep-plunging folds of the belt and adjacent tonalitic gneisses, where fold hinges are bent into parallelism with the trace of the Astridal belt. Note also the major shear zone boundaries with thrust and sinistral strike-slip characters. The map is from [17].

The Palaeoproterozoic deformation structures of the West Troms Basement Complex (Table 1) include a main NNW-SSE striking, mylonitic foliation mostly present in the meta-supracrustal belts (Figure 11a), that formed axial-planar to early-stage isoclinal folds (Figure 11b) with gently-plunging axes, at amphibolite facies conditions. The foliation has a steep

WSW dip and exhibits a dip-slip, west-plunging stretching lineation. Macroscopic NW-SE-trending and mostly upright antiform-synform folds (Figure 11c) are widespread, causing the steep tilt and apparent repetition of most of the supracrustal belts in synformal troughs [17]. Corresponding upright folds also occur in the adjacent gneisses. Younger (Late Palaeo-proterozoic) superimposed structures include tight to vertical folds on all scales (Figure 11d, e) with axial-planar cleavages coeval with an anastomosing network of steeply-dipping, NW-SE to N-S trending sinistral and dextral strike-slip shear zones (Figure 10). The latter zones are mylonitic and retrogressed into greenschist facies. These shear zones caused sub-vertical drag-folding of the surrounding gneisses and became boudinaged and masked by quartz precipitates along the main foliation. Later on the foliation was folded by steeply north-plunging shear folds and cut by oblique-slip crenulation cleavages, sigmoidal clasts and multiple shear bands, all supportive of strike-slip displacements (Figure 11d, e). The youngest set of structures occurs in the northeastern parts of the West Troms Basement Complex, as gently-dipping, SE-directed phyllonitic shear zones (thrusts) with abundant SE-verging folds and thrusts [17].

Figure 11. Outcrop illustrations of Palaeoproterozoic deformation structures in the West Troms Base-ment Complex: (a) Subvertical high-strain mylonites along the eastern contact between the Torsnes belt and the granitoid Kattfjord gneiss (right). (b) Foliation-parallel felsic vein in garnet-mica-schists of the Astridal bel that is isoclinally folded and sinistrally sheared, causing multiple repeatitions of the vein. View is to the NE on near-horizontal surface. (c) Upright asymmetrical NE-verging fold that refolds isoclinal folds in meta-volcanic and siliciclastic rocks of the Svanfjellet belt, Senja. (d) Subvertical phyl-lonitic shear fabric with quartz precipitates along the main foliation.Note dextral subvertical folding of the main fabric and quartz veins. (e) Ersfjord granite pegmatite dyke in mafic Kattfjord gneiss northeast of the Torsnes belt that is folded by subvertical sinistral folds.

The early and middle stages of deformation were associated with prograde metamorphism varying from low grade in the northeast to amphibolite and granulite facies in the central and southern parts and terminated with late stage retrogressive greenschist facies reworking [27]. A migmatitic shear zone in meta-psammites southwest of Ringvassøya displays a garnet granulite facies assemblage succeeded by a two-pyroxene granulite facies assemblage and is dated with the zircon U-Pb method to 1777 ± 12 Ma, the same age obtained for a granitic dike (1776.6 ± 1.1 Ma) cutting the lens [51]. Other critical radiometric ages for Palaeoproterozoic high-strain deformation include metamorphic overprints of Ersfjord granite pegmatite dykes at c. 1.77-1.76 Ga, and an interval of 1774, 1768 ± 4 Ma and 1751 ± 8 Ma for a granitoid pegmatite dyke formed syn-kinematically with late-stage strike slip shear zones in the Astridal belt [16].

2.1.6. Mesoproterozoic reactivation

Late-tectonically deformed granitic pegmatites in the Astridal belt of the Senja Shear Belt yield U-Pb ages of 1725 ±22 Ma and 1562 ± 2 Ma, indicating that formation of the pegmatites occurred after termination of the main orogenic events (Corfu et al. in review). These occurrences are attributed to 'anorogenic' far-field effects reflecting intracratonic strain, possibly caused by the emplacement of A-type massifs in the core of the Fennoscandian Shield.

2.2. The Lofoten-Vesterålen province

This province comprises gneisses and major plutonic suites of Precambrian age [52] that suffered major tectono-magmatic events at 2.8 and 1.8 Ga [53]. The basement complex in western parts of Lofoten and Vesterålen (Figure 3) comprises granulite facies rocks, whose distribution also coincides with a major magnetic and gravity high caused by the presence of dense rocks in the crust and an elevated Moho discontinuity [9, 11, 54, 55]. The latter is the result of differential uplift and extension in the aftermath of the Caledonian orogeny and the subsequent Late Palaeozoic and Mesozoic processes that led to the opening of the North Atlantic. The eastern part of the region consists of various amphibolite facies gneisses, migmatites, greenstone belts and granitic plutons. After the initial geochronological studies more work followed using Pb-Pb, Rb-Sr and Sm-Nd dating [9, 36, 56-58]. The chronology of the basement complex and the anorthosite-mangerite-charnockite-granite suite (AMCG) has now been refined by modern U-Pb geochronology [10].

2.2.1. Neoarchaean basement gneisses

Neoarchaean crust occupies large parts of the islands of Langøya in Vesterålen and Hinnøya farther east (Figure 3), and small remnants are also present at the southwestern tip of Austvågøya [58]. The Neoarchaean rocks of Langøya consist mainly of high-grade gneisses interpreted to represent metasupracrustal rocks of intermediate composition, while Neoarchaean gneisses on Hinnøya define an amphibolite facies metamorphic domain that was migmatized in the Neoarchaean and subsequently deformed and metamorphosed at granulite-facies conditions in the Palaeoproterozoic [9]. The appearance of orthopyroxene to the west of the amphibolite-facies domain on Hinnøya has been interpreted as either a prograde metamorphic transition [9, 54], or an abrupt transition marked by a crustal scale ductile shear zone of presumed Neoarchaean age. The zone east of the metamorphic boundary (Figure 3) is dominated by intrusive rocks of tonalitic to granitic composition, migmatitic domains, and local greenstone belt remnants, considered to be Neoarchaean in age [9, 59].

2.2.2. Palaeoproterozoic supracrustal rocks, deformation and metamorphism

A younger sequence of heterogenous mafic gneisses in Lofoten has been interpreted as a Palaeoproterozoic supracrustal succession of volcanogenic derivation as deduced from geochemical compositions [9]. Metasedimentary rocks consist of fine-grained gneisses, locally quartzitic (Figure 12a), with subordinate graphite schist, banded iron formation and marble. These gneisses have been overprinted by the same granulite-facies metamorphism as the Neoarchaean rocks, and the boundary between the two metamorphic domains corresponds to the eastern limit of the magnetic high and the first appearance of orthopyroxene [54] in eastern Langøya (Figure 3). In the south this boundary is considered a major thrust that juxtaposes the Eidsfjord anorthosite and deformed intrusive mangerite [54, 60]. Some studies [9] suggest that the orthopyroxene isograde is folded and continues southeastward across Hinnøya (Figure 3). The granulite facies gneisses tend to be rather massive, with faint banding, and they equilibrated at 3 to 4 kb and 750 to 780 °C [61]. They are considered to be orthogneisses [59] or migmatized supracrustal rocks of intermediate composition [9].

Figure 12. Outcrop photographs of the AMCG suite and metasupracrustal rocks in western parts of Lofoten (Figure 3)(a) Foliated paragneisses composed of alternating quartz-rich and mafic meta-volcanic rocks. (b) Detail view of mangerite (hypersthene-bearing monzonite) which is the dominant rock type of the Lofoten igneous province. Note phenocrysts of plagioclase and orthoclase. (c) Mangerites with mafic intercalations and dykes aligned parallel to a weak magmatic foliation. From a road cut in eastern Lofoten (Austvågøya). (d) Panorama view of the Hopen pluton mangerite and its boundary to migmatized gneisses and altered mangerites. View is toward the north.

2.2.3. Palaeoproterozoic magmatic rocks, deformation and metamorphism

Plutons of the AMCG suite occupy about 50 % of the Lofoten islands [62]. The suite (Figure 3) is dominated by mangerite and charnockite and their metamorphosed equivalents (Figure 12b-d), with local but important occurrences of gabbro, anorthosite, and granite and associated mafic dykes (Figure 12c). The mafic and felsic phases locally grade into, or mutually cross-cut each other indicating that some of the intrusions are genetically related. In addition, a widespread network of feldspathic veins is present throughout the granulite facies domain [9]. Mangerite and charnockites are widespread throughout the region, in the form of large plutons (Figure 12d) such as in the southwestern Lofoten islands [9, 62, 63]. Anorthosite and gabbro forms plutons in most of the islands[54, 64, 65]. There is one distinctive body of granite-syenite on Langøya [54, 66]. To the east, on Hinnøya, Neoarchaean gneisses are cut by Palaeoproterozoic plutons such as the Lødingen granite with a zircon and titanite age of 1870 Ma [10, 36], whereas the Tysford granite covers a large area on the mainland and has given U-Pb ages between 1.8 and 1.7 Ga [67].

The mangerites and charnockites were shown to have intruded at about 4 kb and >925 to 850 °C, whereas the anorthosite on Flakstadøy records a polybaric crystallization history from 9 kb to about 4 kb at 1180 to 1120 °C. The data imply that anorthosites, mangerites and charnockites in Lofoten were emplaced at the same depth of about 12 km [64]. Later on in the same time span, the Palaeoproterozoic plutonic rocks and surrounding Neoarchaean gneisses were deformed and metamorphosed up to granulite facies conditions, and portions of the plutonic rocks, e.g. anorthosite bodies, were thrusted over mangeritic gneisses (Figure 13a, b) [54]. Structural fabrics include transposed foliations (Figure 13c), intrafolial isoclinal folds (Figure 13d) and local migmatization structures, and upright folds that refolded the earlier shear fabrics (Figure 13e). A younger, very extensive but narrow ductile shear zone network is characterized by greenschist facie retrogressive shear zones that truncate all other structures (Figure 13f) and display both low- and high-angle attitudes. The exact age of the latter is unknown, but presumably, late Palaeoproterozoic.

The structural and isotopic data show that the Neoarchaean crust played an important role in the genesis and deformation of the Lofoten igneous rocks, and both as a source and as a contaminant [58, 68]). The Pb data define a linear trend that may represent mixing between Neoarchaean lower crustal components and late Palaeoproterozoic juvenile additions, whereby the mangerites and charnockites contain more Neoarchaean Pb than the mafic rocks. A multistage evolution with a basaltic parental melt undergoing polybaric crystallization and differentiation to form anorthosites as cumulates and ferrodiorites as residual melts has been proposed [69]. The mangerites and charnockites are inferred to represent feldspar cumulates and residual liquids, respectively, derived from magmas broadly syenitic in composition.

The more recent U-Pb results show that the Lofoten AMCG suite was emplaced during two quite distinct events (Figure 6), the first one at 1.87-1.86 Ga followed by a second and dominant magmatic event at 1.80-1.79 Ga. A concluding period, lasting some 20-30 my, was characterized by local hydration of the dry AMCG rocks, and by the infiltration of pegmatite melts [10]. Local granitic pegmatites belong to a distinct Palaeoproterozoic (ca. 1.77 Ga) generation.

Figure 13. Outcrop features illustrating Palaeoproterozoic deformation fabrics of the Lofoten-Vesterålen province. (a) View of an anorthosite complex in Langøya which is thrusted over granulite facies mangeritic gneisses to the southeast. The ductile thrust zone is c. 50 m thick and made up of mylonitic gneisses [54]. Note the lack of vegetation on the grey-coloured Eidsfjord anorthosite above the shear zone contact. View is toward NE. (b) Thrust in Palaeoproterozoic granitic gneisses on Langøya [54]. (c) Transposed foliation in mangeritic gneiss with granitoid bands and intercalations. Locality: Austvågøya. (d) Ductile shear zone in mangeritic gneiss with felsic intercalations. (e) Open upright fold in mangeritic gneiss. The fold axis trends NW-SE, and view is toward SE. (f) Steep and localized retrogressed ductile shear zone in mangerite in southern part of Vestvågøya.

3. Geological features of the Lewisian Complex of Scotland

The Lewisian Complex in NW Scotland (Figure 1a) is situated to the west of the Caledonian Moine thrust and covers the Outer Hebrides and the NW Scottish mainland of the Inner Hebrides in the south (Figure 4). The Lewisian rocks also form inliers in the Caledonian orogenic belt possibly continuing under the Moine Group up to the Great Glenn Fault in the southeast [18-20], while Mesozoic to early Cenozoic extensional basins bound the Lewisian outcrops offshore Scotland (Figure 1a).

The Lewisian Complex was considered by earlier workers [70, 71] as a continuous crustal block composed of up to three Neoarchaean gneiss domains overlain by Palaeoproterozoic metasedimentary, metavolcanic and intrusive rocks separated by NW-SE trending ductile shear zones. In the last decade it has been proposed that the region consists of distinct terranes [7, 72], although there are disparate views on how these terranes are related [8]. A classical Lewisian nomenclature (Badcallian, Scourian, Inverian, Laxfordian) evolved progressively, originally to designate specific rock forming, deformational or metamorphic

events, and eventually becoming linked to specific dates obtained from forthcoming geochronological studies. Some of these terms, however, have now become problematic since
they have been and can be used to designate different tectonic expressions and times. To
maintain clarity in the following we shall therefore avoid their use.

3.1. Main structures

The Neoarchean rocks of the Lewisian were deformed during an early granulite facies
tectonic event involving crustal shearing and isoclinal folding that formed a gently dipping
gneiss foliation (Figure 14a, b). During a later retrogressive event the gently dipping fabric
was macro-folded into NE and SW-dipping steep attitudes and subjected to steep strike-slip
shearing (Figure 14 c, d) in amphibolite facies [73-75]. A key observation is the local
truncation of these macrofolds by the c. 2.4 Ga Scourie dykes [76], as outlined later (Figure
15), indicating that these early deformation events are likely Archaean in age.

Figure 14. Outcrop features of Neoarchaean tonalitic gneisses in the Lewisian Complex. (a) Tight to
isoclinal intrafolial folds in the TTG gneisses with a presumed 2.7 Ga age, northwest of the Canisp shear
zone in Assynt terrane. (b) Cliff face made up of Neoarchaean TTG-gneisses with a subhorizontal foliation. Height of the cliff is ca. 50 m. Locality: in Assynt terrane. (c) Steeply dipping, alternating banded
tonalitic, granitic and mafic orthogneisses. Locality: north tip of Rhiconich terrane. (d) TTG- gneiss with
an older foliation cut and transposed into a steep ductile shear zone of presumed 2.7 Ga age. Locality:
north of Canisp shear zone.

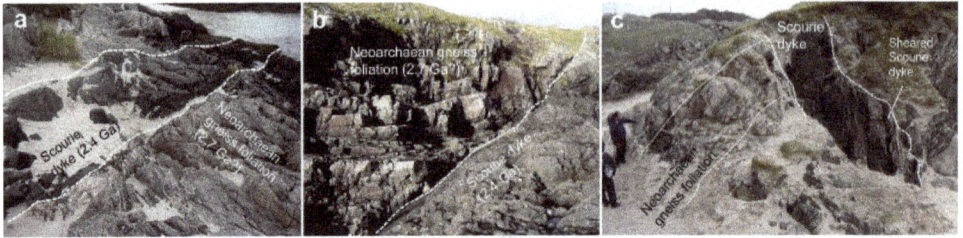

Figure 15. The Scourie dyke swarm in the Assynt terrane (see Figure 4). (a) Steep, Scourie mafic dyke that cuts through Neoarchaean gneiss foliation. Note the very sharp an unaffected intrusive contacts. Dyke thickness is approximately 5 m. Locality, north of Canisp shear zone. (b) Scourie dyke intruded into gently dipping TTG-gneisses. (c) Scourie dyke cutting Neoarchaean gneisses and which is again strongly sheared along steep, presumed 2.49 Ga Palaeoproterozoic shear zones. Locality, near contact to Canisp shear zone.

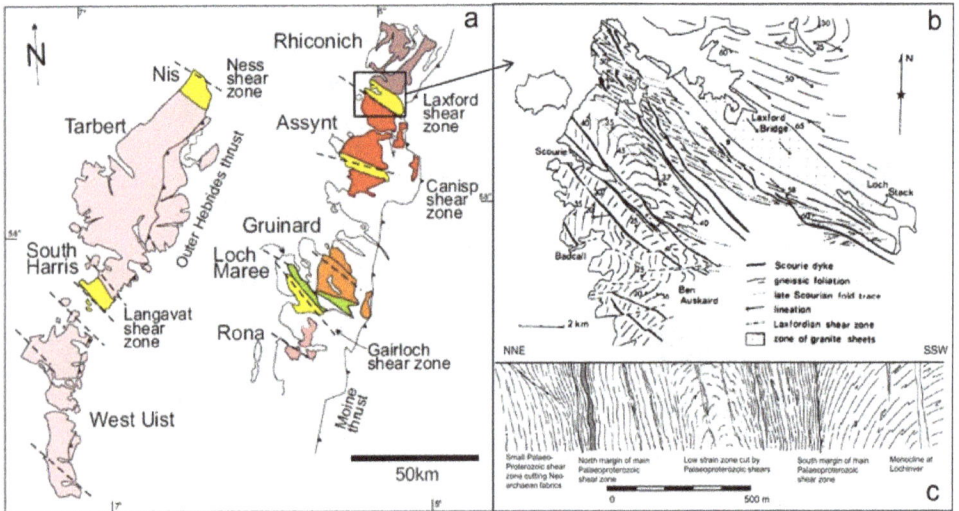

Figure 16. (a) Regional structural map showing major ductile shear zones in the Lewisian Complex, that suffered Palaeoproterozoic deformation and reworking [37]. (b) Map of the Laxford shear zone, with Scourie dykes that cut presumed regional 2.49 Ga folds that are becoming tightened toward the Laxford shear zone [37]. (c) Schematick NNE-SSW cross-section of the Canisp shear zone showing major upright folding of the gneiss foliation and formation of localized, steep, axial-planar shear zones [77].

The main episodes of Palaeoproterozoic crustal deformation in the Lewisian produced major folds and NW-SE striking, dextral-reverse, transpressive shear zones (Figure 16a) that were superimposed on, and largely obliterated the pre-Scourie dykes fabrics, except in some low strain lenses. An early/main stage of deformation involved tight to isoclinal folding of the flat-lying Neoarchaean gneiss foliation and subsequent upright folding leaving the limbs in a steep attitude (Figure 16b, 17a). In addition, localized moderately NE-dipping ductile reverse and dextral oblique shear zones developed by strain partitioning, likely due to reactivation of steep pre-Scourie dyke shear zones [77], and they affected the Scourie dykes

(Figure 17b). This event was associated with upper amphibolite, locally granulite facies metamorphism [37] and intrusion of syn-tectonic pegmatite sheets and veins in e.g. the Laxford shear zone (Figure 4). The major ductile shear zones show complex multiphase strain partitioning, including thrusting and refolding of early-stage subhorizontal shear zones developed parallel to the pre-existing foliation [78] and later development of steep strike-slip shear zones [77]. The late stage of deformation coincides with retrogression to greenschist facies conditions [79] associated with the formation of steep-plunging asymmetric folds and retrogressed cleavage and strike-slip shear zones [37].

From northeast to southwest the main shear zones are the Laxford, Canisp, Gairloch and Diabaig shear zones (Figure 16a). The ca. 8 km wide, SW-dipping *Laxford shear zone* [37] is a major zone of folding (Figure 16b) that reflects thrusting of the gneisses in the Assynt block over the Rhiconich gneisses to the north. This presumed terrane-bounding shear zone [80] evolved from a pre-existing, steep fabric that displayed early stages of sinistral and dextral shearing and subsequent oblique-thrusting and dextral strike-slip movement [81], evidenced by SSE plunging stretching lineations [37]. Numerous granites and pegmatite sheets were injected within this shear zone.

Figure 17. Outcrop features illustrating Palaeoproterozoic (c. 1.9-1.67 Ga) deformation fabrics in the Lewisian Complex. (a) Meso-scale upright folds within low-strain domain of Canisp shear zone. Hammer is parallel to fold axis, ESE-WNW, and the axial-surface dips steeply to the ENE. A steep, SSW-dipping mylonitic shear zone developed on the fold limb (to the right). (b) Steep shear zones that cut and displace a Scourie dyke and the Neoarchaean gneiss foliation near Canisp shear zone. (c) Upright folding of TTG-gneisses outside the Canisp shear zone (left) and refolding by tight sub-vertical folds within the shear zone (central and right). View is toward ESE. (d) Contact between TTG- gneiss and the Canisp shear zone. Note subvertical sinistral drag-folding of the gneisses into the mylonitic shear zone. View is toward WNW. (e) High-strain mylonite in the Canisp shear zone, with retrogressed chlorite-mylonitic schist and quartz veins aligned along the main fabric. View is toward the WNW. (f) Detail from the steep mylonite zone in e, showing asymmetric sinistral folding of the main fabric, including the quartz veins, seen on a horizontal view .

Farther south, the Canisp and Gairloch shear zones define steep oblique crust-internal shear zones of the Assynth and Gruinard terranes (Figure 16a). The *Canisp shear zone* dips steeply SSW [37] and typically truncates the Scourie dykes (Figure 17b) and displaces them into zones of alternating high- and low-strain [77]. Variably dipping high-strain thrust zones evolved in the hinge zone of a major ESE-WNW trending fold (Figure 16c), while steep high-strain mylonitic shear zones truncate the limbs of the macro-fold (Figure 17c, d). Strain partitioning is observed at all scales, and low strain lenses are typically overprinted and transposed into high strain zones. Folds in the high strain zones are asymmetric, tight and have generally steep plunge (Figure 17c). These folds may have been produced by early deformation which partitioned later into the higher strain domain as tighter and steeply plunging folds (Figure 16b), leaving a weak signature in the low strain domain [82]. Fabric-parallel quartz veins are abundant in the high-strain zones, and these veins have been further deformed and folded internally (Figure 17e, f). Stretching lineations in high-strain parts of the *Canisp shear zone* suggest dextral-oblique-reverse movement (south-side up thrust) followed by strike-slip shearing [77, 79, 82]. The *Diabaig shear zone* is thought to be an inclined thrust ramp dipping toward NE. Similar shear zones exist in the Outer Hebrides portion of the Lewisian Complex (see Figures 4 and16a).

3.2. A chronology of events

In the past sixty years the Lewisian Complex has been the subject of very extensive geochronological studies with the application of many different decay systems. Much of the initial work, especially that done with whole rock methods, documented the existence of Archaean and Palaeoproterozoic events, matching the subdivision proposed by [70] based on the pre- and post- Scourie dyke position of the rocks [83-88]. The details of the picture, however, have remained fuzzy due to the complications introduced by the multistage evolution of the rocks. More recent dating of zircon and other minerals with U-Pb has helped to shed more light into the timing and importance of the events in the different domains. This evolution included five major events, interspersed by some minor but tectonically important events (Figure 6): (i) Meso- and Neoarchean orogenic activity, mainly between 3.00 and 2.70 Ga, built the bulk of the Lewisian crust; (ii) an earliest Palaeoproterozoic event at 2.50-2.48 Ga, had a profound influence on the Assynt block, but is not seen elsewhere, except for anorthosite in South Harris; (iii) emplacement of Scourie dyke during at least two episodes at 2418 and 1992 Ma; (iv) deposition of clastic sediments at Loch Maree sometime between 2.0 and 1.9 Ga, leading to, or associated with a localized but intense episode of arc magmatism at 1.90-1.85 Ga at Loch Maree, Laxford Bridge and South Harris; and (v) local migmatization and emplacement of pegmatite dyke swarms at 1.70-1.65 Ga, mainly in South Harris but recorded by sporadic pegmatites and titanite across the entire Lewisian Complex.

i.   Meso- and Neoarchaean evolution

The Lewisian Archaean crust is dominated by banded, felsic to intermediate TTG-gneisses of presumed igneous origin. Meso- and Neoarchaean supracrustal rocks composed of semipelites, calc-silicate schists, meta-arkoses and metavolcanic rocks associated with mafic to ultramafic and anorthositic rocks with a tholeiitic composition are also preserved, and when present, may represent the protoliths of some of the gneisses. The metamorphic grade reached granulite facies in the Assynt block, and varies from granulite to amphibolite facies elsewhere. The geochronology of these gneisses tends to be very complex because of the multiple overprints which are recorded in zircon so that in many cases the U-Pb data are scattered and do not entirely resolve the sequence of events. However, the overall picture permits to distinguish a main pattern typical of the Assynt block and distinct from that of the other segments (Figure 6).

The TTG rocks in the Assynt block preserve some of the oldest zircon ages of 3.04-2.96 Ga and it has been suggested that this is the age of formation of the rock [89-91], but it remains uncertain whether the oldest grains may just be xenocrystic and the gneisses actually formed about 2.85 Ga [92]. The early high grade metamorphism occurred at around 2.80-2.75 Ga, and was followed by the intrusion of trondhjemite at 2.72 Ga [93-95] and emplacement of mafic-ultramafic rocks [88].

The other Archaean segments of the Lewisian rocks had the most prominent period of development between 2.9 and 2.0 Ga, with a peak at 2.84-2.82 Ga [7, 90, 91, 93, 96-101]. An exception is the Rona segment where ages between 3135 and 2880 Ma have been reported for tonalitic gneisses [102]. The Gruinard block underwent granitic to trondhjemitic magmatism and high-grade metamorphism at about 2.73 Ga [91, 96, 97] and in the Richonich block there are some indications for further activity as late as ca. 2.6 Ga [90].

ii.   Earliest Palaeoproterozoic metamorphic progression: 2.50-2.48 Ga

The Assynt block was affected by a second high-grade metamorphic event at 2.50-2.48 Ga. This event caused strong resetting of U-Pb in zircon due to recrystallization and local new-growth [90-94] and is also recorded by Sm-Nd isochrons obtained from garnet and coexisting minerals in ultramfic pods [104], as well as by U-Pb in titanite and monazite [94, 96]. The dry high temperature metamorphism was concluded by re-hydration which caused local formation of granitic pegmatites and leucosome in migmatites [94] and likely led to retrogressive amphibolite facies metamorphism and deformation, termed the Inverian [73, 74]. The only well constrained temporal analogue to these metamorphic events elsewhere in the Lewision Complex is an anorthosite body present in the younger South Harris igneous complex [105].

iii.   Scourie dyke swarms

The Scourie dykes form part of an extensive dyke swarm present throughout the Lewisian Complex [37] but most abundant in the southern region. These dykes display many different geometries and attitudes, mostly steeply dipping, and they cut all Neoarchaean folds and planar fabrics (Figure 15). The composition of the dykes varies from mafic to ultramafic [105], and their trace element geochemistry indicate formation in a marginal continental setting, as island arcs or from a mixtures of crustal and oceanic material [106, 107]. The Scourie dykes intruded during at least two different events at 2418 Ma and 1992 Ma [72, 75, 108, 109]. The youngest of these dykes intruded into hot gneissic and migmatitic rocks syn-kinematically with a late stage of shear zone deformation [79, 107], of the same kind as that shown in Figure 14d.

iv.   Deposition of supracrustal rocks (2.00-1.90 Ga) and arc magmatism (1.90 – 1.85 Ga)

Banded iron-formation, marble, chlorite schists and meta-psammites of the Loch Maree Group were deposited on the Neoarchaean crust and are now arranged in two narrow, NW-SE trending synformal belts [110, 111] in the Gairloch area (Figure 18). Deposition of the Loch Maree Group is constrained between 2.0 Ga, the youngest detrital zircons from a metagreywacke in the Gairloch area [112], and 1903 Ma, the age of the syntectonic Ard gneiss intrusion [102, 110]. The latter was emplaced during the early stages of deformation associated with amphibolite to granulite facies metamorphism and interpreted to be related to the development of a subhorizontal shear zone net work. The Ard gneiss is considered to be a product of arc magmatism and the deformation caused by lateral accretion of oceanic plateaus and primitive island arcs [110].

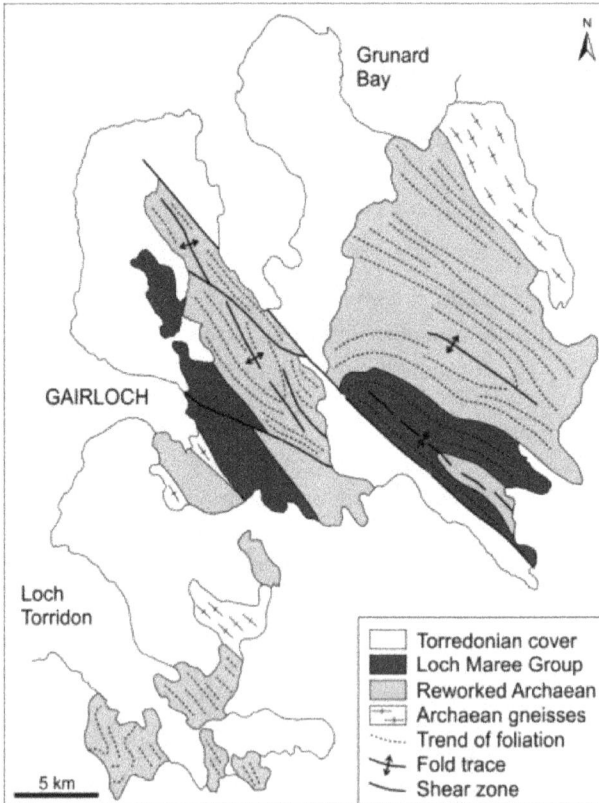

Figure 18. Simplified map of the Palaeoproterozoic Loch Maree Group (shaded) within the Lewisian Complex of the Gairloch area, NW Scotland [110]. This group consists of highly deformed amphibolites and metasedimentary rocks cut by 1.9 Ga granitoid alkaline rocks. Note macro-scale reverse and sinistral folds and related off-sets and lateral shear zone displacement.

The other important tectonic element with similar age and genesis is the South Harris igneous complex in the Outer Hebrides (Figure 4) that comprises a magmatic arc sequence of
mafic to felsic intrusive rocks formed at 1.89-1.88 Ga [98, 100]. Clastic sedimentary rocks in
this region contain abundant detritus of the same age [7, 98]. Crudely coeval high grade
metamorphism strongly overprinted zircon in the 2.5 Ga anorthosite [104] and has also
been dated by Sm-Nd mineral systems [113]. This event also caused deformation and metamorphic zircon growth in a shear belt at the northern tip of the Tarbert block at Nis (Figure 6) [98, 99]. As for the Loch Maree assemblage, the South Harris situation has been interpreted 'to represent a magmatic arc, complete with contemporaneously derived clastic
sediments, developed in a collisional orogen, which culminated in granulite facies metamorphism' [98].

A third important occurrence of intrusive rocks of this age is within the Laxford shear
zone. Granite sheets, with a U-Pb age of 1854 ±13 Ma [72] occur subparallel to the shear
zone boundaries but cutting the pre-Scourie dyke fabric (Figure 19a, b). Synchronous shear
deformation affected the dykes and aligned them as lenses into shear zones that also affected the surrounding tonalitic and mafic gneisses by mostly contractional deformation (Figure 19b-d). There is some consensus that the Laxford shear zone likely represents a terrane
boundary but the timing of juxtaposition is debated, and the role played by the granites
unclear. Recent work [80] concludes that the terranes were probably brought together
sometime between 2.5 and 2.4 Ga, after the earliest Palaeoproterozoic deformation and
retrogression, but before intrusion of the Scourie dykes. Others, however, suggest that the
juxtaposition probably occurred around the time of emplacement of the granites, consistent
with the timing proposed for the Loch Maree group and South Harris igneous complex [7,
8, 114].

v.   Late orogenic events (1.70 – 1.65 Ga)

A metamorphic event at around 1.75 Ga is indicated by a Sm-Nd age of garnet and coexisting metamorphic minerals from a mafic dyke of the Assynt block [109,], and by titanite from
rocks near the Laxford shear zone [90, 93]. Because of the localized occurrence of the rocks,
however, the tectonic significance of this age is uncertain.

By contrast a later event at 1.70 to 1.65 Ga had a much stronger impact across all of the Lewisian Complex (Figure 6). The main expression of this event is the granitic and pegmatitic
migmatite complex associated to, and bordering, the South Harris igneous complex [7, 100,
101, 115, 116]. Felsic dykes and pegmatites of this age have also been found in most other
parts of the Lewisian [7, 98, 110] and they also coincide with the ages of rutile and a younger
titanite generation [76, 93, 102]. The exact significance of this event is still uncertain. One
view is that emplacement of pegmatites of this ages pre-dated, but likely broadly coincided
with the late flexuring, steep shearing and greenschist facies retrogression [110], thus attrib-
uting these events to contractional processes during late-stage collision of the Lewisian
Complex with e.g. a southern block [8].

Figure 19. Outcrop features of Palaeoproterozoic intrusive rocks in the Lewisian complex. (a) Granitic pegmatite intrusion into TTG-gneisses of the Rhiconich terrane. (b) Syn-kinematic felsic pegmatite dykes intruded into mafic gneisses and sheared during the main Palaeoproterozoic tectono-thermal events. Note boudinaged and asymmetric lenses indicating top-NE movement sense (right in photo). View is toward SE. (c) Syn- and post-kinematic granite pegmatite dykes injected into mafic Neoarchaean gneisses. Note lensoidal sheaped mafic bodies in the sheared gneisses. (d) Revers ductile shear zone displaying a duplex geometry, cutting obliquely across steeply dipping TTG-gneisses in the Rhiconich terrane. View is to the NW.

4. Discussion

4.1. Similarities of terranes and terrane juxtaposition

In order to discuss terrane aspects and hypothetic assembly history of the West Troms Basement Complex, the Lofoten-Vesterålen province and the Lewisian Complex as a framework for further correlation, we focus on critical similarities such as the presence of crustal segments with contrasting age, tectono-magmatic and/or metamorphic histories, crustal-scale ductile shear zones (sutures), overlapping nature and character of deformation (convergent and strike-slip) and major metamorphic breaks that may have juxtaposed different crustal levels instead of spatial terranes [8, 72]. The first step would be to locate different crustal segments, then to localize the suture(s) formed by the juxtaposition of terranes, and finally, discussing assembly of different crustal levels in order to explain metamorphic and petrologic differences known from all the studied regions [8, 17].

Although metamorphic and structural characteristics of Precambrian continental crust in general is not uniquely diagnostic for correlation [107], the geotransect in western Troms and Lofoten is underlain by age-equivalent TTG-gneisses and granitoid gneisses of Meso- and Neoarchaean age, with only modest differences in metamorphic grades and histories [32]. This indicates that at the end of the Archaean these rocks were likely assembled as a large single terrane, formed during a prolonged cratonization event (2.92-2.67 Ga), though with considerable complexity in detail [32]. In this scenario, the migmatized ductile shear zones bounding the Dåfjord and Kvalsund gneisses on Ringvassøya [17] and, potentially also the major tectonic and metamorphic boundary in the Lofoten-Vesterålen area [10] may reflect Neoarchaean intra-cratonic terrane boundaries. An argument for multiple terranes in the West Troms Basement Complex at the end of the Neoarchaean, however, appears from the contrasting ages, e.g. 2.8 Ga for the Ringvassøya greenstone belt, 2.4-2.2 Ga for the Vanna group [29] and c. 1.9 Ga for the Torsnes belt [31].

By comparison, the Lewisian Complex was earlier considered to have originated from a single Neoarchaean continent that split up into multiple terranes in the early Palaeoproterozoic and later on were juxtaposed during the 1.90-1.67 Ga events [79]. A revised model of Palaeaoproterozoic juxtaposition in the Lewisian, however, included ten terranes and one block [7, 72, 90]. The Laxford shear zone was considered a terrane boundary between the Rhiconich terrain in the northeast and the Assynth terrane in the southwest (see Figure 4). These two terranes were likely accreted after intrusion of the pegmatite sheets in the Rhiconich terrane at c. 1.855 Ga, since they are absent in the Assynth terrane, but prior to the early-Palaeoproterozoic structures common in both regions. One model suggests that the accretion occurred at 1.74 Ga, synchronously with an amphibolite facies metamorphism recorded in the Rhiconich terrane and a metamorphic retrogression associated with the formation of shear zones in the Assynth terrane [7]. Recently, it was argued that juxtaposition of the Assynth and Rhiconich terranes occurred *prior* to the 1.9-1.75 Ga period, e.g. in late Neoarchaean since Scourie dykes are present on both sides of the Laxford shear zone [80].

A second model [8] involved only two continental plates during the Neoarchaean and Palaeoproterozoic history of the Lewisian Complex. In this model the Lewisian Complex was divided into two blocks classified as *upper-plate* and *lower-plate* blocks (Figure 20) that differed considerably with regards to position of the supracrustal rocks relative to the accreted versus the overriding plates, which is critical for the metamorphic conditions. The *upper-plate* block preserved rare and weak Palaeoproterozoic deformation and amphibolite/greenschist facies retrogression of granulite facies gneisses, while the *lower-plate* portion involved prograde and peak amphibolite facies metamorphism and high-strain assemblages. The *upper-plate* blocks displayed weak deformation and could be located on the low-strain areas above the terrane boundary shear zone in the crustal model for the mainland [8, 37], while the *lower-plate* blocks with strong Palaeoproterozoic structures could correspond to the mid-deep level of the shear zone itself, i.e. as during the emplacement of the Loch Maree Group (Figure 20).

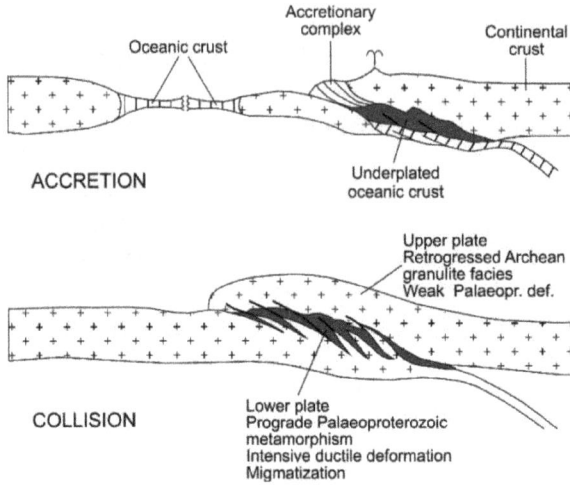

Figure 20. Schematic model of an idealized subduction-accretion collision sequence [8] leading to domains of contrasting deformation and metamorphism, e.g. high-grade/prograde in *lower plate* and low-grade retrogressive metamorphism in *upper plate* settings.

By analogy, in the West Troms Basement Complex, a significant Palaeoproterozoic tectono-metamorphic break is thought to exist southwest of the island of Senja (Figure 2), in a region that separates dominantly amphibolite facies gneisses from granulite facies AMCG-suite rocks of the Lofoten-Vesterålen province [9, 10]. This major boundary is also inferred by contrasting gravity and magnetic characters, and could therefore reflect a Palaeoproterozoic suture [11, 13, 50]. The slightly older ages of the peak Palaeoproterozoic deformation (1.87-1.78 Ga) versus 1.78 Ga in the West Troms Basement Complex (Table 1) suggest progressive southwestward accretion toward an orogenic hinterland near Lofoten (Figure 21), which is also consistent with the observed increase in metamorphic grade. On the other hand, the contrast in U-Pb ages of basement rocks in the southwest (2.7-2.6 Ga) compared to in the north (2.92-2.8 Ga), could indicate a second terrane in the northeast (Figure 21)[10, 17].

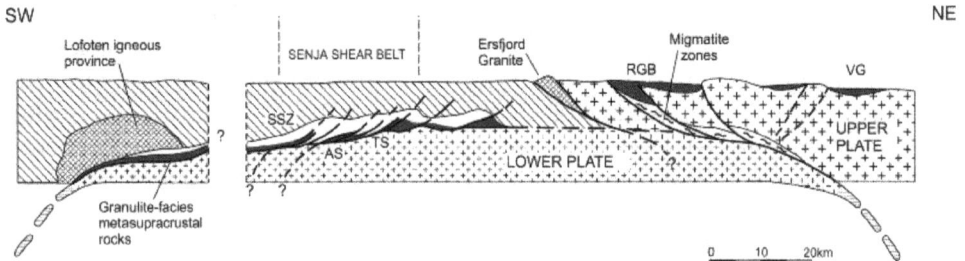

Figure 21. Schematic model of Palaeoproterozoic accretion in the West Troms Basement Complex. Lower – Upper plate model used to explain among others, metamorphic differences [8].

A model involving SW-directed convergence/accretion toward an orogenic front near Lofoten and an oppositely NE-dipping terrane boundary zone in the northeast (e.g. reactivated Neoarchaean migmatite zones in Ringvassøya) may explain both the contrast in deformation styles and metamorphic grades along the geotransect. For example, an accretion/subduction-derived shear zone situated in the Senja shear belt adjacent to eastern Lofoten-Vesterålen would locate them to the *lower plate* of two or more unified colliding terranes (Figure 21, left). This would have formed prograde amphibolite to granulite-facies metamorphic assemblages, granitic melts, migmatite zones and strong ductile deformation, including a detachment in the Senja shear belt that could have become the locus of later-stage partitioned crustal deformation. Conversely, in an upper plate position (Figure 21, right), localized and weakly developed but more likely retrogressed, low-grade metamorphosed cratonic-marginal shear zones and supracrustal platform sequences may appear (as in Vanna). This model would favor assembly of different crustal levels of at least two main crustal segments (terranes) rather than spatially separated smaller terranes [8].

4.2. Comparison between Lewisian, western Troms and Lofoten-Vesterålen

Plate tectonic reconstructions of Precambrian units in the North Atlantic realm with respect to Fennoscandia and Laurentia (Figure 22) have to some extent failed to demonstrate whether these cratons belonged to the same supercraton in the Neoarchean (2.8–2.5 Ga) and Palaeoproterozoic (1.8-1.6 Ga). A tectonic linkage is supported by paleomagnetic reconstructions [8, 119] stating that Fennoscandia was positioned close to the Greenland/Laurentia and Superior supercraton in the late Neoarchean (Figure 22). Geological similarities and differences between domains are important criteria for restoring possible supercontinents, as stated by [3, 4, 119]. In this context, the studied basement outliers west of the Scandinavian Caledonides in North Norway and the Lewisian complex in Scotland both have a pivotal central location within the marginal orogenic belts constituting the presumed Neoarchaen supercontinent (Figure 1) [110, 120]. These units, however, also occupy an interior position of the Caledonian orogen far from the autochthonous shield rocks and are bounded by younger faults. They are, thus, usually not considered part of any shield areas, but instead assigned an uncertain or exotic tectonostratigraphic status [3, 121].

Based on the comparison between the Archaean-Palaeoproterozoic basement suites in North Norway and the Lewisian of Scotland outlined above we can discuss potential correlation of these suites in the context of the North Atlantic realm. Such a correlation can be tested using similarities or dissimilarities in lithology, age, supracrustal units, igneous/ petrogenetic, structural and metamorphic features and evolutionary and tectono-metamorphic history (see Table 1).

4.2.1. Archaean components

Archaean crust forms the backbone of both the western Troms, Lofoten-Vesterålen province and the Lewisian basement complexes. These complexes reveal some broad similarities in terms of lithology and general age patterns but also some differences, which, however, are most pronounced within each of the regions (Figure 6). Thus, we are comparing two heteorgeneous Archaean crustal segments.

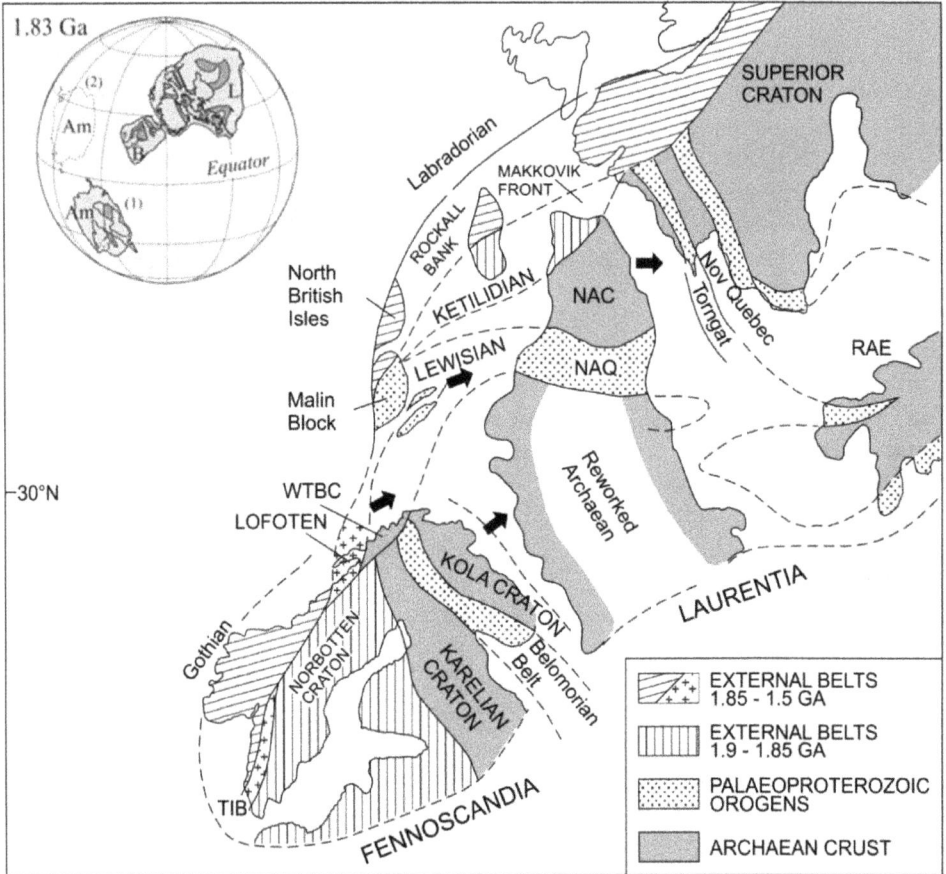

Figure 22. (a) Reconstruction of Laurentia and Fennoscandia during the Palaeoproterozoic based on the palaeomagnetic fit of [118]. Note that the West Troms Basement Complex (including Lofoten) and the Lewisian Complex lies within a continuous Palaeoproterozoic belt extending from the Torngat orogen of Laurentia through the Ketilidian and Nagssugtoqidian orogens to link up with the Kola and Karelian/Belomorian provinces of Fennoscandia. The arrows show inferred movement directions of various crustal segments relative to Laurentia. The map is modified from [4, 110], while the reconstruction of continents at 1.83 Ga is based on [118]. Abbreviations: WTBC =West Troms Basement Complex.

In the West Troms Basement Complex we can distinguish: (i) a Mesoarchaean tonalitic domain formed between 2.9 and 2.8 Ga in Ringvassøya and Vanna, overlain by (ii) the broadly coeval, but tectonically distinct Ringvassøya greenstone belt. These two domains are separated by the (iii) late orogenically active Kvalsund gneiss migmatite zone from (iv) the Neoarchean domain of Kvaløya and Senja farther south, which formed during a short time interval between 2.72 and 2.67 Ga. The Lofoten-Vesterålen province has broadly the same age structure as the southern part of the West Troms Basement Complex, except for the fact that it underwent high-grade metamorphism at 2.63 Ga. The apparent formation of leucosome in southern Senja at around 2.61 Ga [32] may be the expression of the same event, indicating that Senja and Vesterålen may simply represent different crustal levels experiencing metamorphism at different times, a very common pattern of crust construction and maturation [122]. The late Archaean deformation and migmatization that seems to characterize the Kvalsund migmatized zone separating the Dåfjord and Kvalsund gneisses [17, 32] may be another expression of this late orogenic activity.

In the Lewisian there are also some differences in the Archaean history of the contrasting blocks (Figure 6). Several of them appear to have formed mainly in the late Mesoarchaean, between 2.85 and 2.80 Ga. Older ages up to 3150 Ma are recorded in the Rona and Assynt blocks whereas Neoarchean magmatic activity seems to be restricted mainly to the Assynt and Gruinard blocks, although there are local indications of such activity also in the Tarbert-West Uist and Rhiconich blocks.

The single most distint event that is fully missing in the West Troms Basement Complex and Lofoten-Vesterålen is the pervasive high-grade metamorphism and subsequent rehydration and retrogression in the Assynt block at 2.5-2.48 Ga. However, this event is not seen in the other Lewisian blocks except for the anorthosite body in South Harris (Figure 6). One explanation is that it was a specific terrain formed in another orogen, the alternative is that the 2.5 Ga high-grade event reflects a process affecting lower crustal levels but not recorded higher in the crust, in the same way as seen for example in the Superior Province [123].

A comparison of the evolution of these two crustal sectors is difficult because of the inherent internal differences, which may reflect dependencies from crustal level, and probably also the tectonic juxtaposition of different terranes combined with the unequal geochronological coverage in different blocks and the technical difficulties to cleanly date the age of protolith and orogenic tranformations of complex polymetamorphic gneisses. Hence we can conclude that the West Troms Basement Complex, the Lofoten-Vesterålen province and the Lewisian have certain affinities in common (Figure 6), suggesting that they could have been linked to some degree in the Archaean period, but the opposite conclusion is also possible.

2.40 to 1.98 Ga supracrustal rocks and mafic dyke swarms

Intrusion of the huge Ringvassøya mafic dyke swarm in the West Troms Basement Complex (Figure 23b) occurred at c. 2.40 Ga (Figure 6) [34]. This event is part of a major mafic dyke-producing event that affected several Archaean cratons, and as such it does not necessarily represent a unique stitching tool for linking these crustal domains. One argument supporting such a role, however, is the apparent south-westerly shift in age from ca 2.5 Ga events in Kola, to the most widespread phase at 2.45 Ga in Kola and Karelia, and finally to the 2.40 Ga phase in Ringvassøy [34], which is close to the age of the older Scourie dyke generation in the Lewisian. The 2221 Ma dioritic sill intruding meta-sedimentary rocks of the Vanna group [29] is also the expression of a localized but very ubiquitous magmatic phase across northern Fennoscandia [124] and also in Laurentia. No equivalents have so far been described from the Lewisian. However, there is a good temporal correlation, instead, between mafic magmatism at 1.98 Ga in the Mjelde-Skorelvvatn belt of the West Troms Basement Complex [31] and emplacement of the younger generation of the Scourie dykes in the Lewisian, both corresponding to a period of extension and rifting in Laurentia and Fennoscandia.

1.90-1.85 Ga arc magmatism and convergence

The subsequent period of arc magmatism, likely connected to plate convergence and subduction, and final collision was important in the Fennoscandian Shield [4] and also in the Lewisian where it formed the well documented successions at Loch Maree and South Harris in a sequence of events between 1.90 and 1.85 Ga. Granitic sheets of this age are also important along the Laxford shear zone. In Lofoten there was a correlative event at 1.87 Ga, which emplaced granite and local mangerite-charnockite intrusions. Such rocks, however, have so far not been reported from the West Troms Basement Complex, a feature that may reflect a more distal position relative to the orogenic front near Lofoten (Figure 23c; see below). In the Lofoten-Vesterålen province, there is also clear evidence of meta-supracrustal units that post-date the Neoarchaean gneisses [9], but Palaeoproterozoic ages have not yet been documented by radiometric age dating.

1.80 – 1.78 Ga magmatism

The single most important and widespread magmatic event affecting the West Troms Basement Complex and Lofoten-Vesterålen province occurred in a short burst at 1.80 -1.79 Ga. It formed most of the AMCG suite in Lofoten and the major Paleoproterozoic intrusions in Kvaløya and Senja. In Lofoten the event was pre- and post-dated by high grade metamorphism and ductile deformation [9, 10], whereas in western Troms there is no evidence for much activity preceding this magmatic phase. These events can be correlated with a well-defined period of late orogenic magmatism in Fennoscandia [10, 125]. Interestingly, there are no such plutons or strong metamorphic overprint of this age in the northern part of Ringvassøya and Vanna, even though granulite facies metamorphism and partial melting occurred at about 1.78 Ga in Sandøya, just at the edge of this block, supporting an allochthonous origin of the latter [17]. A similar situation is also characteristic of the entire Lewisian which lacks 1.80-1.78 Ga intrusion altogether

1.80-1.75 Ga deformation and metamorphism

Regional deformation and metamorphism are well documented in the West Troms Basement Complex at c. 1.80-1.75 Ga (Table 1). These processes involved high-strain deformation and prograde metamorphism up to granulite facies (1.78-1.768 Ga). The deformation was focused mainly along the boundaries to metasupracrustal belts, e.g. in the Senja Shear Belt (Figure 10, 23c, d), and was probably also superimposed on pre-existing Neoarchaean structures [17, 32, 51]. The deformation started with ENE-directed thrusting and was followed by macroscopic upright folding and combined, late-stage strike-slip shearing and SE-directed thrusting (Figure 24) [17]. The late stages of deformation, not yet documented by age datings (but likely younger than 1.75 Ga), were characterized by partitioned contraction and lateral displacements. In the limbs of the mid-stage macroscopic folds (Figure 24b) the subsequent oblique deformation produced foliation-parallel sinistral strike-slip faults and steeply-plunging folds (Figure 24c), creating a regional lens-shaped structural pattern in the West Troms Basement Complex (Figure 10). The final phase of SE-directed thrusting (Figure 24d) was temporally linked to the strike-slip shearing, thus indicating partitioned transpression as the overall deformation mechanism [17].

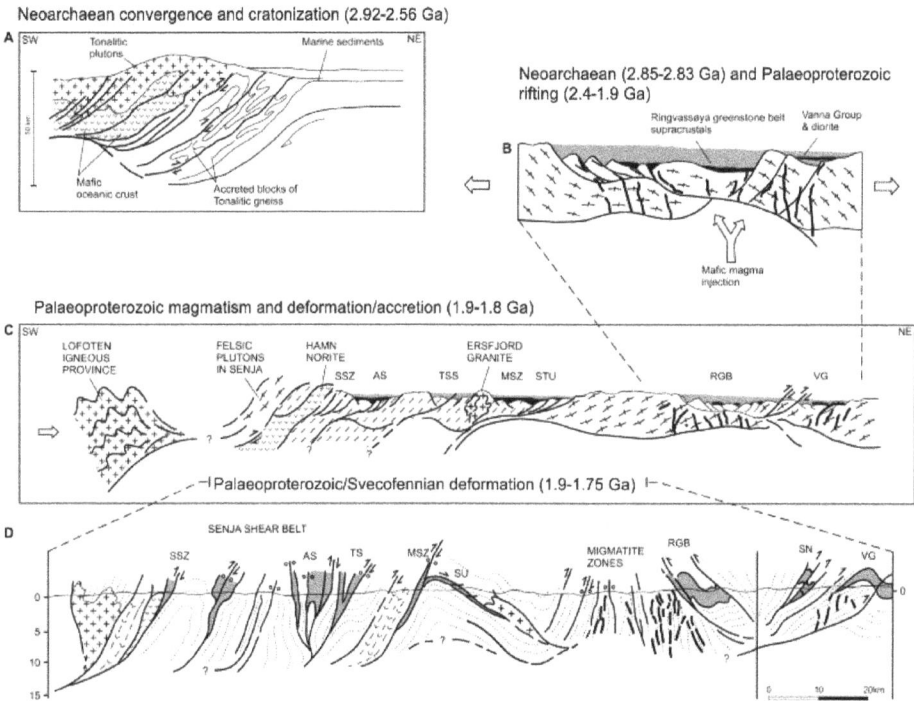

Figure 23. Cartoon sections summarizing the Neoarchaean to Palaeoproterozoic tectonic evolution of the West Troms Basement Complex: (a) Neoarchaean (2.92-2.56 Ga) tonalitic gneiss forming events with crustal accretion and thickening due to underplating. Note position of possible precursory volcanic deposits of the Ringvassøya greenstone belt. (b) Neoarchaean and Early Palaeo proterozoic (2.4-1.9 Ga) crustal extension, basin formation and intrusion of the Ringvassøya mafic dyke swarm. (c) Palaeoproterozoic (1.9-1.8 Ga) continental contraction and probable magmatic arc accretion in the southwest (including the Lofoten AMCG suite). (d) Section illustrating the composite result of Palaeoproterozoic crustal contraction, accretion and continent-continent collision with increasing transpressive deformation through time. For abbreviations, see Figure 2.

In contrast, although the deformation style is similar, there is not full evidence for temporally equivalent deformation events in the Lewisian. The exception is titanite ages of about 1750 Ma near the Laxford shear zone. These ages are considered as the potential expression of a phase of regional metamorphism but the evidence in favor of such an interpretation is dubious. In the Lewisian there are, for example, no datable dykes interspersed with the deformation events like in the West Troms Basement Complex.

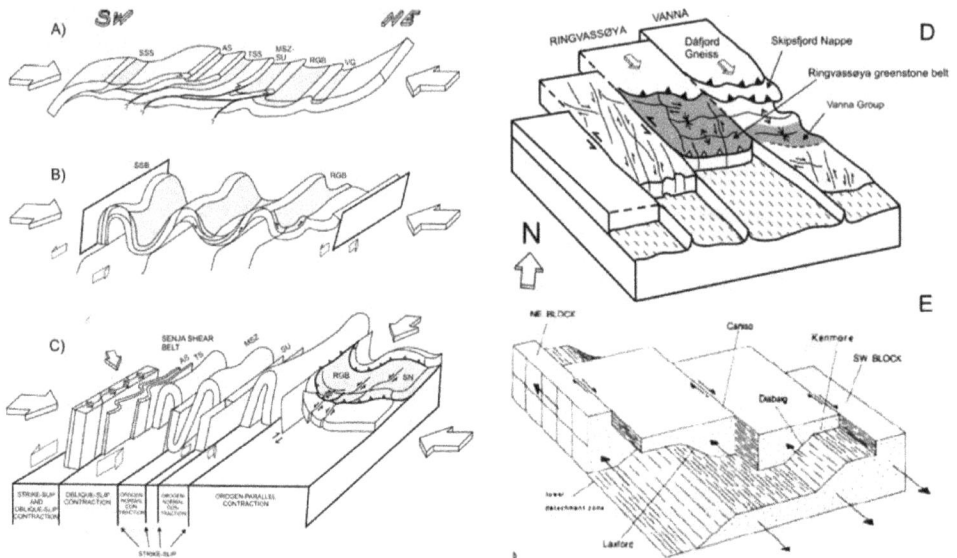

Figure 24. Tectonic model for the Palaeoproterozoic deformation in the West Troms Basement Complex (A-D) [17] compared with the Lewisian (E) [37]. The overall framework is that of NE-SW directed orthogonal shortening and an increasing transpressive component with time. The spatial domains, named in Figure 2, and their kinematic characters are also illustrated. (A) Early-stage formation of NE-directed thrusts and a low-angle main mylonitic foliation in the metasupracrustal belts. (B) Continued orthogonal NE-SW contraction produced upright macro-folds with steep limbs. Note that the main foliation and early thrusts were folded. (C) Late-stage tectonism involved NE-SW orthogonal and/or oblique to orogen-parallel contraction (NW-SE) and mostly sinistral strike-slip reactivation of steep macro-fold limbs, e.g. in the Senja Shear Belt. The eastern, more flat-lying macro-fold hinges (e.g. Ringvassøya greenstone belt) provided the locus for potential low-angle thrust detachments that may have accommodated partitioned NW-SE shortening and SE-directed thrusting. (D) Late-stage Palaeoproterozoic kinematic model for the north-eastern part of the West Troms Basement Complex, where potential low-angle shear zones/detachments accommodated NW-SE directed thrust movements on flats and steep orogen-parallel strike-slip/transfer-type shear zones on ramps [17]. (E) Simplified kinematic model for the Palaeo protero-zoic deformation in the Lewisian Complex involving a combination of thrust and strike-slip movements on flats/detachments and ramps/steep transfer-type shear zone [37].

1.70-1.65 Ga deformation and metamorphism

The Lewisian underwent a very distinct set of events at 1.70-1.65 Ga including deformation, the local development of migmatites, the ubiquitous intrusion of pegmatites, and low grade metamorphic overprints reflected in secondary titanite and rutile ages. These events largely post-date similar pegmatite intrusions affecting the West Troms Basment Complex and the Lofoten-Vesterålen province.

The late stages of magmatism and deformation in the Lewisian at 1.70-1.65 Ga involved localized steep ductile reverse (Figure 17) and dextral oblique shear zones developed by strain partitioning likely due to reactivation of steep pre-existing shear zones [77]. This partitioned deformation was interpreted to have been related to major flat-lying detachment zones (Figure 24e) that generated dip-slip thrust movement on flat portions, i.e. the Laxford shear zone, and strike-slip shear zones on steep oblique ramps, i.e. the Canisp shear zone [37]. The 1.70-1.65 Ga stage corresponds to a retrogression of the amphibolite/granulite facies conditions to greenschist facies, indicating exhumation of the rocks from mid to upper crustal levels [79]. This event was associated with the formation of steep-plunging asymmetric folds, retrogressed cleavage, widespread fluid-flow and quartz vein precipitation, and multiple strike-slip shear zones [37].

A similar crustal model is invoked for the late-stage partitioned deformation in the West Troms Basement Complex [17], which, despite lack of critical age dating may correspond to the 1.70-1.67 Ga event in the Lewisian (Figure 24d). In this model, potential flat-lying thrust detachments (shear zones) are present in the northern part of the region.

In summary, the Palaeoproterozoic deformation in the Lewisian Complex and that of the West Troms Basement Complex and Lofoten-Vesterålen, display obvious similarities in terms of tectonic style and partitioning of the deformation, even though they may not be fully temporally correlative (Figure 23). They show: (i) Long-term protracted deformation character, (ii) presence of crustal scale ductile shear zones (iii) partitioned transpressive crustal deformation, e.g. thrusts and orogen-parallel strike-slip shear zones, and (iv) spatial changes in metamorphic grades, i.e. major tectono-metamorphic breaks such as the regional metamorphic boundary in Lofoten and the Laxford shear zone in the Lewisian. These similarities are all consistent with comparable tectonic assembly processes caused by accretion followed by crustal convergence and orogen-oblique/parallel displacement of crustal segments, and a terminal phase of crustal/differential uplift and reactivation [8, 17].

4.3. Linking Palaeoproterozoic terranes and events in the North Atlantic realm

A major problem when trying to restore Precambrian plate tectonics is the nature and processes of assembly of lower crustal blocks or terranes [3, 4, 7, 8]. In terms of the North Atlantic realm Williams et al. (1991)[126] proposed that at the end of the Neoarchean there was a supercontinent, Kenorland, whose breakup led to the formation of several micro-continents which were reassembled together with juvenile terranes in the Palaeoproterozoic (Figure 22). Others, however, have argued for the existence of several micro-continents at the end of the Neoarchaean instead of one single supercontinent [33].

Most workers agree that the Karelia and Superior cratons of Fennoscandia and Laurentia were in close vicinity to each other or connected in the Neoarchaean [24]. The outline of these cratons (Figure 22) is a result of cycles of collision, granitoid intrusion, extension, rifting and basin formation, and if several of these events can be correlated between various cratons, then it is possible to reconstruct former crustal assemblages or supercratons [33]. In particular, timing of large igneous provinces and associated episodes of continental breakup and supracrustal deposits can be used for such analyses, whereby the most detailed record known is that of the Laurentian cratonic fragments [127]. Similar Palaeoproterozoic configurations have also been discussed in the literature, and various models presented [8, 110, 128, 129, 130]. Following breakup of a potential Neoarchaean supercraton, oceanic arcs started to converge from c. 2.0 Ga, with eventually accretion of the cratons along sutures that follow the grain of the Palaeoproterozoic orogens (Figure 22). The model proposed by [110] suggests a rather familiar configuration of the various Fennoscandia/Baltica and Laurentia cratons at the beginning of the Palaeoproterozoic (Figure 22), and despite being a speculative model it addresses the need for more detailed research within these cratons and especially along their margins.

Recent paleomagnetic reconstruction of the Palaeoproterozoic [8, 128] suggests the presence of several large colliding plates, including the North Atlantic and west Greenland plates, the Central Greenland Craton and the Fennoscandian (Baltic-Kola) plate, with the Lewisian somewhere in between (Figure 25). Most workers link the Lewisian to the Palaeoproterozoic Nagssugtoqidian belt in Greenland [5, 129, 131], and consider that this belt may have counterparts both in North America and/or the Fennoscandian Shield [3, 4, 110]. A link between the Lewisian of NW Scotland and the Lappland-Kola and Karelia craton of northern Fennoscandia would then place the West Troms Basement Complex and Lofoten-Vesterålen province exactly along the line of intersection between these major Palaeoproterozoic orogenic belts (Figure 22). A similar reconstruction [129] supports a correlation of Palaeoproterozoic orogens in Greenland and Fennoscandia at the c. 1.8 Ga supercontinent stage.

The scenario proposed by Park (2005) [8] gives a valid plate setting for the end of the Neoarchaean (Figure 25a) and explains the subsequent Palaeoproterozoic tectono-metamorphic events in the Lewisian Complex and tentatively, also the deformation events in the West Troms Basement Complex and Lofoten-Vesterålen province. At ca. 1.9-1.87 Ga, volcanic arcs were created between North American craton and Central Greenland craton/Kola craton due to the subduction/ accretion of the oceanic crust located between them (Figure 25b). The calk-alkaline plutonic intrusions within the Loch Maree Group, the South Harris Igneous complex, and potentially, the earliest phases of magma intrusions in the Lofoten igneous province and West Troms Basement Complex, and accompanied convergent deformation and granulite facies metamorphism manifest this regional accretionary event [8]. At ca. 1.87 Ga, the Central Greenland craton and Kola-Karelian craton collided and was under-thrusted beneath the the North American craton in a NW-SE direction within the Lapland-Kola belt, and resulting in the main phase of deformation (Figure 25c). Granulite facies metamorphism occurred in the down-going slab due to under-thrusting (lower plate). The line of collision between juvenile terranes was likely oriented in the same direction as the Palaeoproterozoic Nagssugtoqidian belt and the orientation of the collision could be given by the orientation of the main NW-SE trend of the Laxford shear zone [8]. At ca. 1.8 Ga, subduction of oceanic crust to the SW of this new continent may have created a volcanic arc trending NW-SE (Figure 25d), and this arc may have been involved with renewed collision at ca. 1.75 Ga (Figure 25d, e), corresponding to the main stages of deformation in the West Troms Basement Complex. There, the intrusion of the calk-alkaline Hamn norite (1.8 Ga) and the Ersfjord granite (1.79 Ga) may have been related to this phase. Similar calk-alkaline intrusive rocks exist further south of the

Lewisian complex [132, 133], known as the 'Malin block' and this block was thought to be part of a belt comprising the Labradorian of NE Canada, the Ketilidian of South Greenland and the Gothian of Scandinavia (see Figure 22). This belt became tectono-magmatic active at *c.* 1.8 Ga and an event at c. 1.7 Ga could be the result of igneous activity and deformation related both to the latest Palaeoproterozoic events in the Lewisian and in the West Troms Basement Complex.

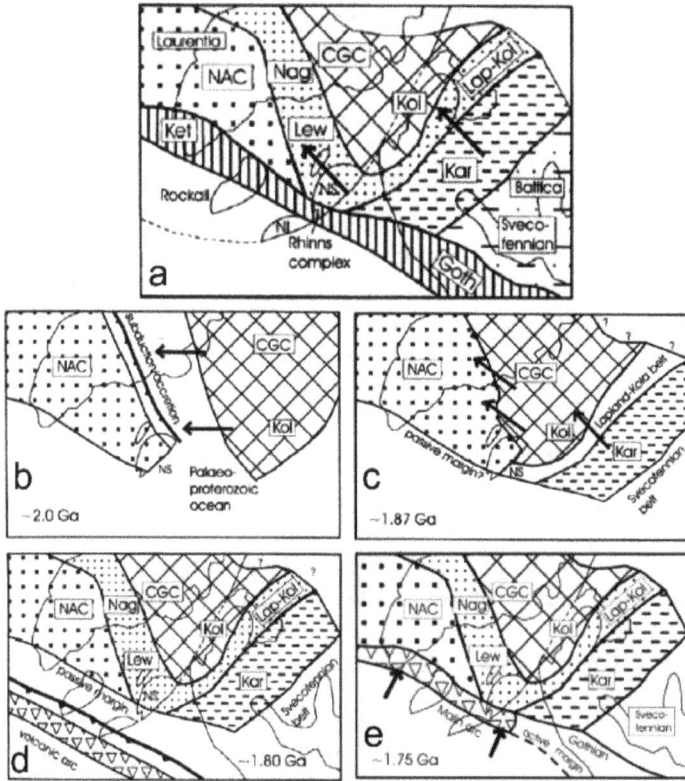

Figure 25. Plate tectonic setting of the Lewisian complex during the Palaeoproterozoic, based on the North Atlantic reconstruction [8, 128]. Abbreviations: CGC=Central Greenland craton, Goth = Gothian belt, Kar = Karelia craton, Ket = Ketilidian belt, Kola= Kola craton, Lap-Kol = Lapland-Kola belt, Lew = Lewisian, NAC = North Atlantic craton, Nag = Nagssugtoqidian belt, NI = north Ireland, NS = north Scotland. (a) Distribution of cratons and orogenic belts during the Mesoproterozoic. (b) 2.0 Ga: Subduction and creation of a volcanic arc in oceanic crust between two continental plates (NAC and CGC/Kol) followed by accretion of oceanic/arc elements along the leading edge of the NAC. (c)1.87 Ga: Collision of the two continents followed by underthrusting of the CGC/Kola craton beneath the NAC, causing the early Palaeoproterozoic deformation and metamorphism. At the same time, collision occurs in the Lapland/Kola belt to the SE caused by collision with the Karelia craton. Note the NW-SE movement direction. (d) 1.80 Ga: Development of a volcanic arc in oceanic crust SW of the amalgamated continent created in b. (e) 1.75 Ga: Collision between the "Malin block" and the continent, causing late-Palaeoproterozoic deformation, metamorphism and granitic melt formation in the Lewisian complex.

5. Conclusions

(1) The West Troms Basement Complex is underlain by Archaean gneisses (2.92-2.56 Ga), metasupracrustal rocks (2.4-1.9 Ga) and mafic dyke swarms (2.4 and 2.2 Ga) that were variably reworked, metamorphosed and intruded by felsic and mafic plutons at c.1.8 Ga. Along strike to the southwest, Neoarchaean high-grade gneisses in the Lofoten-Vesterålen province display magmatic protolith ages of between 2.85 and 2.7 Ga and record a high-grade metamorphic event at c. 2.64 Ga. The Neoarchaean basement rocks have been intruded by a huge 1.8 Ga magmatic suite composed of anorthosites, mangerites, charnockites, gabbros and granites, which corresponds in age with the 1.81–1.77 Ga, NW-trending granitoids in the older part of the Transscandinavian igneous belt of southern Sweden. A similar present structural position of the basement high in the Lofoten-Vesterålen province and the West Troms Basement Complex invokes they are along-strike correlatives.

(2) The Lewisian rocks of NW Scotland comprise a series of Neoarchaean blocks thought to have been amalgamated during a multistage and complex set of Palaeoproterozoic collision events between 1.97 and 1.67 Ga, producing a variety of block-bounding accretional and intrablock shear zones. This province also records Neoarchaean crustal deformation and metamorphism at intervals of 2.7-2.6 Ga and 2.49-2.40 Ga followed by episodes of crustal rifting and mafic dyke intrusion (2.4 and 2.0 Ga), deposition of continental margin-like metasedimentary sequences between 2.0 and 1.9 Ga ago upon the substratum of Neoarchaean gneisses and later on subjected to major orogenic deformation and metamorphism (c. 1.85 and 1.70 Ga).

(3) The West Troms Basement Complex, the Lofoten-Vesterålen province and the Lewisian rocks of Scotland are thus very similar crustal regions in terms of lithology, age, igneous, structural and metamorphic features and tend to share a similar tectono-magmatic and evolutionary history, but there are also sharp differences such as the lack of 1.80 magmatism in the Lewisian and the c.100 m.y. difference in the timing of the latest Palaeoproterozoic deformation overprints in the two regions.

(4) Reconstructing Palaeoproterozoic plate scenarios is a difficult task. Nevertheless, paleomagnetic restorations suggest the presence of several large colliding plates, including the North Atlantic and western Greenland plates, the Central Greenland craton and the Fennoscandian (Baltic-Kola) Shield, with the Lewisian somewhere in between. In this context, the Lewisian has been temporally linked to the Palaeoproterozoic Nagssugtoqidian belt in Greenland and may have its counterpart in North America and/or the Fennoscandian Shield. A link between the Lewisian of NW Scotland and the Lappland-Kola and Karelia craton of northern Fennoscandia would locate the West Troms Basement Complex and Lofoten-Vesterålen province directly along the line of intersection between these major Palaeoproterozoic orogenic belts at the c. 1.8 Ga supercontinent stage.

(5) Tentative similar Palaeoproterozoic terrane models (1.80-1.67 Ga) can be invoked for the basement outliers in northern Norway and the Lewisian Complex. The continental assembly may have involved either multiple small terranes or crustal rejuvenation of one or two large

terranes. The latter model is based on component similarities and metamorphic variations and can be explained by the presence of at least two different crustal blocks and/or depth portions assembled along crustal scale ductile shear zones. The juxtaposition included arc-magmatism and accretion of Neoarchaean continental terranes in the vicinity of the Fennoscandia-Laurentia border, followed by uplift and reworking.

Author details

Steffen G. Bergh[1], Per Inge Myhre, Kåre Kullerud and Erling K. Ravna
Dept. of Geology, University of Tromsø, Tromsø, Norway

Fernando Corfu
Dept. of Geosciences, University of Oslo, Blindern, Oslo, Norway

Paul E.B. Armitage
Consulting Ltd, Rochester, England, UK

Klaas B. Zwaan
Geological Survey of Norway, Trondheim, Norway

Robert E. Holdsworth
Dept. of Earth Sciences, University of Durham, UK

Anupam Chattopadhya
Dept. of Geology, University of Delhi, Delhi, India

Acknowledgement

This paper is based on extensive research in the West Troms Basement Complex over the last decade. The still ongoing work is an interdisciplinary study at the University of Tromsø, the University of Oslo and the Geological Survey of Norway, aimed at resolving Precambrian regional tectonic questions. The work in the Lewisian was undertaken during the first author's sabbatical leave at University of Durham in 2006-07, and the Department of Earth Science is thanked for hosting his visit and providing necessary infrastructural facilities for the research. SB also wishes to thank Dr. R.W. Wilson and Prof. Ken McCaffrey for constructive collaboration during the project.

6. References

[1] Gaal G, Gorbatschev R (1987) An outline of the Precambrian evolution of the Baltic Shield. Prec Res 35: 15-52.
[2] Gorbatschev R, Bogdanova S (1993) Frontiers in the Baltic Shield. Prec Res 64: 3-21.
[3] Höltta P, Balagansky V, Garde A, Mertanen S, Peltonen P, Slabunov A, Sorjonen-Ward P, Whitehouse M (2008) Archaean of Greenland and Fennoscandia. Episodes 31: 13-19.

---

[1] Corresponding Author

[4] Lahtinen R, Garde A, Melezhik VA (2008) Palaeoproterozoic evolution of Fennoscandia and Greenland. Episodes 31: 20-28.

[5] Van Gool JAM, Connelly J N, Marker M, Mengel FC (2002) The Nagssugtoqidian orogen of West Greenland: tectonic evolution and regional correlation from a West Greenland perspective. Can J Earth Sci 39: 665–686.

[6] Sidgren A.-S, Page L, Garde AA (2006) New hornblende and muscovite 40Ar/39Ar cooling ages in the central Rinkian fold belt, West Greenland. Geol Surv Den Greenl Bull 11: 115–123.

[7] Friend PD, Kinny PD (2001) A reappraisal of the Lewisian Gneiss Complex: geochronological evidence for its tectonic assembly from disparate terranes in the Proterozoic. Cont Min Pet 142:198–218.

[8] Park R G (2005) The Lewisian terrane model: a review. Scott J Geol 41:105–118.

[9] Griffin WL, Taylor PN, Hakkinen JW, Heier KS, Iden IK, Krogh EJ, Malm O, Olsen KI, Ormaasen DE, Tveten E (1978) Archaean and Proterozoic crustal evolution in Lofoten-Vesterålen, N. Norway. J Geol Soc Lond 135: 629-647.

[10] Corfu F (2004) U-Pb age, setting and tectonic significance of the anorthosite-mangerite-charnockite-granite suite, Lofoten-Vesterålen, Norway. J Pet 45:1799-1819.

[11] Olesen O, Torsvik T, Tveten E, Zwaan KB, Løseth H, Henningsen T (1997) Basement structure of the continental margin in the Lofoten-Lopphavet area, northern Norway: constraints from potential field data, on-land structural mapping and paleomagnetic data. Norw J Geol 77:15-30.

[12] Blystad P, Brekke H, Færseth RB, Larsen BT, Skogseid J, Tørudbakken B (1995) Structural elements of the Norwegian continental shelf. Part II: The Norwegian Sea Region. Norw Petr Dir Bull 8: 1-45.

[13] Henkel H (1991) Magnetic crustal structures in Northern Fennoscandia. Tectonophysics 192: 57-79.

[14] Gorbatschev R (2004) The Transscandinavian Igneous Belt – introduction and background. In Högdahl K, Andersson UB, Eklund O, editors. The Transscandinavian Igneous Belt (TIB) in Sweden: a review of its character and evolution. Geol Surv Finland Specl Paper 37: 9–15.

[15] Lahtinen R, Korja A, Nironen M (2005) Palaeoproterozoic tectonic evolution of the Fennoscandian Shield. In: Lehtinen M, Nurmi P, Rämö T, editors. The Precambrian Bedrock of Finland- Key to the evolution of the Fennoscandian Shield. Elsevier Science BV: 418–532.

[16] Corfu F, Armitage PEB, Kullerud K, Bergh SG (2003) Preliminary U-Pb geochronology in the West Troms Basement Complex, North Norway: Archaean and Palaeoproterozoic events and younger overprints. Geol Surv Norway Bull 441: 61-72.

[17] Bergh SG, Kullerud K, Armitage PEB, Zwaan KB, Corfu F, Ravna EJK, Myhre PI (2010) Neoarchaean to Svecofennian tectono-magmatic evolution of the West Troms Basement Complex, North Norway. Norw J Geol 90: 21-48.

[18] Bamford D, Nunn K, Prodehl C, Jacob B (1978) Crustal structure of Northern Britain. Geophys J Int 54: 43-60

[19] Park RG, Cliff RA, Fettes DJ, Stewart AD (1994) Precambrian rocks in northwest Scotland west of the Moine Thrust: the Lewisian Complex and the Torridonian. Geol Soc Lond Spec Rep 22:6-22

[20] Woodcock N, Strachan RA (2000) Geological History of Britain and Ireland. Blackwell Science Ltd.

[21] Park RG (1995) Palaeozoic Laurentia-Baltica relationships: a view from the Lewisian. In Coward MP, Ries AC, editors, Early Precambrian processes. Geol Soc Lond Spec Publ 105:299–310.

[22] Bergh SG, Kullerud K, Holdsworth RE, Corfu F, Armitage PEB, McCaffrey K, Ravna E, Wilson RW, Chattopadhyay A (2007) The West Troms Basement Complex, North Norway: A segment of the Lewisian crust assembled by multiple Archaean through Palaeoproterozoic collision events. Abstr, Geol Soc Am, Continental Tectonics and Mountain Building, Peach & Horne Centennial Meeting, NW Scotland.

[23] Bleeker W, Ernst R (2006) Short-lived mantle generated magmatic events and their dyke swarms: The key unlocking Earth's paleogeographic record back to 2.6 Ga. In Hanski E, Mertanen S, Rämö T, Vuollo J, editors. Dyke Swarms - Time Markers of Crustal Evolution. Balkema Publishers, Rotterdam, 3-26.

[24] Mertanen S, Korhonen F (2011) Paleomagnetic constraints on an Archean-Paleoproterozoic Superior-Karelia connection: New evidence from Archean Karelia. Prec Res 186:193-204.

[25] Bridgwater D, Austrheim H, Hansen BT, Mengel F, Pedersen S, Winter J (1990) The Proterozoic Nagssugtoqidian mobile belt of southeast Greenland: a link between the Eastern Canadian and Baltic Shields. Geosci Can 17:305–310.

[26] Connelly JN, Van Gool JAM, Mengel FC (2000) Temporal evolution of a deeply eroded orogen: the Nagssugtoqidian Orogen, West Greenland. Can J Earth Sci 37: 1121-1142.

[27] Armitage PEB, Bergh SG (2005) Structural development of the Mjelde-Skorelvvatn Zone on Kvaløya, Troms: a metasupracrustal shear belt in the Precambrian West Troms Basement Complex, North Norway. Norw J Geol 85:117-132.

[28] Kullerud K, Corfu F, Bergh SG, Davidsen B, Ravna E K (2006) U-Pb constraints on the Archaean and Early Proterozoic evolution of the West Troms Basement Complex, North Norway. Abstr Bull Geol Soc Finl Spec Issue I, pp. 79.

[29] Bergh SG, Kullerud K, Corfu F, Armitage PEB, Davidsen B, Johansen HW, Pettersen T, Knudsen S (2007) Low-grade sedimentary rocks on Vanna, North Norway: a new occurrence of a Palaeoproterozoic (2.4-2.2 Ga) cover succession in northern Fennoscandia. Norw J Geol 87: 301-318.

[30] Myhre PI, Heaman LM, Bergh SG (2010) Svecofennian (c. 1780 Ma) metamorphic zircon ages from the West Troms Basement Complex, northern Norway. In NGF Abstr Proc Geol Soc Norway. pp. 128-129.

[31] Myhre PI, Corfu F, Bergh S (2011) Palaeoproterozoic (2.0–1.95Ga) pre-orogenic supracrustal sequences in the West Troms Basement Complex, North Norway. Precamb Res 186: 89-100.

[32] Myhre PI (2011) U-Pb geochronology along a Meso-Neoarchean geotransect in the West Troms Basement Complex, North Norway. Unpubl PhD thesis, University of Tromsø, Norway.

[33] Bleeker W (2003) The late Archaean record: a puzzle in ca. 35 pieces. Lithos 71: 99-134.

[34] Kullerud K, Skjerlie KP, Corfu F, DeLaRosa J (2006) The 2.40 Ga Ringvassøy mafic dykes, West Troms Basement Complex, Norway: The concluding act of Early Palaeoproterozoic continental breakup. Prec Res 150:183-200.

[35] Tveten E (1978) Geological map of Norway, bed rock map Svolvær 1:250 000, Geological Survey of Norway

[36] Andresen A, Tull JF (1983) The age of the Lødingen granite and its possible regional significance. Norw J Geol 63:269-276.

[37] Coward MP, Park RG (1987) The role of mid-crustal shear zones in the Early Proterozoic evolution of the Lewisian. In Park RG, Tarney J, Editors. Evolution of the Lewisian and Comparable Precambrian High Grade Terrains. Geol Soc Lond Spec Publi 27: 127-138

[38] Zwaan KB (1995) Geology of the Precambrian West Troms Basement Complex, northern Norway, with special emphasis onthe Senja Shear Belt: a preliminary account. Geol Surv Norw Bull 427: 33-36.

[39] Kullerud K, Skjerlie KP, Corfu F, DeLaRosa J (2006) The 2.40 Ga Ringvassøy mafic dykes, West Troms Basement Complex, Norway: The concluding act of Early Palaeoproterozoic continental breakup. Prec Res 150:183-200.

[40] Zwaan KB, Fareth E, Grogan PW (1998) Geologic map of Norway, bed rock map Tromsø, M 1:250.000. Geol Surv Norw.

[41] Zwaan KB, Tucker RD (1996) Absolute and relative age relationships in the Precambrian West Troms Basement Complex, northern Norway (Abstract). 22nd Nord Geol Wint Meet, Finland, pp 237.

[42] Zozulya D, Kullerud K, Ravna EK, Corfu F, Savchenko Y (2009) Geology, age and geochemical constraints on the origin of the Late Archaean Mikkelvik alkaline massif, West Troms Basement Complex in Northern Norway. Norw J Geol 89:327-340

[43] Zwaan KB (1989) Berggrunnsgeologisk kartlegging av det prekambriske grønnsteinsbeltet på Ringvassøy, Troms. Geol Surv Norw Report 89:101

[44] Motuza G, Motuza V, Beliatsky B, Savva E (2001) Volcanic rocks of the Ringvassøya Greenstone Belt (North Norway):Implication for the Stratigraphy and Tectonic Setting. J Conf (Abstract) EUG XI, 6: 578.

[45] Binns RE, Chroston PN, Matthews DW (1980) Low-grade sediments on Precambrian gneiss on Vanna, Troms, Northern Norway. Geol Surv Norw Bull 359: 61-70.

[46]Andresen A (1979) The age of the Precambrian basement in western Troms, Norway. Geol För Stockh Förh 101: 291-298.

[47] Krill AG, Fareth E (1984) Rb-Sr whole-rock ages from Senja, North Norway. Norw J Geol 64:171-172.

[48] Condie KC (2005) TTGs and adakites: are they both slab melts? Lithos 80: 33-44

[49] Henderson I, Kendrick M (2003) Structural controls on graphite mineralisation, Senja, Troms. Geol Surv Norw Report 2003.011: 111pp.

[50] Doré AG, Lundin ER, Fichler C, Olesen O (1997) Patterns of basement structure and reactivation along the NE Atlantic margin. J Geol Soc Lond 154: 85-92.

[51] Gjerløw E (2008) Petrologi og alder av høymetamorfe mafiske bergarter i det vestlige gneiskomplekset i Troms. Unpubl MS thesis, Univ Tromsø, 90 pp.

[52] Herr W, Wolfe R, Kopp E, Eberhardt P (1967) Development and recent applications of the Re/Os dating method. In Radioactive dating and methods of low-level counting. Int Atomic Energy Agency, Vienna, pp 499-508.

[53] Heier KS, Compston W (1969) Interpretation of Rb-Sr age patterns in high grade metamorphic rocks, north Norway. NorGeol Tidsskr 49:257-283.

[54] Heier KS (1960) Petrology and geochemistry of high-grade metamorphic and igneous rocks on Langøy, Northern Norway. Nor Geol Tidsskr 207:1-246.

[55] Mjelde R, Sellevoll MA, Shimamura H, Iwasaki T, Kanazawa T (1993) Crustal structure beneath Lofoten, N. Norway from vertical incidence and wide-angle seismic data. Geoph J Int 114:116-126.

[56] Taylor PN (1975) An Early Precambrian Age for Migmatitic Gneisses from Vikan i Bø, Vesterålen, North Norway. Earth Plan Sci Lett 27:35-42.

[57] Jacobsen SB, Wasserburg GJ (1978) Interpretation of Nd, Sr and Pb isotope data from Archaean migmatites in Lofoten-Vesterålen, Norway. Earth Plan Sci Lett 41:245-253.

[58] Wade SJR (1985) Radiogenic Isotope Studies of Crust-forming Processes in the Lofoten-Vesterålen Province of North Norway. Unpubl PhD Thesis, Univ Oxford, vol. I:1-285, vol. II:1-292.

[59] Taylor PN (1974) Isotope Geology and Related Geochemical Studies of Ancient High-Grade Metamorphic Basement Complexes: Lofoten and Vesterålen, North Norway. Unp PhD thesis, Oxford Univ:1-187.

[60] Markl G (1998) The Eidsfjord anorthosite, Vesterålen, Norway: field observations and geochemical data. Nor Geol Tidsskr434: 51-73.

[61] Olsen KI (1978) Metamorphic petrology and fluid-inclusion studies of granulites and amphibolite-facies gneisses on Langøy and W Hinnøy, Vesterålen, N Norway. Unpubl Cand Real. thesis, Univ Oslo, 214 pp.

[62] Malm O, Ormaasen DE (1978) Mangerite-charnockite Intrusives in the Lofoten-Vesterålen Area, North Norway: Petrography, Chemistry and Petrology. Nor Geol Tidsskr 338: 83-114.

[63] Ormaasen, D.E. 1977: Petrology of the Hopen mangerite-charnockite intrusion, Lofoten, north Norway. Lithos 10, 291-310.

[64] Markl G, Frost BR, Bucher K (1998) The origin of anorthosites and related rocks from the Lofoten Islands, Northern Norway: I. Field relations and estimation of intrinsic variables. J Petrol 39:1425-1452.

[65] Markl G, Frost BR (1999) The origin of anorthosites and related rocks from the Lofoten Islands, Northern Norway: II. Modelling of parental melts for anorthosites. J Petrol 40: 61-77.

[66] Brueckner, H.K. 1971: The age of the Torset granite, Langöy, Northern Norway. Norsk Geologisk Tidsskrift 51, 85-87.

[67] Romer RL, Kjøsnes B, Korneliussen A, Lindahl I, Skyseth T, Stendal M, Sundvoll   B (1992) The Archaean-Proterozoic boundary beneath the Caledonides of northern Norway and Sweden: U-Pb, Rb-Sr and ε Nd isotope data from the Rombak-Tysfjord area. Geol Surv Norw Report 91.225, 67 pp.

[68] Markl G, Höhndorf A (2003) Isotopic constraints on the origin of AMCG-suite rocks on the Lofoten Islands, N Norway. Min Petrol 78:149-171.

[69] Markl G (2001) REE constraints on fractionation processes of massive-type anorthosites on the Lofoten Islands, Norway. Min Petrol 72:325-351.

[70] Sutton J, Watson J (1951) The pre-Torridonian metamorphic history of the Loch Torridon and Scourie areas in the North-west Highlands, and its bearing on the chronological classification of the Lewisian. Q J Geol Soc Lond 106:241-307

[71] Sutton J, Watson J (1962) Further observations on the margins of the Laxfordian Complex of the Lewisian near Loch Laxford, Sutherland. Trans Roy Soc Edin 65: 89-106

[72] Kinny PD, Friend CRL, Love GJ (2005) Proposal for a terrane-based nomenclature for the Lewisian Gneiss Complex of NW Scotland. J Geol Soc Lond 162:175-186.

[73] Evans CR,Tarney J (1964) Isotopic ages of Assynth dykes. Nature Lond 204: 638-641

[74] Evans CR (1965) Geochronology of the Lewisian basement near Lochinver, Sutherland. Nature 207: 54–56.

[75] Park RG (1970) Observations on Lewisian chronology. Scott J Geol 6: 379-399.

[76] Heaman LM, Tarney J (1989) U-Pb baddeleyite ages for the Scourie dyke swarm, Scotland: evidence for two distinct intrusion events. Nature 340: 705-708

[77] Attfield P (1987) The structural history of the Canisp Shear Zone. In: Park RG, Tarney J, editors. Evolution of the Lewisian and Comparable Precambrian High-Grade Terrains. Geol Soc Lond Spec Publ 27: 165–173.

[78] Coward MP (1984) Major shear zones in the Precambrian crust; examples from NW Scotland and southern Africa and their significance. In Kroner A, Greiling SR, editors. Prec Tect Illustr, Stuttgart, pp 207-235.

[79] Park RG, Crane A, Niamatullah M (1987) Early Proterozoic structure and kinematic evolution of the southern mainland Lewisian. In Park RG, Tarney J, editors, Evolution of the Lewisian and Comparable Precambrian High Grade Terrains. Geol Soc Lond Spec Publ 27: 139-151.

[80] Goodenough KM, Park RG, Krabbendam M, Myers J, Wheeler J, Loughlin SC,Crowley QG, Friend CRL, Beach A, Kinny P, Graham R (2010) The Laxford Shear Zone: an end-Archaean terrane boundary? Spec Publ Geol Soc 335:103-120.

[81] Davies FB (1976) Early Scourian structures in the Scourie-Laxford region and their bearing on the evolution of the Laxford Front. J Geol Soc Lond 132: 543–554.

[82] Chattopadhya A (2007) Digital mapping and analysis of continental basement shear zones. Unpubl report, Royal Soc Lond 56pp

[83] Giletti BJ, Moorbath S, Lambert RStJ (1961) A geochronological study of the metamorphic complexes of the Scottish Highlands. Q J Geol Soc Lond 117: 233-272

[84] Moorbath S, Welke H, Gale NH (1969) The significance of lead isotope studies in ancient high-grade metamorphic basement complexes, as exemplified by the Lewisian rocks of northwest Scotland. Earth Planet Sci Lett 6: 245-256

[85] Pidgeon RT, Bowes DR (1972) Zircon U-Pb ages of granulites from the Central Region of the Lewisian, northwesternScotland. Geol Mag 109: 247-258

[86] Chapman HJ, Moorbath S (1977) Lead isotope measurements from the oldest recognised Lewisian gneisses of north-west Scotland. Nature 268: 41-42

[87] Whitehouse MJ, Moorbath S (1986) Pb-Pb systematics of Lewisian gneisses - implications for crustal differentiation. Nature 319: 488-489

[88] Cohen AS, O'Nions RK, O'Hara MJ (1991) Chronology and mechanism of depletion in Lewisian granulites. Contrib Mineral Petrol 106: 142-153

[89] Friend CRL, Kinny PD (1995) New evidence for protolith ages of Lewisian granulites, northwest Scotland. Geology 23: 1027-1030

[90] Kinny PD, Friend CRL (1997) U-Pb isotopic evidence for the accretion of different crustal blocks to form the Lewisian Complex of northwest Scotland. Contrib Mineral Petrol 129:326-340

[91] Love GP, Kinny PD, Friend, CRL (2004) Timing of magmatism and metamorphism in the Gruinard Bay area of the Lewisian Gneiss Complex: comparisons with the Assynt Terrane and implications for terrane accretion. Contr Mineral Petrol 146:620-636

[92] Whitehouse MJ, Kemp AIS (2010) On the difficulty of assigning crustal residence, magmatic protolith and metamorphic ages to Lewisian granulites: constraints from combined in situ U-Pb and Lu-Hf. In Law RD, Butler RWH, Holdsworth RE, Krabbendam M, Strachan RA, editors. Continental Tectonics and Mountain Building: The Legacy of Peach and Horne. Geol Soc Lond Spec Publ 335: 81-101.

[93] Corfu F, Heaman LM, Rogers G (1994) Polymetamorphic evolution of the Lewisian complex, NW Scotland, as recorded by U-Pb isotopic compositions of zircon, titanite and rutile. Contrib Mineral Petrol 117:215-228

[94] Corfu F (2007) Comment to paper: Timing of magmatism and metamorphism in the Gruinard Bay area of the Lewisian gneiss complex: comparison with the Assynt Terrane and implications for terrane accretion by D.J. Love, P.D. Kinny and C.R.L. Friend (Contr Mineral Petrol (2004) 146:620-636). Contr Min Petr 153: 483–488 doi: 10.1007/s00410-006-0157-5

[95] Zhu XK, O'Nions RK, Belshaw NS, Gibb AJ (1997) Lewisian crustal history from in situ SIMS mineral chronometry and related metamorphich textures. Chem Geol 136: 205-218.

[96] Whitehouse MJ, Claesson S, Sunde T, Vestin J (1997) Ionmicroprobe U-Pb zircon geochronology and correlation of Achaean gneisses from the Lewisian complex of Gruinard Bay, northwestern Scotland. Geoch Cos Acta 61: 4429-4438

[97] Corfu F, Crane A, Moser D, Rogers G (1998) U-Pb zircon systematics at Gruinard Bay, northwest Scotland: implications for the early orogenic evolution of the Lewisian Complex. Contrib Mineral Petrol 133: 329–345

[98] Whitehouse MJ, Bridgwater D (2001) Geochronological constraints on Palaeoproterozoic crustal evolution and regional correlations of the northern Outer Hebridean Lewisian Complex, Scotland. Precamb Res 105: 227–245.

[99] Whitehouse MJ (2003) Rare earth elements in zircon: a review of applications and case studies from the Outer Heebrides Lewisian Complex, NW Scotland. In Vance D, Muller

W, Villa JM, editors. Geochronology: Linking the Isotopic Record with Petrology and Textures. Geol Soc Lond Special Publ 220: 49-64.

[100] Mason AJ, Brewer TS (2005) A re-evaluation of a Laxfordian terrane boundary in the Lewisian Complex of South Harris, NW Scotland. J Geol Soc Lond 162: 401–408.

[101] Kelly NM, Hinton RW, Harley SL, Appleby SK (2008) New SIMS U-Pb zircon ages from the Langavat Belt, South Harris, NW Scotland: implications for the Lewisian terrane model. J Geol Soc Lond 165: 967-981.

[102] Love GJ, Friend CRL, Kinny PD (2010) Palaeoproterozoic terrane assembly in the Lewisian Gneiss Complex on the Scottish mainland, south of Gruinard Bay: SHRIMP U-Pb zircon evidence. Precamb Res 183: 89-111.

[103] Humphries FJ, Cliff RA (1982) Sm-Nd dating and cooling history of Scourian granulites, Sutherland. Nature 295: 515-517

[104] Mason AJ, Parrish RR, Brewer TS (2004) U-Pb geochronology of Lewisian orthogneiss in the Outer Hebrides, Scotland: implications for the tectonic setting and correlation of the South Harris Complex. J Geol Soc Lond 161: 45–54.

[105] Tarney J, Weaver BL (1987) Geochemistry of the Scourian complex: petrogenesis and tectonic models. In: Park RG, Tarney J, editors. Evolution of the Lewisian and comparable Precambrian high-grade terrains. Geol Soc Spec Publ 27, pp 45-56

[106] Weaver BL, Tarney J (1981) The Scourie dyke suite: petrogenesis and geochemical nature of the Proterozoic sub-continental mantle. Contrib Min Petrol 78: 175-188.

[107] Park RG, Tarney J (1987) The Lewisian complex: a typical Precambrian high-grade terrain? Geol Soc Lond Spec Publ 27: 13-25 DOI: 10.1144/GSL.SP.1987.027.01.03

[108] Chapman HJ (1979) 2390 Myr Rb-Sr whole-rock for the Scourie dykes of north-west Scotland. Nature 277: 642-643

[109] Waters FG, Cohen AS, O'Nions RK, O'Hara MJ (1990) Development of Archaean lithosphere deduced from chronology and isotope chemistry of Scourie dykes. Earth Planet Sci Lett 97: 241-255

[110] Park RG, Tarney J, Connelly JN (2001) The Loch Maree Group: Palaeoproterozoic subduction-accretion complex in the Lewisian of NW Scotland. Precambr Research 105, 205-226.

[111] Park RG (2002) The Lewisian geology of Gairloch. Geol Soc Lond Mem 26: 76pp.

[112] Whitehouse MJ, Bridgwater D, Park RG (1997) Detrital zircons from the Loch Maree Group, Lewisian complex, NW Scotland: confirmation of a Palaeoproterozoic Laurentia-Fennoscandian connection. Terra Nova, 9: 260-263.

[113] Cliff RA, Gray CM, Huhma H (1983) A Sm–Nd isotopic study of the South Harris Igneous complex, the Outer Hebrides. Contrib. Miner. Petrol. 82, 91–98.

[114] Barooah BC, Bowes DR (2009) Multi-episodic modification of high-grade terrane near Scourie and its significance in elucidating the history of the Lewisian Complex. Scott J Geol 45: 19–41.

[115] Van Breemen O, Aftalion M, Pidgeon RT (1971) The age of the granite injection complex of Harris, Outer Hebrides. Scott. J. Geol. 7: 139–152.

[116] Pidgeon RT, Aftalion M (1972) The geochronological significance of discordant U-Pb ages of oval-shaped zircons from a Lewisian gneiss from Harris, Outer Hebrides. Earth Planet Sci Lett 17: 269-274.

[117] Buchan KL, Mortensen JK, Card KD, Percival JA (1998) Paleomagnetism and U-Pb geochronology of diabase dyke swarms of Minto block, Superior Province, Quebec, Canada. Can J Earth Sci 35: 1054-1069.

[118] Pesonen LJ, Elming S-u, Mertanen S, Pisarevsky S, D'Agrella-Filho MS, Meert JG, Schmidt PW, Abrahamsen,N, Bylund G (2003) Palaeomagnetic configuration of continents during the Proterozoic: Tectonophysics 375: 289-324.

[119] Mertanen S, Vuollo JI, Huhma H, Arestova NA, Kovalenko A (2006) Early Paleoproterozoic-Archean dykes and gneisses in Russian Karelia of the Fennoscandian Shield–New paleomagnetic, isotope age and geochemical investigations: Precambr Res 144: 239–260.

[120] Karlstrom KE, Williams M, McLelland J, Geissman JW, Åhall K-I (1999) Refining Rodinia: Geological evidence for the Australia-Western U.S. connection in the Proterozoic. GSA Today 9: 2-7.

[121] Koistinen T, Stephens M B, Bogatchev V, Nordgulen Ø, Wennerström M, Korhonen J (compilers) (2001) Geological map of the Fennoscandian Shield, scale 1:2 000 000. Espoo, Geol Surv Finl; Geol Surv Norw, Geol Surv Sweden, Ministry of Natural Resources of Russia, Moscow.

[122] Krogh TE (1993) High precision U-Pb ages for granulite metamorphism and deformation in the Archean Kapuskasing structural zone, Ontario: implications for structure and development of the lower crust. Earth Planet Sci Letters 119: 1-18.

[123] Moser DE, Heaman LM (1997) Proterozoic zircon growth in Archean lower-crustal xenoliths, southern Superior craton – a consequence of Matachewan ocean opening, Contrib Min Petrol 128: 164-175.

[124] Hanski E, Huhma H, Vaasjoki M (2001) Geochronology of northern Finland: a summary and discussion. In Vaasjoki M, editor. Radiometric age determinations from Finnish Lappland and their bearing on the timing of Precambrian volcano-sedimentary sequences. Geol Surv Finl Spec Paper 33: 255-279.

[125] Nironen M (1997) The Svecofennian Orogen: a tectonic model. Precambr Res 86: 21-44.

[126] Williams H, Hoffman PF, Lewry JF, Monger JWH, Rivers T (1991) Anatomy of North America: thematic geological portrayals of the continent: Tectonophysics 187: 117–134.

[127] Ernst R, Bleeker W (2010) Large igneous provinces (LIPs), giant dyke swarms, and mantle plumes: significance for breakup events within Canada and adjacent regions from 2.5 Ga to the Present. Can J Earth Sci 47: 695-739.

[128] Buchan KL, Mertanen S, Park RG, Pesonen LJ, Elming S-A, Abrahamsen N, Bylund G (2000) The drift of Laurentia and Baltica in the Proterozoic: a comparison based on key palaeomagnetic poles. Tectonophysics 319: 167-198.

[129] Connelly JN, Van Gool JAM, Mengel FC (2000) Temporal evolution of a deeply eroded orogen: the Nagssugtoqidian Orogen, West Greenland. Can J Earth Sci 37: 1121-1142.

[130] Pesonen LJ, Elming S, Mertanen S, Pisarevsky S, D'Agrella-Filho MS, Meert JG, Schmidt PW, Abrahamsen N, Bylund G (2003) Palaeomagnetic configuration of continents during the Proterozoic. Tectonophysics 375: 289-324.

[131] Connelly JN, Mengel FC (2000) Evolution of Archean components in the Paleoproterozoic Nagssugtoqidian orogen, West Greenland. Geol Soc Amer Bull 112: 747-763.

[132] Muir RJ, Fitches WR, Maltman AJ, Bentley MR (1994) Precambrian rocks of the southern, Inner Hebrides - Malin Sea region: Colonsay, West Islay, Inishtrahull and Iona. In Gibbons W, Harris AL editors. A Revised Correlation of Precambrian Rocks in the British Isles., Geol Soc Lond Spec Rep 22:54-58

[133] Daly JS, Muir RJ, Cliff RA (1991) A precise U-Pb zircon age for the Inishtrahull syenite gneiss, County Donegal, Ireland. J Geol Soc Lond 148: 639-642.

Permissions

The contributors of this book come from diverse backgrounds, making this book a truly international effort. This book will bring forth new frontiers with its revolutionizing research information and detailed analysis of the nascent developments around the world.

We would like to thank Evgenii V. Sharkov, for lending his expertise to make the book truly unique. He has played a crucial role in the development of this book. Without his invaluable contribution this book wouldn't have been possible. He has made vital efforts to compile up to date information on the varied aspects of this subject to make this book a valuable addition to the collection of many professionals and students.

This book was conceptualized with the vision of imparting up-to-date information and advanced data in this field. To ensure the same, a matchless editorial board was set up. Every individual on the board went through rigorous rounds of assessment to prove their worth. After which they invested a large part of their time researching and compiling the most relevant data for our readers. Conferences and sessions were held from time to time between the editorial board and the contributing authors to present the data in the most comprehensible form. The editorial team has worked tirelessly to provide valuable and valid information to help people across the globe.

Every chapter published in this book has been scrutinized by our experts. Their significance has been extensively debated. The topics covered herein carry significant findings which will fuel the growth of the discipline. They may even be implemented as practical applications or may be referred to as a beginning point for another development. Chapters in this book were first published by InTech; hereby published with permission under the Creative Commons Attribution License or equivalent.

The editorial board has been involved in producing this book since its inception. They have spent rigorous hours researching and exploring the diverse topics which have resulted in the successful publishing of this book. They have passed on their knowledge of decades through this book. To expedite this challenging task, the publisher supported the team at every step. A small team of assistant editors was also appointed to further simplify the editing procedure and attain best results for the readers.

Our editorial team has been hand-picked from every corner of the world. Their multi-ethnicity adds dynamic inputs to the discussions which result in innovative outcomes. These outcomes are then further discussed with the researchers and contributors who give their valuable feedback and opinion regarding the same. The feedback is then collaborated with the researches and they are edited in a comprehensive manner to aid the understanding of the subject.

Apart from the editorial board, the designing team has also invested a significant amount of their time in understanding the subject and creating the most relevant covers. They scrutinized every image to scout for the most suitable representation of the subject and create an appropriate cover for the book.

The publishing team has been involved in this book since its early stages. They were actively engaged in every process, be it collecting the data, connecting with the contributors or procuring relevant information. The team has been an ardent support to the editorial, designing and production team. Their endless efforts to recruit the best for this project, has resulted in the accomplishment of this book. They are a veteran in the field of academics and their pool of knowledge is as vast as their experience in printing. Their expertise and guidance has proved useful at every step. Their uncompromising quality standards have made this book an exceptional effort. Their encouragement from time to time has been an inspiration for everyone.

The publisher and the editorial board hope that this book will prove to be a valuable piece of knowledge for researchers, students, practitioners and scholars across the globe.

List of Contributors

Evgenii Sharkov, V.A. Lebedev and A.V. Chugaev
Institute of Geology of Ore Deposits, Petrography, Mineralogy and Geochemistry RAS, Moscow, Russia

A.G. Rodnikov, N.A. Sergeeva and L.P. Zabarinskaya
Geophysical Center RAS, Moscow, Russia

E.V. Sharkov
Institute of Geology of Ore Deposits, Petrography, Mineralogy and Geochemistry (IGEM), Russian Academy of Sciences, Moscow, Russia

Muslim B. Aminu
Adekunle Ajasin University, Akungba-Akoko, Nigeria

Moses O. Olorunniwo
Obafemi Awolowo University, Ile-Ife, Nigeria

A.A. Okiwelu and I.A. Ude
Geophysics Research Unit, Department of Physics, University of Calabar, Nigeria

Silvia Patricia Barredo
Department of Petroleum Engeneering, Instituto Tecnológico de Buenos Aires (ITBA), Buenos Aires, Argentina

Fetheddine Melki and Fouad Zargouni
Department of Earth Sciences, FST, Tunis El Manar University, Tunis, Tunisia

Taher Zouaghi and Mourad Bédir
Georessources Laboratory, CERTE, Borj Cédria Technopole, University of Carthage, Soliman, Tunisia

Mohamed Ben Chelbi
Water Institute of Gabès, University of Gabès, Tunisia

Akın Kürçer
General Directorate of Mineral Research and Exploration, Department of Geology, Çankaya, Ankara, Turkey

Yaşar Ergun Gökten
Ankara University, Faculty of Engineering, Department of Geological Engineering, Tectonic Research Group, Ankara, Turkey

Anupam Chattopadhya
Dept. of Geology, University of Delhi, Delhi, India

Alexandros Chatzipetros, Spyros Pavlides, George Syrides, Kostas Vouvalidis and Özkan Ateş
Aristotle University of Thessaloniki, Faculty of Sciences, Department of Geology, Thessaloniki, Greece

Salih Zeki Tutkun and Süha Özden
Çanakkale Onsekiz Mart University, Department of Geological Engineering, Çanakkale, Turkey

Emin Ulugergerli and Yunus Levent Ekinci
Çanakkale Onsekiz Mart University, Department of Geophysical Engineering, Çanakkale, Turkey

Alessandro Ellero and Giuseppe Ottria
CNR Institute of Geosciences and Earth Resources, Pisa, Italy

Marco G. Malusà
Department of Geological Sciences and Geotechnology, University of Milano Bicocca, Milano, Italy

Hassan Ouanaimi
Université Cadi Ayyad Ecole Normale Supérieure Département de Géologie, Marrakech, Morocco

Jan Golonka
AGH University of Science and Technology, Kraków, Poland

Aleksandra Gawęda
Faculty of Earth Sciences, University of Silesia, Sosnowiec, Poland

Steffen G. Bergh1, Per Inge Myhre, Kåre Kullerud and Erling K. Ravna
Dept. of Geology, University of Tromsø, Tromsø, Norway

Fernando Corfu
Dept. of Geosciences, University of Oslo, Blindern, Oslo, Norway

Paul E.B. Armitage
Consulting Ltd, Rochester, England, UK

Klaas B. Zwaan
Geological Survey of Norway, Trondheim, Norway

Robert E. Holdsworth
Dept. of Earth Sciences, University of Durham, UK